Origination of Organismal Form

The Vienna Series in Theoretical Biology
Gerd B. Müller, Günter P. Wagner, and Werner Callebaut, editors

The Evolution of Cognition, edited by Cecilia Heyes and Ludwig Huber, 2000

Origination of Organismal Form: Beyond the Gene in Developmental and Evolutionary Biology, edited by Gerd B. Müller and Stuart A. Newman, 2003

Origination of Organismal Form
Beyond the Gene in Developmental and Evolutionary Biology

edited by Gerd B. Müller and Stuart A. Newman

A Bradford Book
The MIT Press
Cambridge, Massachusetts
London, England

© 2003 Massachusetts Institute of Technology

All rights reserved. No part of this book may be reproduced in any form by any electronic or mechanical means (including photocopying, recording, or information storage and retrieval) without permission in writing from the publisher.

This book was set in Times Roman by Interactive Composition Corporation and was printed and bound in the United States of America.

Library of Congress Cataloging-in-Publication Data

Origination of organismal form : beyond the gene in developmental and evolutionary biology / edited by Gerd B. Müller and Stuart A. Newman.
    p. cm. — (The Vienna series in theoretical biology)
"A Bradford book."
Includes bibliographical references (p.).
ISBN 0-262-13419-5 (hc. : alk. paper)
  1. Morphogenesis. 2. Evolution (Biology) I. Müller, Gerd (Gerd B.) II. Newman, Stuart. III. Series.

QH491 .O576 2003
571.3—dc21

2002070314

10 9 8 7 6 5 4 3 2 1

Contents

	Series Foreword	vii
	Preface	ix
I	**INTRODUCTION**	1
1	**Origination of Organismal Form: The Forgotten Cause in Evolutionary Theory** Gerd B. Müller and Stuart A. Newman	3
II	**PROBLEMS OF MORPHOLOGICAL EVOLUTION**	11
2	**The Cambrian "Explosion" of Metazoans** Simon Conway Morris	13
3	**Convergence and Homoplasy in the Evolution of Organismal Form** Pat Willmer	33
4	**Homology: The Evolution of Morphological Organization** Gerd B. Müller	51
III	**RELATIONSHIPS BETWEEN GENES AND FORM**	71
5	**Only Details Determine** Roy J. Britten	75
6	**The Reactive Genome** Scott F. Gilbert	87
7	**Tissue Specificity: Structural Cues Allow Diverse Phenotypes from a Constant Genotype** Mina J. Bissell, I. Saira Mian, Derek Radisky, and Eva Turley	103
8	**Genes, Cell Behavior, and the Evolution of Form** Ellen Larsen	119
IV	**PHYSICAL DETERMINANTS OF MORPHOGENESIS**	133
9	**Cell Adhesive Interactions and Tissue Self-Organization** Malcolm Steinberg	137

10	Gradients, Diffusion, and Genes in Pattern Formation H. Frederik Nijhout	165
11	A Biochemical Oscillator Linked to Vertebrate Segmentation Olivier Pourquié	183
12	Organization through Intra-Inter Dynamics Kunihiko Kaneko	195
13	From Physics to Development: The Evolution of Morphogenetic Mechanisms Stuart A. Newman	221
V	**ORIGINATION AND EVOLVABILITY**	241
14	Phenotypic Plasticity and Evolution by Genetic Assimilation Vidyanand Nanjundiah	245
15	Genetic and Epigenetic Factors in the Origin of the Tetrapod Limb Günter P. Wagner and Chi-hua Chiu	265
16	Epigenesis and Evolution of Brains: From Embryonic Divisions to Functional Systems Georg F. Striedter	287
17	Boundary Constraints for the Emergence of Form Diego Rasskin-Gutman	305
	Contributors	323
	Index	325

Series Foreword

Biology promises to be the leading science in this century. As in all other sciences, progress in biology depends on interactions between empirical research, theory building, and modeling. But whereas the techniques and methods of descriptive and experimental biology have dramatically evolved in recent years, generating a flood of highly detailed empirical data, the integration of these results into useful theoretical frameworks has lagged behind. Driven largely by pragmatic and technical considerations, research in biology continues to be less guided by theory than it is in other fundamental sciences. By promoting the discussion and formulation of new theoretical concepts in the biosciences, this series intends to help fill conceptual gaps in our understanding of some of the major open questions of biology, such as the origin and organization of organismal form, the relationship between development and evolution, or the biological bases of cognition and mind.

Theoretical biology is firmly rooted in the experimental biology movement of early twentieth-century Vienna. Paul Weiss and Ludwig von Bertalanffy were among the first to use the term *theoretical biology* in a modern scientific context. In their understanding the subject was not limited to mathematical formalization, as is often the case today, but extended to the general theoretical foundations of biology. Their synthetic endeavors aimed at connecting the laws underlying the organization, metabolism, development, and evolution of organisms. It is this commitment to a comprehensive, cross-disciplinary integration of theoretical concepts that the present series intends to emphasize. A successful integrative theoretical biology must encompass not only genetic, developmental, and evolutionary components, the major connective concepts in modern biology, but also relevant aspects of computational biology, semiotics, and cognition, and should have continuities with a modern philosophy of the sciences of natural systems.

The series, whose name reflects the location of its initiating meetings and commemorates the seminal work of the aforementioned scientists, grew out of the yearly "Altenberg Workshops in Theoretical Biology" held near Vienna, at the Konrad Lorenz Institute for Evolution and Cognition Research (KLI), a private, nonprofit institution closely associated with the University of Vienna. KLI fosters research projects, seminars, workshops, and symposia on all aspects of theoretical biology, with an emphasis on the developmental, evolutionary, and cognitive sciences. The workshops, each organized by leading experts in their fields, concentrate on new conceptual advances originating in these disciplines, and are meant to facilitate the formulation of integrative, cross-disciplinary models. Volumes on emerging topics of crucial theoretical importance not directly related to any of the workshops will also be included in the series. The series editors welcome suggestions for book projects on theoretical advances in the biosciences.

Gerd B. Müller, University of Vienna, KLI
Günter P. Wagner, Yale University, KLI
Werner Callebaut, Limburg University Center, KLI

Preface

Concepts concerning the evolution of biological form are undergoing ferment. The vast progress made in the last two decades in characterizing the genetic mechanisms involved in embryonic development has demonstrated unexpected degrees of functional redundancy in these processes and unanticipated discordances between conserved forms and conserved genes. At the same time, evolutionary studies have revealed the surprising extent of homoplasy and other forms of parallel morphological evolution in disparate lineages; they have found evidence of extensive morphological diversity much earlier in the evolution of multicellular life than previously thought. Finally, theoretical models have suggested that certain regimes of natural selection can lead to extensive "rewiring" of genetic circuitry with no overt change in organismal phenotype. These findings have raised new questions concerning the relationship between gene content and activity and the generation of biological form.

The present volume is motivated by the conviction that the origination of morphological structures, body plans, and forms should be regarded as a problem distinct from that of the variation and diversification of such entities (the central theme of current neo-Darwinian theory) and that the generative determinants of organismal phenotype must be included in any productive account of the evolution of developmental systems and organismal form in our postgenomic era. It is an outgrowth of the 1999 Altenberg workshop in theoretical biology "The Origins of Organismal Form," organized by the Konrad Lorenz Institute for Evolution and Cognition Research (KLI). The workshop brought together scientists in fields ranging across paleontology, developmental biology, developmental and population genetics, cancer research, physics, and theoretical biology whose work has in various ways attempted to supply the missing generative element in standard accounts of the development and evolution of biological form. Despite the wide diversity of the participants' fields of research, three days of discussion only strengthened an initial sense that gene-level descriptions and analyses are just part of the story in development and evolution, that increased attention must be paid to the generative mechanisms, and that computational models will play an important role in the analysis and understanding of the properties and potentialities of generative systems, thus paving the way for formal integration into evolutionary theory.

We thank the staff of the KLI for their enthusiastic assistance in organizing the workshop. The institute's beautiful setting and a workshop format conducive to the exchange of ideas and controversial views combined to create a uniquely satisfying intellectual experience. As before, it was our great pleasure to work with MIT Press in preparing this volume, and we particularly thank Robert Prior for his continued support of the series.

I INTRODUCTION

1 Origination of Organismal Form: The Forgotten Cause in Evolutionary Theory

Gerd B. Müller and Stuart A. Newman

Evolutionary biology arose from the age-old desire to understand the origin and the diversification of organismal forms. During the past 150 years, the question of how these two aspects of evolution are causally realized has become a field of scientific inquiry, and the standard answer, encapsulated in a central tenet of Darwinism, is by "variation of traits" and "natural selection." The modern version of this tenet holds that the continued modification and inheritance of a basic genetic tool kit for the regulation of developmental processes, directed by mechanisms acting at the population level, has generated the panoply of organismal body plans encountered in nature. This notion is superimposed on a sophisticated, mathematically based population genetics, which became the dominant mode of evolutionary biology in the second half of the twentieth century. As a consequence, much of present-day evolutionary theory is concerned with formal accounts of quantitative variation and diversification. Other major branches of evolutionary biology have concentrated on patterns of evolution, ecological factors, and, increasingly, on the associated molecular changes. Indeed, the concern with the "gene" has overwhelmed all other aspects, and evolutionary biology today has become almost synonymous with evolutionary genetics.

These developments have edged the field farther and farther away from the second initial theme: the origin of organismal form and structure. The question of why and how certain forms appear in organismal evolution addresses not what is being maintained (and quantitatively varied) but rather what is being generated in a qualitative sense. This causal question concerning the specific generative mechanisms that underlie the origin and innovation of phenotypic characters is probably best embodied in the term *origination,* which will be used in this sense throughout this volume. That this causal question has largely disappeared from evolutionary biology is partly hidden by the semantics of modern genetics, which purports to provide answers to the question of causation, but these answers turn out to be largely restricted to the proximate causes of local form generation in individual development. The molecular mechanisms that bring about biological form in modern-day embryos, however, should not be confused with the causes that led to the appearance of these forms in the first place. Although the forces driving morphological evolution certainly include natural selection, the appearance of specific, phenotypic elements of construction must not be taken as being caused by natural selection; selection can only work on what already exists. Darwin acknowledges this point in the first edition of *The Origin of Species,* where he states that certain characters may have "originated from quite secondary causes, independently from natural selection" (Darwin, 1859, 196), although he attributes "little importance" to such effects. In a modified version of the same paragraph in the sixth edition (Darwin, 1872, 157), he concedes that "we may easily err in attributing importance to characters, and in believing that they have been developed through natural selection."

It is the aim of the present volume to elaborate on this distinction between the origination (innovation) and the diversification (variation) of form by focusing on the plurality of causal factors responsible for the former, relatively neglected aspect, the origination of organismal form. Failure to incorporate this aspect represents one of the major gaps in the canonical theory of evolution, it being quite distinct from the topics with which population genetics or developmental genetics is primarily concerned. As a starting point, we will briefly outline the central questions that arise in the context of origination. We have identified four areas (represented by parts II–V of the book) from which the most important open questions arise: (1) the phenomenology of organismal evolution (phylogenetics); (2) genotype-phenotype relationships; (3) physical determinants of morphogenesis; and (4) the structure of the evolutionary paradigm. It will be noted that the questions that arise in each of these areas are often similar or overlapping. Indeed, the presence of recurrent themes across quite disparate subdisciplines is one important indication of the lacuna with regard to origination in the field as a whole.

Questions Arising from Phylogenetics

The evolution of organismal forms—morphological evolution—consists of the generation, fixation, and variation of structural building elements. Cell masses form microscopic structures such as spheres, cones, tubes, rods, plates, and coils. These are often branched and connected by attachments, fusions, or articulations. Such units assemble to form higher-level, macroscopic building elements, again connected to one another, resulting in the body plans of organisms that evolve further by progressive modification. This scenario raises a number of questions that relate specifically to the macroscopic features of morphological evolution. Why, for instance, did the basic body plans of nearly all metazoans arise within a relatively short time span, soon after the origin of multicellularity? Assuming that evolution is driven by incremental genetic change, should it not be moving at a slow, steady, and gradual pace? And why do similar morphological design solutions arise repeatedly in phylogenetically independent lineages that do not share the same molecular mechanisms and developmental systems? And why do building elements fixate into body plans that remain largely unchanged within a given phylogenetic lineage? And why and how are new elements occasionally introduced into an existing body plan?

Many of the phenomena on which these questions are based bear classical names (table 1.1; most "why" questions are also "how" questions here and in table 1.2), giving the issues a seemingly old-fashioned aura. But hardly any of the problems specified by this traditional terminology are explained in the modern theory of evolution. Whereas the

Table 1.1
Open questions concerning morphological evolution

Burgess shale effect	Why did metazoan body plans arise in a burst?
Homoplasy	Why do similar morphologies arise independently and repeatedly?
Convergence	Why do distantly related lineages produce similar designs?
Homology	Why do building elements organize as fixed body plans and organ forms?
Novelty	How are new elements introduced into existing body plans?
Modularity	Why are design units reused repeatedly?
Constraint	Why are not all design options of a phenotype space realized?
Atavisms	Why do characters long absent in a lineage reappear?
Tempo	Why are the rates of morphological change unequal?

classical questions refer to phenomena at the organismal level, most can also be applied to the microscopic and even to the molecular level. All are linked by one common, underlying theme: the origin of organization. The nature of the determinants and rules for the organization of design elements constitutes one of the major unsolved problems in the scientific account of organismal form. The chapters of part II explore some of the most important aspects of this problem.

Questions Arising from Genetics

A second set of open questions relates to the role of genes in the origination of biological form (table 1.2). Organismal evolution is nowadays almost exclusively discussed in terms of genetics. But are genes the determinants of form? Is it true that complete knowledge of the genetic-molecular machinery of an organism also explains how it was brought into being? That is, if we were to know all components and functions of an anonymous genome, would we be able to compute the form of its organism? And is it correct to assume that morphological evolution is driven solely by molecular evolution? Comparative evidence indicates substantial incongruences between genetic and morphological evolution, and the same genotypes do not necessarily correspond with identical phenotypes (Lowe and Wray, 1997). On the one hand, genetic and developmental pathways can change over evolutionary time even when morphology remains constant (Felix et al., 2000); on the other, similar gene expression patterns can be associated with different morphologies.

These questions converge in the second major unsolved problem of organismal form: the genotype-phenotype relationship. Now that entire genomes are mapped out and the genomic approach is seen to be unable to explain biological complexity, this problem will be a central concern of future research. Recognizing that the origination of biological form

Table 1.2
Open questions concerning the genotype-phenotype relationship in development and evolution

Jurassic Park scenario	Does the genetic code contain the complete information of organismal form?
Novelty	Do new structural elements arise from mutations?
Polyphenism	Why can identical genetic content be associated with very different morphological phenotypes?
Redundancy and overdetermination	Why are there multiple genetic and biochemical pathways to the realization of biological forms?
Discordance	Why do morphological and genetic evolution proceed at different paces?
Epigenesis	How is the genotype-phenotype relationship mediated in development?

cannot be understood solely from genetic analysis will necessarily stimulate investigation of the processes that actually construct the phenotype from materials provided, in part, by the genotype. Also, to analyze, interpret, and predict the genotype-phenotype relationship, mathematical model building and computer simulation will be essential, representing a new research approach that has been called "phenomics" (Palsson, 2000). The chapters of part III provide viewpoints on several of the problems that will have to be taken into account in future modelling approaches.

Questions Arising from Development

Two causal processes interact in the generation of organismal form: development and evolution. The new field of evolutionary developmental biology acknowledges this fact, but much work in this area proceeds under the assumption that the only important link between the two processes lies in genetics—as if the individual generation of form were merely a reading out of evolved genetic programs. However, development does not appear to behave like any program known to computer science—phenotypic outcomes persist despite extensive derangement in lines of "program code" (i.e., gene expression levels and interactions) induced by such evolutionarily unprecedented manipulations as experimental "knockouts" (Shastry, 1995) and nuclear transfer (Humpherys et al., 2001). Moreover, that genetic circuitry involved in development can undergo evolutionary "rewiring" without overt changes in the phenotype (Szathmary, 2001) suggests that phenotypes have autonomy that can trump that of the programs they supposedly express.

Epigenesis, the sum of processes that determine the transformation of a zygote into an adult phenotype poses a number of unanswered questions regarding the generation of individual forms (table 1.3). Among the most fundamental but least understood class of epigenetic factors are the physical properties of biological materials that participate in

Table 1.3
Open questions concerning epigenesis and its role in morphological evolution

Programs	Does the developmental generation of organismal form result from deterministic programs?
Context	How are developmental processes modulated by epigenetic context?
Generic Properties	What is the role of the physicochemical properties of biological materials?
Environment	What is the role of the external environment in development?

morphogenesis. How do the generic, physical properties of cell aggregates and tissues shape the constructional outcomes of development (segmentation, multilayering, body cavity formation, and so forth), and, equally important, to what extent are these same properties relevant to the origin of these forms in evolution? Although the properties are paradigmatic of the determinants that generate form, these determinants may take on different importance at different stages of evolution. The chapters of part IV deal with them individually and collectively.

Questions Arising from Evolutionary Theory

The neo-Darwinian paradigm still represents the central explanatory framework of evolution, as exemplified by recent textbooks (e.g., Mayr, 1998; Futuyma, 1998; Stearns and Hoekstra, 2000). This refined and canonical theory concerns the variational dynamics and adaptation of existing forms. It is a gene-centered, gradualistic, and externalistic theory, according to which all evolutionary modification is a result of external selection acting on incremental genetic variation. The resulting adaptations lead to successive replacement of phenotypes and hence to evolution.

Although this theory can account for the phenomena it concentrates on, namely, variation of traits in populations, it leaves aside a number of other aspects of evolution, such as the roles of developmental plasticity and epigenesis or of nonstandard mechanisms such as assimilation (table 1.4). Most important, it completely avoids the origination of phenotypic traits and of organismal form. In other words, neo-Darwinism has no theory of the generative. As a consequence, current evolutionary theory can predict what will be maintained, but not what will appear. Although recent years have seen attempts to extend evolutionary theory to organism-environment interactions (Oyama, 2000; Johnston and Gottlieb, 1990; Sober and Wilson, 1998) and self-organizing processes (Kauffman, 1993), what is still lacking is an evolutionary theory that specifically addresses the morphological aspects of evolution and integrates the interactional-epigenetic aspects with the genetic. The missing generative dimension in evolutionary theory is the subject of part V, whose

Table 1.4
Open questions concerning the theory of morphological evolution

Origination	What generative mechanisms are responsible for the origin and innovation of phenotypic characters?
Plasticity	Are developmental response capacities specifically evolved, or is plasticity a primitive property?
Epigenesis	Do the rules of developmental transformation shape evolution?
Evolvability	Is the evolutionary potential of a lineage associated with the capacity of its developmental system to respond to the environment?
Assimilation	What is the role of genetic co-optation and assimilation in the evolution of organismal form?

chapters illustrate, with specific examples across a range of morphogenetic systems, the ways in which epigenetic processes are beginning to take their place in a more complete and comprehensive evolutionary theory.

Elements of a Postgenomic Synthesis

If, as we suggest, the failure of the current theory of evolution to deal with the problem of origination is the major obstacle to a scientific understanding of organismal form, it is incumbent on us to provide at least a sketch of an alternative view. In fact, it is our contention that a synthetic, causal understanding of both the development and the evolution of morphology can be achieved only by relinquishing a gene-centered view of these processes (Newman and Müller, 2000).

Processes of natural selection can lead to morphological novelty by unleashing new epigenetic relationships (Müller, 1990; Müller and Wagner, 1996). Alternatively, they can consolidate the expression of a morphological phenotype that was previously dependent on developmental or environmental conditionalities (Johnston, Barnett, and Sharpe, 1995). In neither case does an understanding of changes in gene frequencies shed light on the evolution of forms—only on the evolution of genes. And even though hierarchical programs of gene expression often govern the sequential mobilization of morphogenetic processes in modern-day organisms, the mobilized processes are distinct from these triggering events. Again, detailed information at the level of the gene does not serve to explain form.

In the framework we propose, epigenetic processes—first, the physics of condensed, excitable media represented by primitive cell aggregates and, later, the conditional responses of tissues to each other and to external forces—replace gene sequence variation and gene expression as the primary causal agents in morphological origination. These determinants and their outcomes are considered to have set out the original, morphological templates during the evolution of bodies and organs, and to have remained, to varying extents, effective causal factors in the development of all modern, multicellular organisms (Newman and Müller, 2000).

Genetic evolution is highly suited for enhancing the reliability and inheritance of forms originally brought about by conditional processes: promotor duplication and diversification, metabolic integration, and functional redundancy can all add parallel routes to the same endpoint (Newman, 1994). By such means, the morphogenetic outcomes originated by epigenetic propensities become captured and routinized, "assimilated" (Waddington, 1961), by genetic circuitry over the course of evolution. In this view, morphological plasticity, and much of evolvability are primitive properties—the phylogenetic retention of the conditionality of the originating, epigenetic processes. At the end of long evolutionary trajectories, organisms come to embody a species-characteristic mix of conditional and programmed modes of development. Finally, in any given species the ratio of conditional to programmed determinants of morphogenesis may vary at different stages and developmental subsystems.

The view described here emphasizes the distinction between the mechanisms underlying origination and those underlying variation in morphological evolution and hence the necessity to account for that distinction in evolutionary theory. It clearly suggests that the relationship between genotype and phenotype in the earliest metazoans was different from that in their modern counterparts and that the present relationship between genes and form is a derived condition, a product of evolution rather than its precondition.

Although not all contributors to this volume would accept the most radical implications of this view, which challenges major tenets of neo-Darwinism, including its incrementalism, uniformitarianism, and genocentricity, all were invited to participate in this project because their work explicitly influenced the development of the ideas behind it. Readers will evaluate each chapter on its own terms; we hope they will also recognize a coherence that transcends the disciplinary boundaries of the contributors.

References

Darwin C (1859) The Origin of Species by Means of Natural Selection. London: John Murray.

Darwin C (1872) The Origin of Species by Means of Natural Selection (6th ed). London: John Murray.

Felix MA, De Ley P, Sommer RJ, Frisse L, Nadler SA, et al. (2000) Evolution of vulva development in the Cephalobina (Nematoda). Dev Biol 221: 68–86.

Futuyma DJ (1998) Evolutionary Biology. Sunderland: Sinauer.

Humpherys D, Eggan K, Akutsu H, Hochedlinger K, Rideout WM III, et al. (2001) Epigenetic instability in ES cells and cloned mice. Science 293: 95–97.

Johnston CM, Barnett M, Sharpe PT (1995) The molecular biology of temperature-dependent sex determination. Philos Trans R Soc Lond B Biol Sci 350: 297–303; discussion 303–304.

Johnston TD, Gottlieb G (1990) Neophenogenesis: A developmental theory of phenotypic evolution. J Theor Biol 147: 471–495.

Kauffman SA (1993) The Origins of Order. New York: Oxford University Press.

Lowe CJ, Wray GA (1997) Radical alterations in the roles of homeobox genes during echinoderm evolution. Nature 389: 718–721.

Mayr E (1998) This Is Biology: The Science of the Living World. Cambridge, Mass.: Harvard University Press.

Müller GB (1990) Developmental mechanisms at the origin of morphological novelty: A side-effect hypothesis. In: Evolutionary Innovations (Nitecki MH, ed), 99–130. Chicago: University of Chicago Press.

Müller GB, Wagner GP (1996) Homology, Hox genes, and developmental integration. Am Zool 36: 4–13.

Newman SA (1994) Generic physical mechanisms of tissue morphogenesis: A common basis for development and evolution. J Evol Biol 7: 467–488.

Newman SA, Müller GB (2000) Epigenetic mechanisms of character origination. J Exp Zool (Mol Dev Evol) 288: 304–317.

Oyama S (2000) The Ontogeny of Information: Developmental Systems and Evolution (2d ed). Durham, N.C.: Duke University Press.

Palsson B (2000) The challenges of in silico biology. Nat Biotechnol 18: 1147–1150.

Shastry BS (1995) Genetic knockouts in mice: An update. Experientia 51: 1028–1039.

Sober E, Wilson DS (1998) Unto Others: The Evolution and Psychology of Unselfish Behavior. Cambridge, Mass.: Harvard University Press.

Stearns SC, Hoekstra R (2000) Evolution: An Introduction. Oxford: Oxford University Press.

Szathmary E (2001) Developmental circuits rewired. Nature 411: 143–145.

Waddington CH (1961) Genetic assimilation. Adv Genet 10: 257–293.

II PROBLEMS OF MORPHOLOGICAL EVOLUTION

The three chapters of part II highlight the fact that, despite the plethora of variable genetic and developmental mechanisms that emerged in the course of metazoan evolution, only a limited number of constructional themes were realized in the phenotypic realm. In particular, the millions of extant animal species are all elaborations of the thirty-seven presently known basic body designs: the same—or similar—structural elements are used over and over again. Before considering why this is so, however, and what factors and processes are responsible for the development and evolution of these structures, we must first lay out the phenomena that define the subject matter of any consideration of evolution of organismal form.

Simon Conway Morris (chapter 2) reviews the stunning burst of metazoan forms at the beginning of the Cambrian. Over a relatively short geological time period, aggregates of cells yielded a repertoire of macroscopic body plans that encompassed all later metazoan designs. Conway Morris critically analyzes the controversial issues of timescales and speed of change. In his highly original discussion of the factors that plausibly contributed to the Cambrian "explosion," he points out that this proliferation of body designs amounts to a triploblastic event, whose roots, and those of an earlier diploblastic radiation, must lie in the preceding Ediacaran period. He also addresses the possibly significant role of non-fossilized larvae in metazoan evolution.

Pat Willmer (chapter 3) introduces a genuine but puzzling feature of morphological evolution, namely, the recurrence of similar design solutions in different phylogenetic lineages, despite their absence in a common ancestor. This pervasive phenomenon is variously referred to as "convergence," "parallelism," and "homoplasy," each designating a different aspect of the encompassing concept of analogy. The questions raised by these phenomena of similarity are numerous and have been controversially debated since the time of Richard Owen (see, for example, Sanderson and Hufford, 1996). With regard to the themes of this volume, the major issue is whether such morphological similarity results from similar external constraints and contingencies, through natural selection, or from intrinsic properties of tissue masses and the inherent, generative features of developmental systems. Using segmentation, appendages, lophophores, and larvae as specific examples, Willmer discusses convergence within a genetic and developmental framework, underlining the flexibility of developmental mechanisms. She proposes that the frequency of convergence is related to the degree of phenotypic plasticity in a phylogenetic lineage.

Gerd Müller (chapter 4) argues that, despite persistent difficulties with its conceptualization, homology represents the key problem that needs to be addressed by a theory of morphological evolution. After analyzing the reasons why homology fell into undeserved disrepute in recent years, he briefly reviews the quantifiable dimensions of homology in organismal designs. He identifies three steps in the origination of homology requiring causal explanation: (1) the generation of initial parts and innovations; (2) the fixation of such new elements in the body plan of a phylogenetic lineage; and (3) the autonomization of homologues as process-independent elements of organismal design. According to his proposed "organizational homology concept," homology is not merely the passive result of genetic evolution: homologues play an active role as organizers of genetic, developmental, and phenotypic order. This reconceptualization of homology can provide a starting point for new empirical research projects into the causal mechanisms underlying the three- and four-dimensional processes of form generation.

All three contributions remind us that a number of distinct questions about the morphological phenomena of evolution remain unanswered. Notably, how did homoplasy, homology, and particular structural themes, including entire body plans, originate? These questions are among the theoretically most challenging, but there are many others not explicitly addressed in these chapters. How, for example, can we account for morphological trends and stasis, the various kinds of vestigialization, and the occurrence of atavisms? All of these questions are related in one way or another to the three major issues discussed in part II. Coming from different perspectives, these three chapters emphasize that the discordances between genetic and morphological evolution are more prevalent than generally appreciated, and that, to understand these characteristic features of morphological evolution, we must consider processes and mechanisms beyond the realm of genetics.

2 The Cambrian "Explosion" of Metazoans

Simon Conway Morris

Few would disagree that the fossil record provides a genuine narrative of the broad sweep of the history of life: from stromatolites, to dinosaurs, to us. The many details that evolutionary biologists crave, however, are commonly regarded as elusive, if not unattainable. It is received wisdom, for example, that higher taxonomic categories, most notoriously the phyla, appear abruptly and cryptically. The apparent absence of intermediary forms has been explained by appealing to derivation from animals with a very low preservation potential, such as larval forms, or to rapid morphological transitions, and the failure of the rock record to preserve the appropriate fossils (e.g., van Tuinen, Sibley, and Hedges, 2000; but see Foote et al., 1999; Benton, 1999). Such explanations—perhaps even excuses—coalesce into a broad assumption that the emergence of new body plans can only be explained by mechanisms of macroevolution, mechanisms that as often as not differ from those familiar to neo-Darwinians. Another common assumption is that fossils with unfamiliar anatomies are best interpreted as offshoots into regions of otherwise sparsely populated morphospace. However much they may emphasize the degree of faunal disparity, such fossils therefore play no significant part in either the elucidation of phylogenies or the documentation of transitions between supposedly different body plans.

Although cryptic originations, stratigraphic hiatuses, geologically instantaneous originations, and problematic fossils are not to be dismissed, this chapter will argue that, in the context of the early metazoan radiations or the Cambrian "explosion," the fossil record remains a fruitful, and historically unique, source of insights not only into the origin and evolution of organismal form but, if interpreted correctly, also into the phylogenies based on molecular data and the possible role of genomic reorganization in the origin of body plans. How we interpret the Cambrian "explosion"—whether as a slow-fuse detonation, a megatonblast, or even as an artifact (see Conway Morris, 2000a)—has a bearing on how we go about disentangling cause and effect. And it is perhaps here that we begin to realize that, however well we understand some aspects of evolution, in this area of science there exist profound problems that remain remarkably recalcitrant.

Some Problems of Metazoan Complexity

Seeing an elephant demolish a small tree, a bird build a nest, or a bombardier beetle defend itself may seem to make the question of metazoan definition superfluous. Motility, nervous action, sophisticated behavior, and a dependence on heterotrophic sources of energy provide a unique combination of characters. Although life on earth would no doubt cease in the absence of bacteria, or even of plants, animals stamp their imprint (sometimes literally) on the biosphere. Yet, clearly, metazoans evolved from "simpler" eukaryotes, which it seems likely were also closely akin to the ancestors of fungi (e.g., Baldauf, 1999; see also Atkins, McArthur, and Teske, 2000). As to the nature of these transitional forms, we can only speculate: apart from the shared presence in fungi and metazoans of the protein collagen, which may in any event represent convergence (see Celerin et al., 1996), the similarities between these two massively diversified kingdoms are largely identified on the basis of molecular sequences.

Looking at even what are purportedly the most primitive metazoans, the sponges, and their close allies, the protistan choanoflagellates, may not tell us much. It is no coincidence that venerable speculations on the earliest metazoans by zoologists such as Hadzi, Haeckel, Hand, Jägersten, and Steinböck all have a decidedly theoretical air about them, nor that they remain largely untested against either molecular phylogenies or the fossil record. If sponges are the most primitive living representatives of the Metazoa, and at present this seems reasonable (e.g., Gamulin, Müller, and Müller, 2000), then presumably most of the hallmarks of this kingdom (e.g., muscles, neurons) were acquired after the sponges first appeared. Even so, with the surprising discovery that a number of metazoan phyla, notably, Platyhelminthes (Balavoine, 1997), Mesozoa (Kobayashi, Furuya, and Holland, 1999), Myxozoa (e.g., Siddall et al., 1995), and Xenoturbellida (Israelsson, 1999), are highly derived and simplified, it is important to keep an open mind as to the supposed primitiveness of any particular group. Even if we accept the sponges as the most primitive of the metazoans, however, without evidence of transitional (and almost certainly extinct) forms between sponges and coelenterates, or between coelenterates and triploblasts, we can understand the acquisition of complex features only in outline. No doubt much will become clearer as the number of genomic surveys grows; even so, the battery of metazoan genomic and molecular systems already in place in sponges and cnidarians alike (Shenk and Steele, 1993; see also Ono et al., 1999) suggests that the solution to the problem of emerging metazoan complexity will lie as much in redeployment, reorganization, and duplication as in genetic novelty. This, too, is fine in theory. Yet from what we already know, the description of these new genetic pathways will probably appear to us as arbitrary, haphazard, and unsatisfactory. Notwithstanding the potentially potent mechanisms of gene duplication (Lundin, 1999), which may play a key role in vertebrate diversification (e.g., Holland, 1997), at present there is a serious lack of precision so far as it applies to metazoan evolution (Shimeld, 1999). In conclusion, the old divide (and sometimes rivalry) between morphology and molecules—although bridged in places—is not only real, but reinforces our notion that there is no simple series of connections. Life is more complicated than some scientific reductionists are prepared to concede.

There is another aspect to the problem of the uniqueness of the Metazoa and our difficulty in deciding how this came about, which concerns the too often unacknowledged degrees of complexity in the other kingdoms. It is worth reminding the more metazoan-centered readers of the sophistication of single-celled protistans possessing statocyst-like "organs" (Fenchel and Finlay, 1986) or an "eye" (Foster and Smyth, 1980) and, indeed, of the multicellularity of certain bacteria (Shapiro and Dworkin, 1997), perhaps most spectacularly manifestated in myxobacteria such as *Chondromyces*, which produces a treelike fruiting body. The boundary between at least the smaller metazoans and protistans becomes particularly blurred when we look at such complex ciliates as *Ephelota, Ophrydium,* and *Stentor,* which, despite being single-celled, showed a remarkable repertoire of activities, and whose shapes are strongly convergent on various metazoans. It is scarcely surprising that there is at least one case whereby two supposed species of a rotifer turned out to be ciliates (Turner, 1995).

The point of the above remarks is not to suggest that metazoans are necessarily polyphyletic, although this cannot yet be ruled out (Conway Morris, 1998a). It is rather to remind us of two important facts. First, the connection between phenotypic complexity and underlying genomics is far from straightforward, no matter what the reductionist fervor in molecular biology may suggest to the contrary (see Goodwin, 1994; and Gilbert, chapter 6, Newman, chapter 13, Wagner and Chiu, chapter 15, this volume). Second, the earliest metazoans emerged in an already complex milieu, thickly populated by sophisticated protistans, of which the fossil record gives us only a fragmentary glimpse (e.g., Porter and Knoll, 2000). Indeed, it is not clear that a hypothetical observer 700 million years ago would have noticed the incipient metazoan clade as anything other than a group of multicellular eukaryotes.

Metazoans in the History of Life

Life originated at least 3.5 billion years ago. It is widely assumed that prokaryotes, and the especially primitive thermophiles among them, preceded eukaryotes. Although this assumption is treated as orthodoxy, the evidence is more ambiguous than is generally imagined (e.g., Forterre, 1997). Eukaryotes may be very ancient, even if endosymbioses (notably, the acquisition of mitochondria) occurred after their appearance. So far as the fossil record is concerned, the earliest evidence for eukaryotes is in the form of sterane biomarkers (Brocks et al., 1999). The first known example of what may be eukaryotic multicellularity occurs some 600 million years later (Han and Runnegar, 1992), in the form of ribbonlike fossils, although it is by no means impossible these are prokaryotes. By about 1.8 billion years ago, however, the fossil record yields reasonably convincing eukaryote body fossils (as opposed to the much earlier geochemical biomarkers) in the form of algal structures known as acritarchs.

There is a consensus that about 1 billion years ago there was a "Big Bang" in eukaryote evolution, although the reasons for this massive diversification are unclear. The evidence for this event, which saw the emergence of many of the advanced eukaryotes, is largely molecular (e.g., Sogin, 1994), although there is also some degree of support from the fossil record (e.g., Butterfield, Knoll, and Swett, 1990; Butterfield, 2000). Nested somewhere within this remarkably diverse plexus (e.g., Cavalier-Smith, 1998) are the Metazoa, which, as noted above, are linked to the Fungi at the molecular level.

The first convincing evidence for metazoan body fossils appears no earlier than 600 million years ago, which raises the question: Were the Metazoa indeed late entrants in the story of eukaryotic diversification? Or, alternatively, are the first roughly 400 million years of metazoan evolution paleontologically cryptic? How can we decide? Proponents of the merits of "molecular clocks" (e.g., Wray, Levinton, and Shapiro, 1996; Bromham et al., 1998), perhaps predictably, have been vociferous and confident in claiming a remote ancestry for metazoans. Even so, some of their claims verge on the incredible. Bromham and colleagues (1998), for example, infer possible metazoan occurrences in the Archaean, that is, in excess of 2.5 billion years ago. More nuanced views of the data from molecular "clocks" are available, however (e.g., Ayala, Rzhetsky, and Ayala, 1998; Bromham and Hendy, 2000; Cutler, 2000); appropriate statistical techniques (e.g., Huelsenbeck, Larget, and Swofford, 2000), careful consideration of the fossil record (e.g., Norman and Ashley, 2000), and a more realistic evolutionary framework (see Conway Morris, 2000b) may lead to a more coherent and believable view. Nevertheless, these molecular data do not rule out the possibility of a cryptic interval in the evolution of metazoans before their appearance in the fossil record as Ediacaran faunas (e.g., Narbonne, 1998; Waggoner, 1998, 1999). Indeed, sensibly employed, the data may help narrow the likely time interval of origination and thus suggest the most fruitful strategy for paleontological discovery.

Claims that very ancient sedimentary structures are metazoan trace fossils (e.g., Breyer et al., 1995; Seilacher, Bose, and Pflüger, 1998) are not only dubious in their own right, but carry with them a baggage of uniformitarian thinking typified by the general, half-articulated assumption that somehow "worms are primitive." Even if such trace structures are biogenic (however implausible that may be), they need not indicate metazoans. For one thing, it is far from clear how a billion-year-old worm could have evolved and failed to colonize the seafloor, thereby ushering in the Cambrian "explosion" four hundred million years ahead of schedule. Indeed, the general absence of *any* convincing trace fossils before about 570 million years ago and the controversial nature of earlier structures such as those from India (Seilacher, Bose, and Pflüger, 1998) have persuaded most students of this problem that the first metazoans cannot have been larger than about a millimeter. Bona fide millimeter-sized trace fossils, made by similarly sized animals, are readily preserved in fine-grained sediments such as siltstones and are well known from Ediacaran sediments.

Although the most popular candidates for the role of cryptic, pre-Ediacaran metazoans are either the meiofauna (e.g., Fortey et al., 1997) or ciliated larvae (e.g., Peterson, Cameron, and Davidson, 1997), for different reasons, both of these hypotheses face difficulties (Conway Morris, 1998b, c, d, 2000c; Budd and Jensen, 2000). There is little evidence that the extant meiofauna (see Giere, 1993) is phylogenetically primitive. Whereas the cryptic pre-Ediacaran fauna may have adopted (at least in part) an interstitial life, it is unlikely that the denizens of these Neoproterozoic sediments were simply miniaturized versions of familiar metazoans, with organ systems and anatomies poised to "inflate" to macroscopic size as soon as the Cambrian "explosion" got under way. Skepticism about such a body plan "inflation" revolves around the apparent implausibility of such features as a water-vascular system, articulated skeleton, coelom, or complex sensory organs evolving at the lilliputian scale of the meiofauna (Budd and Jensen, 2000). The larval hypothesis harks back to a Haeckelian theme of the supposed primitiveness of metazoan larvae, perhaps best exemplified by trochophore larvae, and to Jägersten's notion (1972) of a biphasic life cycle that alternates between a minute ciliated larva occupying a planktonic niche and a macroscopic, typically benthic adult. In the context of the Cambrian "explosion," this hypothesis has been taken forward vigorously by Peterson, Cameron, and Davidson, (1997; see also Peterson, Cameron, and Davidson, 2000), with an appeal to developmental biology, specifically focusing on the "set-aside cells" that give rise to the adult tissues during the catastrophic metamorphosis separating the stages of this biphasic cycle. Despite various difficulties explained at length elsewhere (Conway Morris, 1998b, c, d; see also Wolpert, 1999; Hughes, 2000), Peterson, Cameron, and Davidson (2000) continue to argue that the evolutionary novelty of "set-aside cells" ushered in the appearance of macroscopic body plans. It seems as parsimonious, however, to argue that, though the primitive metazoan may have had a ciliated larva, the complex biphasic life cycle with "set-aside cells" was introduced on multiple occasions. In this scenario, it is hardly surprising that an adult rudiment is present in the larva, and that such diagnostic features as Hox gene expression are effectively confined to this rudiment and its descendant cells.

If we want to know what the earliest metazoans looked like, the above comments can hardly be regarded as encouraging. What, if neither ciliated larvae poised to invent their "set-aside cells" nor interstitial meiofauna, should we be looking for? Our problem is compounded if we move away from the firm ground of uniformitarian assumptions and try to conjure up the reality of a pre-Ediacaran metazoan. What might we find? For many years, it has been proposed that the decline in stromatolite diversity initiated about 1 billion years ago in deeper water and 800 million years ago in shallower water stemmed from the appearance of metazoans (Walter and Heys, 1985; see also Awramik and Sprinkle, 1999). This proposal, in itself, is controversial, and some workers (e.g., Grotzinger, 1990) look to changes in ocean chemistry as a more plausible mechanism. But, even if the decline in

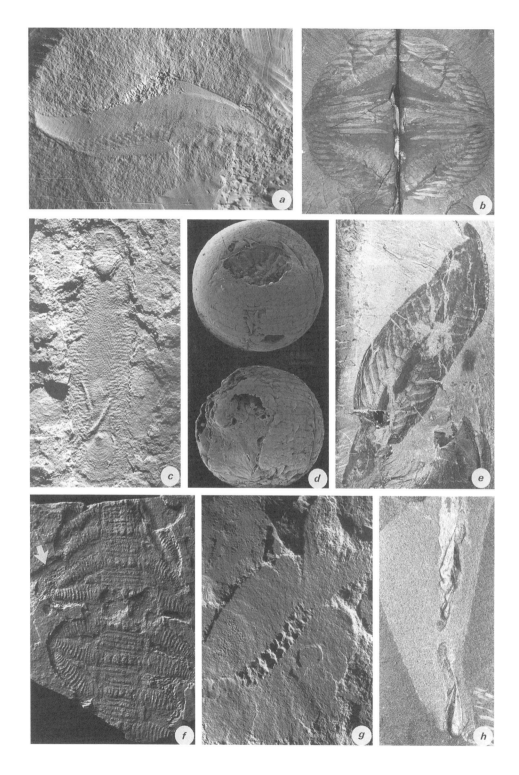

Neoproterozoic stromatolite diversity resulted from disruption of the microbial mats by grazing and burrowing, as Cao (1999) suggests, the organisms responsible may have been unrelated to metazoans.

Given the likelihood that pre-Ediacaran metazoans were minute, consisting of perhaps only a few thousand cells, the chances of finding any fossil representatives would seem decidedly slight. The recent discovery in phosphorites of both Cambrian (figure 2.1d; Bengtson and Yue, 1997; Yue and Bengtson, 1999; Kouchinsky, Bengtson, and Gershwin, 1999) and late Neoproterozoic (Xiao and Knoll, 1999a, 2000) fossil embryos may have much improved those chances, however, opening the way for investigation of earlier phosphorites. The claim that at least some of the Neoproterozoic fossils are metazoan embryos is controversial (e.g., Xue, Zhou, and Tang, 1999; see also Xiao and Knoll, 1999b; Xiao, Yuan, and Knoll, 2000), and any announcement of yet earlier metazoan embryos is likely to generate even greater controversy. Nevertheless, as a starting point, it seems sensible to examine phosphorites from beneath the widely distributed Neoproterozoic tillites. These latter units are interpreted, if somewhat controversially, as the product of very widespread, if not global, glaciation (e.g., Hoffman et al., 1998; Kempf et al., 2000; Kerr, 2000), of which there have been at least two major episodes (e.g., Brasier et al., 2000). The earliest Ediacaran fossils, simple discs of perhaps cnidarian grade, have been found in a unit below the younger of two tillites (Ice Brook Formation) in the Mackenzie Mountains (Narbonne and Aitken, 1995); all other Ediacaran assemblages appear to be firmly posttillite.

Figure 2.1
Representative Lower Cambrian (a, c, d, f, g) and Middle Cambrian (b, e, h) fossils, from the Burgess shale (b, e, h) and from similar deposits in Chengjiang (a) and Sirius Passet (c, f, g), as well as from the phosphatized material of the Petrosvest Formation of east Siberia (d). Together, these fossils encompass some of the most characteristic body plans of the Cambrian "explosion." (a) The agnathan chordate *Myllokunmingia fengiao* (anterior is to the right), showing the gill pouches, dorsal fin, and myomeres. (b) The ctenophore *Fasciculus vesanus*, the holotype and only known specimen, is unusual both in the large number of comb rows it possesses and their division into long and short sets. (c) The halkieriid *Halkieria evangelista* is an armored sluglike organism with a scleritome of some 2,000 sclerites and two large shells, one located at either end of the body. The anterior shell appears to be in a retracted position. The prominent ridges toward the posterior represent superimposed trace fossils and are otherwise unrelated to the halkieriid. (d) Embryos referred to *Markuelia secunda*, possibly derived from the halkieriids, showing both transverse divisions and possible incipient sclerites. (e) The frond *Thaumaptilon walcotti*, an Ediacaran survivor that most probably belongs to the pennatulacean cnidarians and whose body consists of a thick stalk and a blade with prominent cushionlike structures arising either side of the axis. (f) The lobopodian *Hadranax augustus*, showing the prominent lobopods, trunk tubercles, and a possible anterior appendage (arrow). (g) The stem group arthropod *Kerygmachela kierkegaardi*, showing the anterior giant appendages, trunk lobes, and gut trace. (h) The priapulid *Ottoia prolifica*, showing the posterior trunk with intestinal strand containing three hyoliths, two adults and a juvenile. Note all the hyoliths are pointing in the same direction (anterior backward), and that the most anterior has its operculum in place and presumably was swallowed alive. Magnifications are $a: \times 2.0$; $b: \times 0.3$; $c: \times 0.8$; $d: \times 55$; $e: \times 0.3$; $f: \times 0.8$; $g: \times 0.9$; $h: \times 1.9$.

Originally published as figure 2 in Conway Morris 2000a, this figure is reproduced with the permission of the *Proceedings of the National Academy of Sciences, USA,* and also with the specific permission of the following: a, D. Shu, North-West University, Xi'an; $b, c, e,$ and h S. Conway Morris, copyright 2000; d, S. Bengtson, and Z. Yue, copyright 1997, and S. Bengtson; f and g, G. Budd, Uppsala, Sweden.

There are also significant problems of stratigraphic correlation. For example, a well-documented and classic series of tillites exposed in the Neoproterozoic strata of northwest Scotland may turn out to be substantially older than previously thought (Prave, 1999), which would cast doubt on a purported trace fossil from immediately above the tillites (Brasier and McIlroy, 1998), and could also explain my failure (Conway Morris, 1999) to discover convincing Ediacaran fossils in ostensibly suitable lithologies (Jura quartzite) yet higher in this Scottish sequence. An additional problem is that pretillite phosphorites are relatively uncommon. Even so, my preliminary results based on an examination of some pretillite phosphatized sediments exposed in a late Neoproterozoic sequence near Lake Mjøsa, southern Norway, are not encouraging. Although this unit is already known for some very well preserved acritarchs (Speldjnaes, 1967; Vidal, 1990), to date, nothing convincingly metazoan has been found.

The Cambrian "Explosion" and Phyla

The still hypothetical nature of the pretillite metazoans, and the phylogenetically refractory nature of the Ediacaran assemblages give us an undeniably lopsided view of early metazoan evolution. This is simply because of the remarkable riches of the Cambrian fossil record, most notably the ever-growing harvest from the fossil Lagerstätten, especially of faunas of the Lower-Middle Cambrian Burgess shale type, in particular, of those in the Burgess shale itself, as well as in Chengjiang (south China) and Sirius Passet (north Greenland), and of faunas in the Middle-Upper Cambrian Orsten (e.g., Müller, Walossek, and Zakharov, 1995). Continuing discoveries and reinterpretations from these remarkable faunas have led to a series of reassessments and opened up exciting possibilities, touched on below, of dialogues with molecular biology and revitalization of old phylogenetic schemes.

Even so, it is still widely supposed that the Cambrian "explosion" effectively marks the origination, not only of all the metazoan phyla, but also of a swathe of now-extinct body plans, swept from the theatre of life by contingent happenstance rather than by disastrous maladaptation. Such an oversimplification verges on parody; more seriously, it undermines any coherent attempt to establish an evolutionary framework. This Cambrian scenario is littered not only with supposedly extinct phyla but also with the mysterious mechanisms of macroevolution supposedly necessary to power this burst of diversification. At first glance, to be sure, a survey of the Cambrian seas seems to reveal both familiar and extinct phyla. For example, even the nonspecialist has little difficulty in identifying forms characteristic of a wide range of metazoans, from sponges, cnidarians (figure 2.1e), ctenophores (figure 2.1b), and priapulids (figure 2.1h) to arthropods (figures 2.1f, g), brachiopods, annelids, molluscs, hemichordates, echinoderms, and chordates, in the last case including even fish (figure 2.1a; Shu et al., 2000). To this rich array can be added a motley selection of variously strange taxa, the so-called problematica (figure 2.1c). We should keep in mind, however, that defining phyla, whether extant or extinct, tends to discourage even speculation as to how seemingly immutable body plans might be transformed or might otherwise evolve.

To counter this tendency, an effective strategy is now in place with the proposal that, contrary to received wisdom, the bulk of the Cambrian fossil record, especially its earlier stages, consists of representatives, not of phyla per se, but rather of the various stem groups (Budd and Jensen, 2000; see also Conway Morris, 1998d, 2000a, c). In other words, what we see is a series of body plans in the process of construction, with some, but significantly not all, of the character states (in fossils these are necessarily anatomical) that will define the end product, that is, the crown group (defined as the last common ancestor of all living species and all its descendants). This approach yields insights into the supposedly problematic forms, making the idea of extinct phyla largely superfluous. It also encourages a more coherent analysis of how functional transitions are achieved in a realistic ecological milieu. Here the emphasis has been on locomotion and feeding, with the clear implication of necessary changes in both sensory systems and neurology. Despite the seeming welter of body plans emerging in the Cambrian, it is difficult to escape the conclusion that the processes and products, far from requiring a radical revision of existing theory, fit comfortably into the standard neo-Darwinian framework.

Did Anything Happen?

Just as the Great Wall of China is said to be visible from outer space, so no paleontologist could miss seeing the dramatic change in the fossil record across the Vendian-Cambrian boundary. The most obvious manifestation of this change is the appearance of hard parts. Cambrian faunas are dominated by skeletal remains of trilobites and brachiopods, with a number of other groups, such as molluscs, echinoderms, and sponges, being of variable importance. The first skeletal assemblages are unfamiliar; to a certain extent their phylogenetic position remains uncertain. They include tubes, platelike shells, various spicules and sclerites, and slender toothlike objects. The acquisition of skeletal parts is clearly polyphyletic, and most probably represents co-option of proteins (and mucopolysaccharides) as templates for mineral precipitation. Given the precision of biomineralization processes, it seems less likely that changes in seawater chemistry, once a popular explanation for the sudden acquisition of Cambrian skeletons, were the trigger for this event. At present, it seems more likely that the primary stimulus behind the appearance of diverse skeletons, both in terms of their construction and distribution across the Metazoa, was ecological: the need for protection from predators.

It is now realized, of course, that deciphering the Cambrian "explosion" will require more than documenting the fossil record of skeletons. Most famously, deposits of the Burgess shale type (Conway Morris, 1998e; Hou et al., 1999) demonstrate that the proportion of animals with skeletons, preservable under normal conditions for fossilization, is very small (see figures 2.1a–c, e–h). Although such discoveries have generated considerable controversy, this has been largely for ideological reasons, most blatantly by Stephen Jay Gould, who, in support of a strange materialist agenda, has argued that a contingently happenstance origin of humans leads to certain ethical consequences. On the other hand, controversy also surrounds some of the scientific conclusions. Notably, can the range of taxa already described be fitted into a series of credible evolutionary frameworks, compatible with proposed phylogenies? What mechanisms are appropriate for such a seemingly rapid event? These questions are still an active focus of research; in addition, and alongside the main series of Burgess shale campaigns (e.g., Conway Morris, 1998e; Hou et al., 1999), there have been continuing developments along two fronts: trace fossils and the Ediacaran assemblages.

To a first approximation, the diversification of body fossils in the Cambrian is paralleled in the trace fossil record (Bottjer, Hagadorn, and Dornbos, 2000). Particularly obvious features include the development of deep burrow systems, effectively exploiting all three dimensions of the seafloor (e.g., Droser and Li, 2000), and scratch marks that can be attributed with some confidence to the rapid and dramatic diversification of Cambrian arthropods, a group that shows increasingly sophisticated methods of feeding (Budd, 1998). Given that many of the trace fossils were produced by animals with a minimal preservation potential, being either soft-bodied "worms" or having delicate and unmineralized skeletons, this diversification of burrows and trackways would seem to echo the story told by faunas of the Burgess shale type. In contrast, diversity is highly restricted in the Precambrian trace fossil record. Most traces are simple and two-dimensional: there is no effective colonization of the third dimension of the seafloor. Recent restudy of supposedly more complex Ediacaran traces has cast serious doubt on their animal origin (S. Jensen, personal communication). Moreover, considerable doubt still hovers around very ancient structures interpreted as trace fossils, and a cautious investigator would find it difficult to identify any convincing evidence of sediment disturbance by a metazoan earlier than about 570 million years ago (see Martin et al., 2000).

Even though the trace fossil record is congruent with the notion of a Cambrian "explosion," associated with the simple, latest Neoproterozoic traces is a relatively diverse assemblage of Ediacaran organisms, which have been the subject of extensive and rather inconclusive debate. In recent years, Seilacher's daring hypothesis (1989), that these fossils represent a separate excursion in eukaryote multicellularity, has lost ground, although his concept of the Vendobionta may yet apply to some of the more enigmatic taxa, such as *Ernietta* and *Pteridinium,* which are typically built on a body plan of repeated saclike chambers. Along with related taxa, these could be some sort of protistan analogues to the more complex foraminiferans. In general, however, in addressing the problems of understanding these Ediacaran fossils, there has been a more sustained effort in trying to accommodate these admittedly strange-looking fossils into one or another of the schemes of metazoan phylogeny presently under consideration (e.g., Conway Morris, 2000b; Fedonkin and Waggoner, 1997; Dewel, 2000). Even so, ambiguities of interpretation and phylogenetic paradoxes remain. These should not be ignored, nor should alternative outlooks closer to Seilacher's formulation (e.g., Narbonne, 1998; see also Narbonne, Saylor, and Grotzinger, 1997).

What is the way forward? Because there seems to be some degree of congruence with the Cambrian fossil record, the incorporation of molecular data into new, and sometimes unexpected, molecular phylogenies (see Conway Morris, 2000b) has proved fruitful. In the case of all three superclades of triploblasts—deuterostomes, ecdysozoans, and lophotrochozoans—it is now possible to argue that apparently enigmatic fossil animals provide at least a historical glimpse of the actual events. Thus, among the deuterostomes, forms such as *Yunnanozoon* (Shu, Zhang, and Chen,1996) and the very similar *Haikouella* (Chen, Huang, and Li, 1999) may represent animals close to the hemichordate (echinoderm)–chordate divergence. Among the ecdysozoans, the priapulid-lobopodian plexus (see figures 2.1f, g) may be similarly basal, whereas the *Kimberella*-halkieriid clade (see figure 2.1c) may possess comparable importance in the initial stages of lophotrochozoan diversification. Together, the molecular phylogenies and Cambrian fossil record, especially as recovered from faunas of the Burgess shale type, indicate that the Cambrian "explosion" is effectively a triploblastic event (see also Adoutte et al., 1999). The Ediacaran faunas, therefore, are presumed to represent both the roots of the triploblast radiation and as important, the preceding diploblastic diversifications, of which present information is largely dependent upon our knowledge of the cnidarians and ctenophores. It is an exciting prospect—if only the phylogenetic scales would fall from our eyes—to be able to identify intermediates within the diploblasts, as well as plausible connections both upward, toward the triploblasts, and deeper into the metazoan tree, toward the sponges and perhaps still more primitive organisms that ultimately will provide links to other groups of eukaryotes.

Causes and Effects

The Cambrian "explosion," though real, falls into a wider phylogenetic context that probably began a billion years ago, with the eukaryote "Big Bang." The "explosion" itself encompasses not only the dramatic events in triploblast evolution that occurred in the latest

Neoproterozoic and earliest Cambrian, but extends into the Lower Palaeozoic as the various body plans of the marine fauna were assembled (Budd and Jensen, 2000).

In this concluding section, I wish to address three questions: What accounts for the delay in metazoan diversification? Was the trigger for the Cambrian "explosion" extrinsic, such as rising levels of atmospheric oxygen, or intrinsic and perhaps genomic? And how do new ideas, such as light reception by complex eyes, figure into attempts to explain this event?

The normal formulation of Darwinian evolution is that the processes of natural selection depend on an unending supply of variation, which produces a constant and unavoidable selection pressure whereby "any advantage, however slight" is preferred. This view presupposes that (1) the necessary "building blocks" are available; (2) a given structure (e.g., an eye) is in fact adaptively superior to its predecessors; and (3) the structure is in a functional context that makes sense, for example, it is connected to a brain that can interpret the electrical signals in a coherent fashion. For our example of the eye, the first two requirements seem to pose no significant problems. The key building blocks of the eye, notably, proteins in the form of crystallins for the lens and other transparent tissues and opsins for the retina, are phylogenetically quite ancient and, at least in the case of the crystallins, classically exemplify co-option for new functions. In addition, as Nilsson and Pelger (1994) have demonstrated, the transformation of a simple eye spot into a complex camera eye, which has occurred independently at least four times (in vertebrates, cephalopods, alciopid polychaetes, and strombid gastropods), is geologically almost instantaneous ($< 500,000$ years). Knowing this, we might be surprised to learn that metazoans failed to appear by the end of the Archaean. Some proponents of the molecular clock data believe this failure lies in the rock record rather than in our understanding of evolutionary processes. However, if we reject the hypothesis of a very ancient origin for metazoans, then the apparent delay in their emergence might be attributed to some constraint, which in turn might be either extrinsic or intrinsic.

In terms of possible extrinsic triggers, it is clear that there were major environmental changes and crises, although as yet it is far from clear whether any of these can be directly linked to the early evolution of Metazoa. Significant elevation in the levels of atmospheric oxygen during the Neoproterozoic (e.g., Canfield and Teske, 1996) would seem to provide an almost too obvious control for metazoan diversification (Knoll and Holland, 1995); yet establishing a direct correlation between oxygen and evolution is much more difficult. Living metazoans occupying dysaerobic environments, and, by implication, now-extinct taxa from the Phanerozoic, when levels of atmospheric oxygen were significantly depressed (Berner, 2000), might be less diverse and ecologically stressed, but they still manage or managed. Moreover, in a low-oxygen world, one might expect metazoans to adopt strategies such as modifying respiratory proteins or incorporating photosynthetic symbionts to circumvent the most obvious problems of such an existence.

Oxygen is thus not the only candidate for constraining biological diversification; other candidates, perhaps more suited to the deep Precambrian past, are oceans that are either too salty (Knauth, 1998) or hot (Schwartzman, McMenamin, and Volk, 1993). Kirschvink and colleagues (2000) have speculated on the possible biological consequences of a global glaciation some 2 billion years ago. It is certainly intriguing that, closer to the time of major metazoan diversification, there is even firmer evidence for global glaciations. Even so, despite the attention paid to the evidence for at least two Neoproterozoic "snowball Earth" episodes (Hoffman et al., 1998) and their possible evolutionary consequences, the claim that these events were accompanied by biological catastrophe may need to be qualified. In its most extreme version, the oceans are clogged with pack ice, perhaps a kilometer thick, that effectively destroys the photic zone. Surface temperatures tumble to well below 0°C. The proposal for such global refrigeration is, however, inconsistent with some climatic models (e.g., Hyde et al., 2000). As has also been pointed out, the preglacial fossil record indicates the presence of an assemblage of eukaryotes; in even the worst scenario, some must have been able to survive in refugia, such as hot springs oases. This does not rule out a mass extinction, but attempts to link the amelioration of environment after such superglaciations to the metazoan diversifications are difficult to reconcile with a hiatus lasting tens of millions of years.

The mention of mass extinctions will remind the reader of the importance of extraterrestrial impacts, notable as the trigger for the end-Cretaceous extinctions. Such collisions as a factor in driving the early diversification of the metazoans have received sporadic mention in the scientific literature, albeit unsupported by any convincing evidence. Culler and colleagues (2000), in their refinement of the lunar bombardment history, have suggested evidence for an overall decline in impacts from the cataclysmic episode early in the history of the Solar System. They have found evidence, however, of an increase in impacts in the more recent geological past, which they note is "roughly coincident with the 'Cambrian explosion.'" Intriguing as these data are, they are in need of both further refinement and terrestrial evidence, such as shocked quartz, from one or more horizons spanning the interval from approximately 580 to 530 million years ago.

There is, however, another style of extrinsic control, which although equally difficult to test, looks to the ecological milieu. Earlier I noted that any discussion of the earliest metazoans might do well to look at likely ecological analogues, if not avatars, such as the ciliate protistans. In this dynamic, microscopic world, the eventual advantage to the metazoans might have depended on some genomic reorganization, and it is to this topic we need briefly to turn.

In the sponges, we see a wide array of molecules that also play crucial functions in more advanced animals (see Gamulin, Müller, and Müller, 2000). In some ways, this is even more apparent in the cnidarians, notably *Hydra* (and the related *Hydractinia*), which, for experimental convenience, has been the focus of attention. Here we see a somewhat confusing picture. That the cnidarians are diploblastic and radially symmetrical makes correlations with the triploblast Bilateria far from straightforward. Is, for example, the so-called "head" of *Hydra* in any way equivalent to the bilaterian head? The rapidly emerging molecular evidence is somewhat ambiguous. In certain instances, expression patterns are strikingly similar, pointing to a conservation of function, whereas, in others, the gene clearly has one function in *Hydra* but has been subsequently redeployed in the higher animal for another (see, for example, Gauchat et al., 1998; Hassel, 1998; Mokady et al., 1998; Smith et al., 1999; Galliot, 2000; Hobmayer et al., 2000). A hand-waving answer to the dilemma, that is, how to reconcile conservation as against co-option, is to appeal to new networks, gene duplications, and so forth. There are, however, at least two problems with this view. The first, admittedly from very limited information, is that there is evidence even within a particular group, such as the nematodes (e.g., Eizinger and Sommer, 1997), for considerable lability in developmental pathways. Felix (1999, 15) remarks that, "although developmental genes and molecular cascades are well conserved, the developmental context in which they play a role can vary extremely at the cellular level." (See also Steinberg, chapter 9, and Bissell et al., chapter 7.) A similar conclusion is also evident in the frequent switching of larval types in, for example, echinoderms where very similar adults emerge by radically different embryological pathways that, by implication, have substantially altered genetic architecture. Reconciling this genomic lability with the conservation of body plans is one of the major challenges for evolution. But the reverse is also the case, whereby phenotypic diversity emerges from a conserved genomic framework. A striking example comes from the arthropods. Averof (1997) reminds us that the identical complement of Hox genes, which in arthropods underpins their axial reorganization, seems to have no obvious bearing on the widely varying degrees of tagmoses and segment organization. Thus, although genome arrangements and duplications must provide an important basis for metazoan diversifications, the fundamental patterns continue to elude us.

Are time-honored appeals to agencies such as atmospheric oxygen or the evolution of Hox clusters perhaps incidental to the problem? Should we be looking elsewhere? Two proposals certainly merit attention. Stanley (1992) drew attention, albeit briefly, to the possibility that one cell type, the neuron, might have been the trigger for the Cambrian "explosion." Although clearly neurons, and most probably brains, evolved in the late Neoproterozoic, there might still have been a lag time as neural and behavioral systems assembled themselves. This question of the time required for self-assembly of complex biological systems is yet another significant problem in evolution. That protracted, geologically lengthy intervals are necessary for reassembly is evident from a rather different area of evolution. This is the response of Phanerozoic faunas to mass extinction events, where the postcatastrophe assemblages show a remarkably sluggish recovery despite the manifest ecological opportunities (see Kircher and Weil, 2000). As such, this behavior might provide an analogy to the timing and rate of the latest Neoproterozoic–earliest Cambrian radiations: barrels refill, but not instantaneously.

The role of behavioral changes may provide other insights into the Cambrian diversifications. Parker (1998), for example, has identified diffraction gratings in the cuticle of a number of Burgess shale animals, the presence of which implies iridescent color production by such organisms as *Wiwaxia.* Building an elaborate hypothesis on this basis, he argues effectively that the Cambrian "explosion" was driven by optical activity, including display and warning. It can hardly be doubted that optical activity is a hitherto unrecognized factor in the emergence of Cambrian complexity (see also Marcotte, 1999), but it may be simplistic to view it as the principal, let alone sole, mechanism. To start with, metazoans communicate via a variety of other means (e.g., olfactory, vibration). A somewhat deeper question is to inquire: what are the underlying neurological similarities between these different reception mechanisms? And how did they come about in the first place? Thus it may yet transpire that, having moved away from the influential view, at least in some quarters, that the Cambrian "explosion" is only an artifact of our minds, we are now facing the possibility that it would not have happened without a new set of behavioral, and by implication neural, frameworks that ultimately set the mental stage for us to speculate on this remarkable event.

Were Darwin to return, would he feel that his suspicion, articulated in chapter 9 of *On the Origin of Species,* that the seemingly abrupt appearance of skeletons near the beginning of the Cambrian might undermine his notion of evolution proceeding by slow and steady change could now be laid to rest? I believe so. In prospect, we have the emergence of a reasonably coherent phylogeny of early metazoans that supports the view that the fossil record, far from being a disastrously fragmented collage, is instead a jigsaw puzzle sufficiently complete to provide genuine historical insights into how body plans evolve without being functionally compromised. There remains a paradox, however, inasmuch as although molecular biology is to be thanked for largely revitalizing our views of metazoan phylogeny, when it comes to developmental biology, it may transpire that the real evolutionary action is in the realm of functional morphology and ecology, and particularly the realm of neurology and behavioral sophistication. When combined with new insights from epigenesis (Newman and Müller, 2000; see also Gray, 1992), the study of evolution seems poised to break free from a steady and increasingly arid reductionism. And those all-powerful genes? Perhaps their importance is better pursued if we view them as a necessary tool kit, to be used as and when required, than as some sort of master template upon which evolution is meant both to act and unfold (see Budd, 1999). Genes are essential, certainly, but by no means the whole story.

Acknowledgments

I thank Gerd Müller and the staff of the Konrad Lorenz Institute for Evolution and Cognition Research for much encouragement and support, and Sandra Last for typing the manuscript. My research is supported by the Leverhulme Trust, Natural Environment Research Council, and St. John's College, Cambridge.

References

Adoutte A, Balavoine G, Lartillot N, de Rosa R (1999) Animal evolution—the end of the intermediate taxa? Trends Genet 15: 104–108.

Atkins MS, McArthur AG, Teske AP (2000) Ancyromonadida: A new phylogenetic lineage among the Protozoa closely related to the common ancestor of metazoans, fungi and choanoflagellates (Opisthokonta). J Mol Evol 51: 278–285.

Averof M (1997) Arthropod evolution: Same *Hox* genes, different body plans. Curr Biol 7: R634–R636.

Awramik SM, Sprinkle J (1999) Proterozoic stromatolites: The first marine evolutionary biota. Hist Biol 13: 241–253.

Ayala FJ, Rzhetsky A, Ayala FJ (1998) Origin of the metazoan phyla: Molecular clocks confirm paleontological estimates. Proc Natl Acad Sci USA 95: 606–611.

Balavoine G (1997) The early emergence of Platyhelminthes is contradicted by the agreement between 18S rRNA and *Hox* genes data. C R Acad Sci Paris, Sci Vie 320: 83–94.

Baldauf SL (1999) A search for the origins of animals and fungi: Comparing and combining molecular data. Am Nat 154: S178–S188.

Bengtson S, Yue Z (1997) Fossilized metazoan embryos from the earliest Cambrian. Science 277: 1645–1648.

Benton, MJ (1999) Early origins of modern birds and mammals: Molecules vs. morphology. BioEssays 21: 1043–1051.

Berner RA (2000) Isotopic fractionation and atmospheric oxygen: Implications for Phanerozoic O_2 evolution. Science 287: 1630–1633.

Bottjer DJ, Hagadorn JW, Dornbos SQ (2000) The Cambrian substrate revolution. GSA Today 10: 1–7.

Brasier MD, McCarron G, Tucker R, Leather J, Allen P, Shields G (2000) New U-Pb zircon dates for the Neoproterozoic Ghubrah glaciation and for the top of the Huqf Supergroup, Oman. Geology 28: 175–178.

Brasier MD, McIlroy D (1998) *Neonereites uniserialis* from c. 600 M year old rocks in western Scotland and the emergence of animals. J Geol Soc Lond 155: 5–12.

Breyer JA, Busbey AB, Hanson RE, Roy AC (1995) Possible new evidence for the origin of metazoans prior to 1 Ga: Sediment-filled tubes from the Mesoproterozoic Allamoore Formation, Trans-Pecos, Texas. Geology 23: 269–272.

Brocks JJ, Logan GA, Buick R, Summons RE (1999) Archean molecular fossils and the early rise of eukaryotes. Science 285: 1033–1036.

Bromham LD, Hendy MD (2000) Can fast early rates reconcile molecular dates with the Cambrian explosion? Proc R Soc Lond B Biol Sci 267: 1041–1047.

Bromham L, Rambaut A, Fortey R, Cooper A, Penny D (1998) Testing the Cambrian explosion hypothesis by using a molecular dating technique. Proc Natl Acad Sci USA 95: 12386–12389.

Budd GE (1998) Arthropod body-plan evolution in the Cambrian with an example from anomalocaridid muscle. Lethaia 31: 197–201.

Budd GE (1999) Does evolution in body patterning genes drive morphological change—or vice versa? BioEssays 21: 326–333.

Budd GE, Jensen S (2000) A critical appraisal of the fossil record of the bilaterian phyla. Biol Rev 75: 253–295.

Butterfield NJ (2000) *Bangiomorpha pubescens,* n. gen., n. sp.: Implications for the evolution of sex, multicellularity and the Mesoproterozoic/Neoproterozoic radiation of eukaryotes. Paleobiology 26: 386–404.

Butterfield NJ, Knoll AH, Swett K (1990) A bangiophyte red alga from the Proterozoic of Arctic Canada. Science 250: 104–107.

Canfield DE, Teske A (1996) Late Proterozoic rise in atmospheric oxygen concentration inferred from phylogenetic and sulphur-isotope studies. Nature 382: 127–132.

Cao R-j (1999) Evidence for possible metazoan activity in Neoproterozoic aberrant stromatolites and thrombolite structures. Acta Palaeont Sinica 38: 291–304.

Cavalier-Smith T (1998) A revised six-kingdom system of life. Biol Rev 73: 203–266.

Celerin M, Ray JM, Schisler NJ, Day AW, Stetter-Stevenson WG, Laudenbach DE (1996) Fungal fimbriae are composed of collagen. EMBO J 15: 4445–4453.

Chen J-y, Huang D-y, Li C-w (1999) An early Cambrian craniate-like chordate. Nature 402: 518–522.

Conway Morris S (1998a) The question of metazoan monophyly and the fossil record. Prog Mol Subcell Biol 21: 1–19.

Conway Morris S (1998b) Metazoan phylogenies: Falling into place or falling to pieces? A palaeontological perspective. Curr Opin Genet Dev 8: 662–667.

Conway Morris S (1998c) Eggs and embryos from the Cambrian. BioEssays 20: 676–682.

Conway Morris S (1998d) Early metazoan evolution: Reconciling paleontology and molecular biology. Am Zool 38: 867–877.

Conway Morris S (1998e) The Crucible of Creation: The Burgess Shale and the Rise of Animals. Oxford: Oxford University Press.

Conway Morris S (1999) I fossili Ediacarani del Vendiano d'Europa. In: Alle Radici della Storia Naturale d'Europea (Pinna G, ed). Milan: Jaca.

Conway Morris S (2000a) The Cambrian explosion: Slow-fuse or megatonnage? Proc Natl Acad Sci USA 97: 4426–4429.

Conway Morris S (2000b) Evolution: Bringing molecules into the fold. Cell 100: 1–11.

Conway Morris S (2000c) Nipping the Cambrian "explosion" in the bud? BioEssays 22: 1053–1056.

Culler TS, Becker TA, Muller RA, Renne PR (2000) Lunar impact history from $^{40}Ar/^{39}Ar$ dating of glass spherules. Science 287: 1785–1788.

Cutler DJ (2000) Estimating divergence times in the presence of an overdispersed molecular clock. Mol Biol Evol 17: 1647–1660.

Dewel RA (2000) Colonial origin for Eumetazoa: Major morphological transitions and the origin of bilateral complexity. J Morphol 243: 35–74.

Droser ML, Li X (2000) The Cambrian radiation and the diversification of sedimentary fabrics. In: The Ecology of the Cambrian Radiation (Zhuravlev A Yu, Riding R, eds), 137–169. New York: Columbia University Press.

Eizinger A, Sommer RJ (1997) The homeotic gene *lin-39* and the evolution of nematode epidermal cell fates. Science 278: 452–455.

Fedonkin MA, Waggoner BM (1997) The late Precambrian fossil *Kimberella* is a mollusc-like bilaterian organism. Nature 388: 864–871.

Felix M-A (1999) Evolution of developmental mechanisms in nematodes. J Exp Biol (Mol Dev Evol) 285: 3–18.

Fenchel T, Finlay BJ (1986) The structure and function of Müller vesicles in loxodid ciliates. J Protozool 33: 69–76.

Foote M, Hunter JP, Janis CM, Sepkoski JJ (1999) Evolutionary and preservational constraints on origins of biologic groups: Divergence times of eutherian mammals. Science 283: 1310–1314.

Forterre P (1997) Archaea: What can we learn from their sequences? Curr Opin Genet Dev 7: 764–770.

Fortey RA, Briggs DEG, Wills MA (1997) The Cambrian evolutionary "explosion" recalibrated. BioEssays 19: 429–434.

Foster KW, Smyth RD (1980) Light antennas in phototactic algae. Microbiol Rev 44: 572–630.

Galliot B (2000) Conserved and divergent genes in apex and axis development of cnidarians. Curr Opin Genet Dev 10: 629–637.

Gamulin V, Müller IM, Müller WEG (2000) Sponge proteins are more similar to those of *Homo sapiens* than to *Caenorhabditis elegans*. Biol J Linn Soc 71: 821–828.

Gauchat D, Kreger S, Holstein T, Galliot B (1998) *prdl-a,* a gene marker for hydra apical differentiation related to triploblastic *paired*-like head-specific genes. Development 125: 1637–1645.

Giere O (1993) Meiobenthology: The Microscopic Fauna in Aquatic Sediments. Berlin: Springer.

Goodwin B (1994) How the Leopard Changed Its Spots: The Evolution of Complexity. New York: Simon and Schuster.

Gray R (1992) Death of the gene: Developmental systems strike back. In: Trees of Life: Essays in Philosophy of Biology (Griffiths P, ed), 165–209. Dordrecht, Kluwer.

Grotzinger JP (1990) Geochemical model for Proterozoic stromatolite decline. Am J Sci 290A: 80–103.

Han TM, Runnegar B (1992) Megascopic eukaryotic algae from the 2.1 billion-year-old Negaunee Iron-Formation, Michigan. Science 257: 232–235.

Hassel M (1998) Upregulation of a *Hydra vulgaris* CPKC gene is tightly coupled to the differentiation of head structures. Dev Genes Evol 207: 489–501.

Hobmayer B, Rentzsch F, Kuhn K, Happel CM, von Laue CC, Snyder P, Rothbächer U, Holstein TW (2000) WNT signalling molecules act in axis formation in diploblastic metazoan *Hydra*. Nature 407: 186–189.

Hoffman PF, Kaufman AJ, Halverson GP, Schrag DP (1998) A Neoproterozoic snowball Earth. Science 281: 1342–1346

Holland PWH (1997) Vertebrate evolution: Something fishy about Hox genes. Curr Biol 7: R570–R572.

Hou X-g, Bergström J, Wang H-f, Feng X-h, Chen A-l (1999) The Chengjiang Fauna. Yunnan: Science and Technology Press.

Huelsenbeck JP, Larget B, Swofford D (2000) A compound Poisson process for relaxing the molecular clock. Genetics 154: 1879–1892.

Hughes NC (2000) The rocky road to Mendel's play. Evol Dev 2: 63–66.

Hyde WT, Crowley TJ, Baum SK, Peltier WR (2000) Neoproterozoic "Snowball Earth" simulations with a coupled climate/ice-sheet model. Nature 405: 425–429.

Israelsson O (1999) New light on the enigmatic *Xenoturbella* (phylum uncertain): Ontogeny and phylogeny. Proc R Soc Lond B Biol Sci 266: 835–841.

Jägersten G (1972) Evolution of the Metazoan Life Cycle. New York: Academic Press.

Kempf O, Kellerhals P, Lowrie W, Mattes A (2000) Paleomagnetic directions in late Precambrian glaciomarine sediments of the Mirlat Sandstone Formation, Oman. Earth Planet Sci Lett 175: 181–190.

Kerr RA (2000) An appealing Snowball Earth that's still hard to swallow. Science 287: 1734–1736.

Kircher JW, Weil A (2000) Delayed biological recovery from extinctions throughout the fossil record. Nature 404: 177–180.

Kirschvink JL, Gaidos EJ, Bertani LE, Beukes NJ, Gutzmer J, Maepa LN, Steinberger RF (2000) Paleoproterozoic Snowball Earth: Extreme climatic and geochemical global change and its biological consequences. Proc Natl Acad Sci USA 97: 1400–1405.

Knauth LP (1998) Salinity history of the Earth's early ocean. Nature 395: 554–555.

Knoll AH, Holland HD (1995) Oxygen and Proterozoic evolution: An update. In: Biological Responses to Past Environmental Changes, 21–33. Washington, D.C.: National Academy Press.

Kobayashi M, Furuya H, Holland PWH (1999) Dicyemids are higher animals. Nature 401: 762.

Kouchinsky A, Bengtson S, Gershwin L-A (1999) Cnidarian-like embryos associated with the first shelly fossils in Siberia. Geology 27: 609–612.

Lundin L-G (1999) Gene duplications in early metazoan evolution. Cell Dev Biol 10: 523–530.

Marcotte BM (1999) Turbidity, arthropods and the evolution of perception: Toward a new paradigm of marine Phanerozoic diversity. Mar Ecol Prog Ser 191: 267–288.

Martin MW, Grazhdankin DV, Bowring SA, Evans DAD, Fedonkin MA, Kirschvink JL (2000) Age of Neoproterozoic bilaterian body and trace fossils, White Sea, Russia: Implications for metazoan evolution. Science 288: 841–845.

Mokady O, Dick MH, Lackschewitz D, Schierwater B, Buss LW (1998) Over one-half billion years of head conservation? Expression of an *ems* class gene in *Hydractinia symbiolongicarpus* (Cnidaria: Hydrozoa). Proc Natl Acad Sci USA 95: 3673–3678.

Müller KJ, Walossek D, Zakharov A (1995) "Orsten" type phosphatized soft-integument preservations and a new record from the middle Cambrian Kuonamka Formation in Siberia. N Jb Geol Paläont Abh 197: 101–118.

Narbonne GM (1998) The Ediacaran biota: A terminal Neoproterozoic experiment in the evolution of life. Geol Soc Am Today 8: 1–6.

Narbonne GM, Aitken JD (1995) Neoproterozoic of the Mackenzie Mountains, northwestern Canada. Precambrian Res 73: 101–121.

Narbonne GM, Saylor BZ, Grotzinger JP (1997) The youngest Ediacaran fossils from southern Africa. J Paleont 71: 953–967.

Newman SA, Müller GB (2000) Epigenetic mechanisms of character origination. J Exp Zool (Mol Dev Evol) 288: 304–317.

Nilsson D-E, Pelger S (1994) A pessimistic estimate of the time required for an eye to evolve. Proc R Soc Lond B Biol Sci 256: 53–58.

Norman JE, Ashley MV (2000) Phylogenetics of Perissodactyla and tests of the molecular clock. J Mol Evol 50: 11–21.

Ono K, Suga H, Iwabe N, Kuma K-I, Miyata T (1999) Multiple protein tyrosine phosphatases in sponges and explosive gene duplication in the early evolution of animals before the parazoan-eumetazoan split. J Mol Evol 48: 654–662.

Parker AR (1998) Colour in Burgess Shale animals and the effect of light on evolution in the Cambrian. Proc R Soc Lond B Biol Sci 265: 967–972.

Peterson KJ, Cameron RA, Davidson EH (1997) Set-aside cells in maximal indirect development: Evolutionary and developmental significance. BioEssays 19: 623–631.

Peterson KJ, Cameron RA, Davidson EH (2000) Bilaterian origins: Significance of new experimental observations. Dev Biol 219: 1–17.

Porter SM, Knoll AH (2000) Testate amoebae in the Neoproterozoic era: Evidence from vase-shaped microfossils in the Chuar Group, Grand Canyon. Paleobiology 26: 360–385.

Prave AR (1999) The Neoproterozoic Dalradian Supergroup of Scotland: An alternative hypothesis. Geol Mag 136: 609–617.

Schwartzman D, McMenamin M, Volk T (1993) Did surface temperatures constrain microbial evolution? BioScience 43: 390–393.

Seilacher A (1989) Vendozoa: Organismic construction in the Proterozoic biosphere. Lethaia 22: 229–239.

Seilacher A, Bose PK, Pflüger F (1998) Triploblastic animals more than 1 billion years ago: Trace fossil evidence from India. Science 282: 80–83.

Shapiro JA, Dworkin M (eds) (1997) Bacteria as Multicellular Organisms. New York: Oxford University Press.

Shenk MA, Steele RE (1993) A molecular snapshot of the metazoan "Eve." Trends Biochem 18: 459–463.

Shimeld SM (1999) Gene function, gene networks and the fate of duplicated genes. Cell Dev Biol 10: 549–553.

Shu D-g, Luo H-l, Conway Morris S, Zhang X-l, Hu S-x, Chen L, Han J. Zhu M, Li Y, Chen L-z (2000) Lower Cambrian vertebrates from South China. Nature 402: 42–46.

Shu D, Zhang X, Chen L (1996) Reinterpretation of *Yunnanozoon* as the earliest known hemichordate. Nature 380: 428–430.

Siddall ME, Martin DS, Bridge D, Desser SS, Cone DK (1995) The demise of a phylum of protists: Phylogeny of Myxozoa and other parasitic Cnidaria. J Parasitol 81: 961–967.

Smith KM, Gee L, Blitz IL, Bode HR (1999) *CnOtx*, a member of the *Otx* family, has a role in cell movement in *Hydra*. Dev Biol 212: 392–404.

Sogin ML (1994) The origin of eukaryotes and evolution into major kingdoms. In: Early Life on Earth (Bengtson S, ed). Nobel Symposium, vol 84, pp 181–192. New York: Columbia University Press.

Stanley SM (1992) Can neurons explain the Cambrian explosion? Geol Soc Amer Abstr 24: A45.

Turner DN (1995) Rotifer look-alikes: Two species of *Colurella* are ciliated protozoans. Invert Biol 114: 202–204.

van Tuinen M, Sibley, CG, Hedges SB (2000) The early history of modern birds inferred from DNA sequences of nuclear and mitochondrial ribosomal genes. Mol Biol Evol 17: 451–457.

Vidal G (1990) Giant acanthomorph acritarchs from the Upper Proterozoic in southern Norway. Palaeontology 33: 287–298.

Waggoner B (1998) Interpreting the earliest metazoan fossils: What can we learn? Am Zool 38: 975–982.

Waggoner B (1999) Biogeographic analyses of the Ediacaran biota: A conflict with paleotectonic reconstructions. Paleobiology 25: 440–458.

Walter MR, Heys GR (1985) Links between the rise of the Metazoa and the decline of stromatolites. Precambrian Res 29: 149–174.

Wolpert L (1999) From egg to adult to larva. Evol Dev 1: 3–4.

Wray GA, Levinton JS, Shapiro LH (1996) Molecular evidence for deep Pre-Cambrian divergences among metazoan phyla. Science 274: 568–573.

Xiao S-h, Knoll AH (1999a) Fossil preservation in the Neoproterozoic Doushantuo phosphorite Lagerstätte, South China. Lethaia 32: 219–240.

Xiao S-h, Knoll AH (1999b) Embryos or algae? A reply. Acta Micropaleont Sinica 16: 313–323.

Xiao S-h, Knoll AH (2000) Phosphatized animal embryos from the Neoproterozoic Doushantuo Formation at Weng'an, Guizhou, South China. J Paleont 74: 767–788.

Xiao S-h, Yuan X-l, Knoll AH (2000) Eumetazoan fossils in terminal Proterozoic phosphorites? Proc Natl Acad Sci USA 97: 13684–13689.

Xiao S-h, Zhang Y, Knoll AH (1998) Three-dimensional preservation of algae and animal embryos in a Neoproterozoic phosphorite. Nature 391: 553–558.

Xue Y-s, Zhou C-m, Tang T-j (1999) "Animal embryos," a misinterpretation of Neoproterozoic microfossils. Acta Micropalaeont Sinica 16: 1–4.

Yue Z, Bengtson S (1999) Embryonic and post-embryonic development of the Early Cambrian cnidarian *Olivooides*. Lethaia 32: 181–195.

3 Convergence and Homoplasy in the Evolution of Organismal Form

Pat Willmer

Convergent evolution is prevalent at all levels of organismal design—from cell chemistry and microstructure to cell types, organ systems, and whole body plans (Willmer, 1990; Sanderson and Hufford, 1996). Indeed, it may be sufficiently common to undermine methodologies (whether morphological, paleontological, or molecular) for determining animal relationships (Willmer, 1990; Willmer and Holland, 1991; Moore and Willmer, 1997). Yet, even though detecting convergence depends on knowing your taxonomy, methods of establishing taxonomy, particularly the now almost ubiquitous cladistic methods, tend to rely on a parsimonious assumption of minimum convergence. Or, as Foley (1993, 197) put it: "The best phylogeny is essentially the one that has the least convergence. And yet if cladistics is itself showing that convergence is rife in the real world of evolution, then the very assumptions of cladistics are open to question."

Convergence and its prevalence also have a major interaction with the subject of this volume, the development of organismal form. Within a phylum, convergence is particularly likely when animals with different ancestry have been selected for survival in similar and especially demanding environments; thus it might be rare in some circumstances but common in others (see also Wake, 1991). But, at the "higher" level of animal design, between phyla, we know very little about ancestral states; instead, we see a diversity of overlapping sets of characters, which cannot map cleanly onto any single classification. This may lead us to conclude that such characters have evolved repeatedly (Willmer, 1990) and thus that convergent evolution has been very widespread, extending to whole body plans, or perhaps even the whole of the Metazoa.

Two major developments in the last decade may have affected our view of convergent evolution. First, molecular evidence, almost entirely accruing since 1987, suggests we are reaching a consensus on monophyly and a rather traditional-looking tree or cladogram for the Metazoa (e.g., Lake, 1990; Christen et al., 1991; Adoutte and Philippe, 1993). Second, accumulating developmental evidence of similar genes across the animal kingdom controlling shape, form, patterns, and the location of key features also supports monophyly (e.g., Fortey, Briggs, and Wills, 1996; Erwin, Valentine, and Jablonski, 1997; Conway Morris, 1998; Knoll and Carroll, 1999). This chapter addresses these new developments, and their implications for the definition and detection of both convergent evolution and the origins of organismal form.

Definitions

Classical definitions of convergence or homoplasy come from traditional morphology and embryology, and lead many to conclude that the phenomenon is widespread (e.g., Cain,

1982). But defining convergent evolution is a matter of controversy, intimately linked with defining homology (e.g., Doolittle, 1994; Abouheif et al., 1997), the hierarchical basis of comparative biology (Coddington, 1994; Hall, 1994) and the central ordering principle of biological characters. Classical homology was based on comparative morphology and only later came to be applied to evolutionary origins, usually as seen through ontogeny. Different definitions still tend to emphasize either common descent or individual development. Thus Mayr (1969; see also Mayr, 1994) stipulated that characters are homologous in two or more organisms if they can be traced back to the same character in a common ancestor; and cladistic analysis equates homology with a "shared derived character" (synapomorphy; e.g., Patterson, 1982). McKitrick (1994) suggested that synapomorphies are really just hypotheses of homology, and therefore that ontogenetic studies are the most reliable studies for revealing homology. Van Valen (1982) usefully defined homology as "resemblance caused by a continuity of information." This informational continuity may be developmental or historical (Minelli, 1993); some would argue that, far from being necessarily continuous back to the origin of a particular morphology, it may have been added and canalized ("routinized") at a later, postepigenetic stage (Newman and Müller, 2000).

Thus our definition of homology bears directly on our definition of convergence. Elsewhere (Moore and Willmer, 1997), I have argued that convergence needs to be distinguished from parallelism. Mayr (1969) has defined parallel evolution as the development of similar characters separately in two or more lineages sharing common ancestry. Although these characters do not appear in the ancestor, descendants have inherited the potential to express them, and lineages can then change in similar ways when faced with similar problems. In what may properly be called "parallelism," the separately evolved descendants are as similar to each other as were their ancestors. Convergent evolution, by contrast, occurs when distantly related animals evolve separately, yet produce similarity: the descendants are *more* alike than were their ancestors. This distinction is only relative, however; on a sufficiently long timescale, all animals share a common ancestor, and describing that ancestor for any two species as "ancient" or "recent" is arbitrary. Hence the term *convergent evolution* can be used inclusively. The alternative and currently preferred term *homoplasy* implies the independent acquisition of similar attributes in distinct lineages; it includes not only parallelism and convergence but also apparent reversals (see Sanderson and Hufford, 1996; Hodin, 2000). From its use in cladistic analyses, however, where it is not an empirical observation but a post hoc conclusion derived from a specific analysis (e.g., Sanderson and Donoghue, 1989; Friday, 1994), it has acquired other connotations. For present purposes, therefore, *convergence* is the more useful—and less "loaded"—term.

Convergence needs to concentrate on detailed resemblances that might be mistaken for homology, which leads back to the suggestion of different kinds or levels of convergence (Wake, 1991; McShea, 1996). Convergence between animals within relatively low-order taxa experiencing similar selective pressures, though pertinent to the simple prevalence and methodology debate, is perhaps less interesting from the theoretical perspective, having been added on quite late in the course of evolution. Here I want to concentrate more on the possibility of early and extensive convergence of whole body plans, at the phyletic level or higher, where we are inevitably dealing with altogether fundamental characteristics.

Convergence and Developmental Processes

There are a number of reasons why animals may show similarity in very broad fundamental characteristics, but many of them could be grouped in the category of "developmental constraints" (see Maynard Smith et al., 1985; Hall, 1992; Raff, 1996). Developmental mechanisms are part of the legacy of evolution, as many chapters in this volume discuss, and operate by making it difficult for certain kinds of phenotypes (or easy for certain other kinds) to be produced. This constrains the relationships between particular epigenetic events, and in turn between particular kinds of cells and tissues. Organisms thus consist, not of independently selected characters, but rather of interacting developmental pathways. Until recently, it has been presumed that earlier developmental stages are more conservative for this reason (von Baer's rule), with only the later events readily changeable through evolution. However, we are now more aware that embryological characters themselves are extremely flexible, sometimes even arbitrarily so, and that exceptions to von Baer's rule are rather frequent (e.g., Thomson, 1988; Horder, 1994). It is evident that there may be only a limited number of "standard parts" for animals in Riedl's sense (1978) and that there are only a few ways of moving cell sheets and cell masses around to shape an early embryo into a three-layered and three-dimensional animal (Løvtrup, 1974; Wolpert, 1994; Steinberg, chapter 9, this volume).

The current emphasis on the interface of development, evolution, and genetics provides a new appreciation of the nature of some "developmentally constrained" pathways of morphogenesis. In the last fifteen years, analyses beginning with the homeotic genes have seemed to reveal a remarkable conservatism on the part of mechanisms that control developmental processes (and perhaps positional information) available to dividing cells in the embryo, with an apparent conservation of function in genes from worms to fruit flies to mammals. A small number of molecules, controlling mostly cell-cell interactions, giving few possible outcomes, and, conserved across large parts of the animal kingdom, have come to be regarded as key controllers in morphogenesis.

But models of how such processes could work are also appearing (e.g., Wolpert, 1990, 1994; Erwin, 1993; Newman, 1993, 1994). One of the most intriguing possibilities, highly relevant to the development of organismal form, is that morphological evolution may be initially generated by slight variations in intrinsic physical properties of cells and cell aggregations; intense selection may then act to favor biochemical (and heritable) fixation of just a few viable morphologies (Newman, 1994; Newman and Müller, 2000). Even at the very origin of multicellularity, it is not hard to envisage simple physical forces (especially differential adhesivity) producing similar morphologies that may therefore evolve repeatedly. These inevitable physical forces and self-organizing patterning systems may operate even *without* selective forces pushing morphologies to the same end. This suggests various causal links between ontogeny and phylogeny, and could account for a whole range of fundamental (or at least "classical") taxonomic features including gastrulation, tissue layering, cavity formation, and segmentation (see Moore and Willmer, 1997; Newman and Müller, 2000). Gastrulation, which occurs by several quite distinct mechanisms such as epiboly, involution, and delamination, all leading to the same physical outcome, is a particularly good example of the key issues, although gastrulation-related features such as blastopore fate have persisted as key features on which superphyletic groupings (Protostomia and Deuterostomia) are grounded. Løvtrup (1974) specifically linked gastrulation type to the relative positions of cell sheets, and their tendency to move in two dimensions over extracellular matrices. More recently, gastrulation mechanisms have been associated with the inherent tendency of cell sheets to take up equilibrium shapes (Newman, 1994; chapter 13, this volume; Steinberg, chapter 9, this volume). Only the end result matters, as a key means of establishing layering in an embryo; but gastrulation becomes a convergent phenomenon and a poor taxonomic character.

Such findings have forced us to reconsider the old idea that earlier stages in development are highly conservative, indeed, to reexamine the very concept of homology (which some would say should now be sought at a developmental level); they should also lead us to reanalyze fundamental body plans (see Wagner and Misof, 1993; Minelli and Schram, 1994; McKitrick, 1994; Knoll and Carroll, 1999). Perhaps most important, they agree with predictions, not only that major innovations of design occurred very early on in phylogeny, but also that they are *likely* to have arisen more than once, whether by the same or somewhat different routes. Inevitable pattern-forming systems are fully consistent with, but in key senses separate from, classical gene-dependent biological processes using a few controller genes; the two may commonly come together in what Newman calls "generic-genetic coupling" to provide a "belt and suspenders" organization with a greater security of final functional outcome. This again both underlines and, more important, helps to *predict* and *explain* the ubiquity of convergence. Genetic changes that are stabilizing in their interaction with physically induced change in morphology are most likely to be selected for. In this way, convergence becomes pervasive precisely because so much of the simple morphology of organisms is constrained by basic physical laws and processes.

As we now know, there are "developmental networks" that are used differently in a wide range of tissues even within the same organism, which suggests that such networks are modular, and that modules can be snapped in or out and can be redeployed when not needed for a particular task, with widespread redundancy of genetic pathways in developmental processes (Tautz, 1992). Common developmental modules that occur in the genome of different organisms, and in common ancestors, do not necessarily indicate shared morphogenetic processes or true homology. Although similarities in function may be "merely" parallelisms (an issue expertly dissected by Hodin, 2000), it is also evident that *different* developmental pathways can lead to the same adult structures. Wagner and Misof (1993) suggested that homologies, and adult morphology in general, must depend in part on postdevelopmental stabilizing patterns, which could again be determined by a small suite of possible cell interactions; thus a definition of homology (and thus of convergence) based solely on development remains difficult (see Müller, chapter 4, this volume).

Convergence and Developmental Genes

That convergent evolution is rife may appear to conflict with the now widespread concept of a "genetic tool kit," according to which animals are assembled by a set of fixed and extremely influential genes, for example, those controlling the dorsoventral axis, segmental (serially repeating) patterning, and so on. It is common to see lists of the homologous genes in vertebrates and in *Drosophila*. Indeed, the Hox gene clusters may seem to provide new homologies for the Metazoa as a whole (Slack, Holland, and Graham, 1993; Minelli and Schram, 1994) or, in the view of some authors, to be almost the defining character of "animalness." On the other hand, there is now a move to restrict the Hox terminology not simply to genes with sequence similarity (which are legion) but to genes clustered in the right fashion (Holland, 1992); the arrangement of the Hox gene clusters in the genome might then give clues to phylogenetic patterns within the metazoans (e.g., Garcia-Fernandez and Holland, 1994). This raises interesting problems if some "lower" invertebrates have only one Hox cluster, as seems to be the case for cnidarians and ctenophores. Certainly, we would need to know about Hox genes in sponges, mesozoans, and different groups of protozoans before we could assert that Hox clusters really defined an animal. Nevertheless, the idea of "genes for animalness" has been developed, with additional "fundamental designs" such as "Urbilateria" and subdivisions such as "Eutrochozoa" and "Ecdysozoa" emerging from a combination of developmental and molecular analyses (Aguinaldo et al., 1997; Knoll and Carroll, 1999).

This becomes even more worrisome when we are presented with apparently homologous "genes for eyes," "genes for legs," "genes for tails," or even "genes for hearts." Does this undermine the case for convergence? Certainly, arthropods and chordates, once seen as widely separated in the animal kingdom, are now seen by some as sharing multiple homologous features. Arthropods themselves, which many had thought made a good case for multiple convergent origins (e.g., Manton, 1977; Anderson, 1979; Willmer, 1990; Fryer, 1997), are rarely now seen as convergently similar (e.g., Wheeler, Cartwright, and Hayashi, 1993). Rather, the insects, myriapods, crustaceans, and chelicerates are said to share common genes that produce segmentation, limbs, antennae, and so on. By the mid-1990s, we were entirely redefining the meaning of homology and convergence, such that homology became a property of genes, and much more prevalent, whereas convergence became much rarer and seemed to lose ground in importance.

But the apparent clarity of that position from developmental studies is becoming clouded again; instead of common genes for common structures in fruit flies, worms, and mammals, we have arrived at genes that influence or promote the development of a particular "type" of structure, often at a particular spatial site in the embryo—controller genes that are "in the right place at the right time." Take the structure of eyes, much cited as a classical case of multiple convergent origins (Salvini-Plawen and Mayr, 1977), with the clearly convergent similarities between tetrapod and cephalopod eyes being most often mentioned. Eye development has been subject to renewed scrutiny; researchers have found a similar "master switch" gene (the *Pax-6* gene and its "homologues") in vertebrates, *Drosophila*, squid, and even flatworms (Quiring et al., 1994; Gehring, 1996). Although this could imply a common starting point for all eyes, it is more likely an example of the universality of positional and pattern-forming determination systems in animals. Eye evolution may occur in a predetermined spatial (anterior) hot spot, but the eye that evolves still represents a profoundly convergent phenomenon, often remarkably similar in basic structure while varying in neural connection mechanisms, retinal configuration, lens proteins, focusing systems, and most other fine details. Note also that while *Pax-6* in vertebrates is homologous to the *Drosophila* gene *eyeless,* other genes related to eye formation in vertebrates match bizarrely with genes involved in appendage formation and with muscle formation in fruit flies (see Manak and Scott, 1994; Raff, 1996); and that *Pax-6* also regulates the unrelated phenomenon of nasal placode formation in vertebrates.

Detailed Examples of Convergence in an "Evo-Devo" Context

Segmentation

It used to seem evident from virtually all the traditional perspectives that the character "segmentation" had to have arisen at least twice because it occurs in the protostome annelid-arthropod grouping and again in the very distant deuterostome chordates, but not in any of their possible common ancestors (see Willmer, 1990; De Robertis, 1997). Yet this serial repetition of body structure, and especially of mesodermal tissues and organs, can be quite precisely matched in details in these two groupings. For a while, it seemed that this might be attributed not to convergence after all but to possession of a genetic instruction package, the homeobox sequence, common to all essentially segmented animals (e.g., Holland, 1992). Continuing this trend, Holland and colleagues (1997) suggested that because the gene *engrailed* was expressed in both *Drosophila* and chordate metameres, segmentation must have been present in their common ancestor ("Urbilateria") at least 500 million years ago; homologies of *hairy* and *her-1* genes led Kimmel (1996) to a similar conclusion.

This now seems an overinterpretation. Although homeobox proteins function as transcription factors for other genes, the genes they regulate are often quite unrelated to segmentation. Furthermore, this same Hox gene sequence appears in a far greater range of animals, including unsegmented nematodes and echinoderms (see Slack, Holland, and Graham, 1993), than formerly thought. Perhaps all the Hox genes tell us in our present context is that there is an altogether fundamental inheritance of the ability to encode relative position along the axis during development. Thus what we see manifested as segmentation can easily be—and is likely to be—a convergent feature (or a parallelism in Hodin's sense). That common patterns of axial organization and development exist and may in some sense underlie segmentation also explains the varying degrees of less precise serial repetition of structures (especially obvious in some of the pseudocoelomate groups). Or, as Newman (1993) has suggested, "segmentation" is almost a generic property of metazoan organization, an inevitable consequence of cellular molecules' autoregulatory tendencies. "Segmental" repetition of structure becomes a built-in developmental potential of all animals (see also Newman and Müller, 2000), some of which have co-opted Hox genes as developmental executors. Furthermore, the boundaries between groups of segments can then become "hot spots" for expression of new groups of structural genes, leading to a cascade of additional and more complex kinds of serial repetition. But actual morphological manifestation of the character "segmentation" has clearly evolved independently several times from within this framework of developmental possibilities. It is noteworthy that in some phyla the Hox genes appear to have no relation to axial patterning but have been co-opted for other functions.

In the last few years, gene knockout studies have shown that many genes are related to segment formation (or perhaps to periodicity) from Hox onward, acting in cascade or network fashion and still best understood in fruit flies (Lawrence, 1992). But we should keep in mind that segmentation is essentially a rational form of morphology in a body that is motile and thus logically elongate. From an epigenetic perspective, segments are no more "fundamental" than cells or groups of cells. Certainly in many situations, segmentation has a huge functional value and is readily (thus perhaps repeatedly) selected for (see Willmer, 1990); outside of these situations, however, it is very readily reduced or lost.

Note also that segmentation patterns differ fundamentally in different groups. In some insects (long-band species) segments appear synchronously, whereas in other insects, in annelids, and in vertebrates they normally appear in anteroposterior sequence. In insects, segmentation is chiefly ectodermal; in vertebrates, chiefly mesodermal. Primary segmentation may be retained or may be highly modified, with some authors holding that segments have more effect on complexity when highly modified (often reduced) and collected into tagmata, as in arthropods, than when homonomous, as in annelids. Minelli (2000) has argued that there are two "kinds" of segmentation and that the failure to recognize this has clouded the developmental story still further.

Appendages

Where "true" limbs occur, in tetrapods and in arthropods, their changing morphology appears to be linked with changes in specific pattern formation genes, from which some have inferred that the systems were established in a common ancestor (Shubin, Tabin, and Carroll, 1997). In arthropods, not only the Hox genes but also the *distal-less* gene are always involved in leg production, once to make a uniramous insect leg and twice to make a biramous crustacean one; the same *distal-less* gene is also operating in annelids and in starfish and vertebrates. But, here again, the similarity of genes linked with chick limb and insect limb (or wing) formation, often collectively termed *fringe gene,* may lie in processes rather than in real homology. The limb structures cannot themselves be homologous in any meaningful sense, not least because known phylogenetic intermediates (on virtually *any* proposed phylogeny) do not possess such structures. A fringe protein is a boundary-determining factor, triggering other genes that cause growth and proliferation, giving an outgrowth at a specific point. Because the same genes appear to operate in eye and gill proliferation, in branchial arches, and in the notochord (Raff, 1996), they cannot be considered reliable homology markers. It is more logical to suppose that genes operating in one way on one set of tissues have been co-opted serendipitously to operate in other ways elsewhere in the body. In most of these cases, it is feasible that "appendage formation" first extends down to variation of cell surface adhesivity in groups of cells that are then subject to particular physical effects (see again Newman and Müller (2000); Steinberg, chapter 9, this volume).

Because a gene operates to facilitate an outgrowth that becomes a limb in several different taxa of animals thus does not mean that legs evolved only once, much less that all legs are homologous. Rather, it indicates that genes and gene cascades/networks have been recruited convergently as organizers of limb development.

Lophophores

The multiple occurrence of structures collectively called "lophophores" (tentaculate coelomate feeding structures) is perhaps the clearest example of the "power" of convergence for simple selective functional reasons; it has the charm of being a powerful convergence whichever way one chooses to read the phylogeny. The phylum Bryozoa once included both entoprocts and ectoprocts, but now normally only refers to the latter, a group of coelomate lophophore-bearing animals sometimes included along with phoronids and brachiopods within the even larger phylum "Lophophorata." Bryozoans share a number of rather clear characters (both developmental and adult) with these latter two phyla, including detailed similarities of ciliation patterns and current flow over the lophophore itself. That they used to "belong" next to the entoprocts (now commonly regarded as a pseudocoelomate group, possibly akin to rotifers) attests to the marked similarity of form of these two groups, such that at least one modern authority (Nielsen, 1977, 1995) still regards entoprocts and ectoprocts as sister taxa. If ectoprocts belong with phoronids and brachiopods, then their clear similarity to entoprocts is convergent; if they actually belong with the entoprocts, then the detailed shared features of the lophophore itself are convergent with the other two "proper" lophophorate groups.

The position of the lophophorate phyla in relation to the whole animal kingdom, based on new molecular analyses, has added new intrigue. Although brachiopods may have had one or several origins (Valentine, 1975; Wright, 1979; but see also Rowell, 1982), in virtually all recent schemes based on morphology, they have been convincingly placed among the lophophorate phyla and close to the deuterostomes (see Schaeffer, 1987). Molecular sequence analysis, for its part, has consistently placed the brachiopods alongside or even right in the middle of the traditional protostomes, close to annelids. Recently, the ribosomal DNA (rDNAs) of bryozoans and of phoronids have also been sequenced, again indicating affinities with protostome groups, particularly the molluscs, rather than with deuterostomes; indeed, the bryozoans were found to be more distant than the molluscs from the other two groups of lophophorates (Cohen, Gawthrop, and Cavalier-Smith, 1998). If the molecules are "right," this undermines one rather convincing (and once almost universally accepted) view, that the trimeric enterocoelic coelomic cavities of the various groups of deuterostomes and their allies are good evidence of shared ancestry. The fossil record has yielded evidence in support of a new protostome-related lophophorate status for the halkieriids, a group of Cambrian organisms (Conway Morris, 1993, Conway Morris and Peel, 1995) that may be related to both annelids and brachiopods. The presence of almost identical chitinous chaetae in both phyla (see Storch, 1979) might be explained by shared relationships with such an ancestor. More recently still, Rosa and colleagues (1999) report a common pattern of Hox genes in brachiopods and annelids that is distinct from other protostome and pseudocoelomate patterns.

This leaves us with a considerable reevaluation of morphological evidence on our hands. However, it still retains the very precise and detailed lophophore structure as convergent across phyla, because the lophophore occurs not just in traditional "lophophorates" but also in the indisputably deuterostome pterobranch group (part of Hemichordata), where it functions as an almost identical feeding apparatus (Halanych, 1996). It also raises the issue of convergent evolution of deuterostome-like embryologies in the lophophorate groups. Knoll and Carroll (1999) argue that such convergence may arise from early divergence within the Bilateria of groups having both ancestral and derived characters, with the lophophorates lying at the base of the protostome grouping but "before" spiral cleavage and schizocoely were added to the genetic and developmental repertoire.

Larvae

This example, chosen for being different in "kind" from the preceding ones, has its own peculiar controversy. There is an unresolved theoretical argument as to whether developmentally early characters "should" be phylogenetically more informative than adult characters. In relation to larvae, this is augmented by an uncertainty as to whether larvae are primary or secondary in animal life cycles anyway (Strathman, 1988, 1993; McHugh and Rouse, 1998). Larvae have often been seen as key steps along the way of invertebrate evolutionary radiation, especially in the Germanic tradition deriving from Haeckel and represented in Remane 1963, Jägersten 1972, and Nielsen 1985, 1995, all of which invoke phylogenetically significant larval intermediates. It has also been proposed that there is an underlying pattern of embryogenesis (type 1) in all Bilateria that gives rise to a life history with larval stages, linked with increasing expression of Hox gene clusters in particular "set-aside" cells (Davidson, Peterson, and Cameron, 1995; Peterson and Davidson, 2000). But the fundamental controversy is really about whether larvae *can* be a primitive feature of life histories at all, from the ecological (Olive, 1985; Ax, 1989) or developmental (Wolpert, 1994) perspective. To be functional, planktotrophic larvae depend on reliable external fertilization, requiring stored gametes and thus needing relatively big bodies with cavities as storage spaces. The argument goes that they therefore cannot have been present in the life cycles of what must have been tiny ancestral metazoan adults. If planktotrophic larvae are a secondary phenomenon, then of course many of the similarities of larvae *must* be convergent phenomena anyway.

In practice, it is therefore unsurprising that the "defining characters" of larvae are actually rather unclear, and almost certainly not always independent (Willmer, 1990; Popkov, 1993). Larvae show forms suited to their function, converging on the small ciliated ball format of necessity if they are marine and planktonic, although clearly expressing diversity within the constraints of function. Electron microscopy reveals that many larval forms once called "modified trochophores" are no more like the classical polychaete trochophore than they are like other planktonic larvae. Similarly, there is accumulating evidence of frequent and reversible transitions of larval type within some taxa (McEdward and Janies, 1997), with loss and reappearance of planktotrophic forms being quite commonplace (Haszprunar, Salvini-Plawen, and Rieger, 1995; McHugh and Rouse, 1998; and, controversially, Williamson, 1992).

On the other hand, evidence from the paleontologists presents us with the counterview that larvae have been extraordinarily conserved in some taxa, with Jurassic pluteus larvae and even more extreme cases of Cambrian nauplii very like modern forms (see Raff, 1996). Wray's analysis (1992) of the echinoderm pluteus shows conservation of basic form for some 250 million years, albeit with multiple separate transitions to direct development. This might be worrisome in terms of developmental lability were it not for equally clear evidence that larval body plans have on occasion undergone very radical shifts (e.g., in ascidians, and in amphibians, see Raff, 1996) without greatly affecting adult morphology.

Convergence, Development, Molecular Taxonomy, and Body Plans

One of the most important developments in zoology over the last fifteen years has been the application of molecular taxonomy to the "big" problems of phyletic relationships, altering many of our views on where animals belong and how they must have evolved. It now seems clear (Kobayashi et al., 1993; Wainright, Hinkle, and Sogin, 1993; Erwin, 1991; Raff, Marshall, and Turbeville, 1994; but see also Adoutte and Philippe, 1993) that metazoans are monophyletic, with sponges basal and allied to choanoflagellate protistans, and with fungi as a sister group to this clade. Sponges and other animals perhaps took the key step of adding extracellular matrices to their body organization, and thus set off the chain of events of cell-cell and cell-matrix interactions that led to epithelia and movements of sheets of cells in three dimensions to produce complex anatomies, cavities, and coelomate triploblasts.

Far from resolving the problems of convergence, however, molecular taxonomy has in some ways underscored our need to understand patterns of divergence and convergence. There are still major disputes over the status of diploblastic groups, of nemertines, molluscs, pseudocoelomate phyla, and priapulids; and there are suggestions from molecular taxonomy that taxa previously supposed to be primitive may in fact be secondarily simplified (e.g., Balavoine, 1997; Aguinaldo et al., 1997). The problems have arisen partly because, even with molecules, when diversification is rapid, it is almost impossible to resolve precise branching orders; metazoan phyletic diversification, at least of all the triploblasts in the late Precambrian, seems to have been rather rapid (Erwin, 1993; Raff, 1996; Conway Morris, 1998, and chapter 2, this volume). Most recent molecular taxonomies, whatever their methodological assumptions, have found it particularly difficult to resolve the relationships of key protostome groups such as myriapods, which regularly map out as distinct from other "uniramian" arthropods (Abele, Kim, and Felgenhauer, 1989; Ballard et al., 1992; Philippe, Chenuil, and Adoutte, 1995; Raff, 1996). Furthermore, we still do not know enough about molecular genomic variation within phyla to be sure the very few representatives chosen thus far are typical or "normal" for a particular taxon; Maley and Marshall (1998) point out the very large errors that can arise from choosing a single aberrant species. If, however, the phylogeny of "articulate" animals is indeed to be settled as molecular taxonomy is currently suggesting, it of course yields up yet more convergence in all those morphological features that for a century and more have led most zoologists to set myriapods next door to insects.

As discussed earlier, the Hox gene family has also been interpreted as telling us about fundamental monophyly of the whole animal kingdom, an idea that perhaps began in earnest with the "zootype" of Slack, Holland, and Graham (1993) and the idea of a phylotypic stage, and later picked up additions such as "arthrotype" and "trimerotype" by Minelli and Schram (1994). Raff (1996) has argued that the phylotypic stage, although the most conserved evolutionary stage of development, is also attainable independently through nonconserved developmental processes, giving rise to the metaphor of a developmental "hourglass." Part of the answer to this seeming paradox may lie with gene duplications, which are extremely widespread in many or most animal genomes and which may allow structural modifications or additions by freeing up gene copies to accumulate small mutations, to become available for new functions, or both. This phenomenon, well documented for actin genes and for lens crystallins (Raff, 1996), also underlies the history of Hox genes (Holland et al., 1994). Indeed, Minelli (1998) argues persuasively that factors such as gene duplication and exon shuffling should lead us away from a hierarchical view of genes, development, and morphology.

Although, eventually, on the grounds of molecular taxonomy and developmental "homologies," it may become impossible to maintain that convergence extends to polyphyly of the animal kingdom, the jury is still out. Certainly, a conservative explanation of similarities in mechanism across broad sweeps of the animal kingdom is that they represent a common inheritance; in the case of complex structures such as eyes or hearts, perhaps the "stem bilaterians" had primitive and only slightly differentiated forerunners of these structures, from which radiation could occur (Knoll and Carroll, 1999). On the other hand, some of the genes and regulatory pathways involved here may be very much earlier in origin, predating animal organization and having been inherited from unicellular ancestors—either once or convergently many times—for canalizing epigenetically generated morphologies. Wolpert (1990) and Erwin (1993) similarly argue that most of the requirements for development are possessed by "protists," and that many supposed synapomorphies of Metazoa are in fact plesiomorphies shared with a variety of unicellular organisms.

Conclusions

Analyses of animal relationships based on molecules, genes, or a combination of the two present both intriguing new insights and problems that do not apply to morphologically based trees. These include linkage, possible horizontal transfer of genes (deemed to be commonplace in prokaryotic evolution by Doolittle, 1999), and, especially relevant in the present context, the concerted evolution of common multigene families. Many genes affect more than one character; many characters depend on more than one gene. Changes on the chromosomal scale, for example gene duplications and rearrangements, may in practice be much more important than point mutations in setting the trends for animal evolution. There are thus many sources of change in the genome that may invalidate traditional assumptions about homology and the independence of characters (Li and Graur, 1991; Minelli, 1998).

At the level of genes, it is certainly becoming clear that there is considerable evolutionary conservation of developmentally acting spatial genes, and that genes shaping development are often rather similar across whole kingdoms. But now we see that such genes are by no means constrained to produce similar morphologies. The mere presence of shared regulatory genes in distantly related organisms does not guarantee that these genes perform the same role in development; in fact, they may be put to very different uses to fashion divergent body plans (e.g., Lowe and Wray, 1997): the mapping of genes against body plans is far from straightforward and decidedly nonlinear. A substantial number of major morphology-regulating genes are shared among all bilateral animals at least, sometimes with functional similarity and sometimes with homologous genes recruited to entirely different functions, in what Roth (1988) has called "genetic piracy." In this context, it is particularly significant that the Cambrian explosion of diversity of form clearly postdates the origins of the genetic tool kit, with bilaterian radiation and protostome-deuterostome divergence occurring in the late Proterozoic.

Regulatory systems residing in genes evolve at highly variable rates, and morphological change largely depends on the regulatory effect of some genes on others (Raff and Kaufmann, 1983; Raff, 1996), with developmental processes either buffering or amplifying this relationship (e.g., Levinton, 1988; Wray, 1992). Raff (1996) has summarized the possible pathways of evolution of regulatory genes themselves. Master regulator genes can be conserved between phyla and even across whole kingdoms, while producing very different morphologies (sometimes but not always in conserved spatial sites); on the other hand, new master regulators can also evolve without much morphological evolution. Developmental mechanisms that are flexible and sometimes convergent probably underlie many examples of "phenotypic plasticity" (Schlichting and Pigliucci, 1998; Hodin, 2000); lineages with strong phenotypic plasticity are likely to be characterized by multiple occurrences of convergent evolution.

References

Abele LG, Kim W, Felgenhauer B (1989) Molecular evidence for inclusion of the phylum Pentastomida in the Crustacea. Mol Biol Evol 6: 685–691.

Abouheif E, Akam M, Dickinson WJ, Holland PWH, Meyer A, Patel NH, Raff RA, Roth VL, Wray GA (1997) Homology and developmental genes. Trends Genet 13: 432–433.

Adoutte A, Philippe H (1993) The major lines of metazoan evolution: Summary of traditional evidence and lessons from ribosomal RNA sequence analysis. In: Comparative Molecular Neurobiology (Pichon Y, ed), 1–30. Basel: Birkhäuser.

Aguinaldo AMA, Turbeville JM, Linford LS, Rivera MC, Garey JR, Raff RA, Lake JA (1997) Evidence for a clade of nematodes, arthropods and other moulting animals. Nature 387: 489–493.

Anderson DT (1979) Embryos, fate maps, and the phylogeny of arthropods. In: Arthropod Phylogeny (Gupta AP, ed), 59–105. New York: Van Nostrand Reinhold.

Ax P (1989) Basic phylogenetic systematization of the Metazoa. In: The Hierarchy of Life (Fernholm B, Bremer K, Jornvall H, eds), 229–245. Amsterdam: Elsevier.

Balavoine G (1997) The early emergence of Platyhelminths is contradicted by the agreement between 18S rRNA and Hox genes data. CR Acad Sci Paris, Sci Vie 320: 83–94.

Ballard JWO, Olsen J, Faith DP, Odgers WA, Rowell DM, Atkinson PW (1992) Evidence from 12S rRNA sequences that Onychophorans are modified arthropods. Science 258: 1345–1348.

Cain AJ (1982) On homology and convergence. In: Problems of Phylogenetic Reconstruction (Joysey KA, Friday AE, eds), 1–19. London: Academic Press.

Christen R, Ratto A, Baroin A, Perasso R, Grell KG, Adoutte A (1991) Origins of metazoans: A phylogeny deduced from sequences of the 28S rRNA. In: The Early Evolution of Metazoa and the Significance of Problematic Taxa (Simonetta A, Conway Morris S, eds), 1–9. Cambridge: Cambridge University Press.

Coddington JA (1994) The roles of homology and convergence in studies of adaptation. In: Phylogenetics and Ecology (Eggleton P, Vane-Wright RI, eds), 53–78. London: Academic Press.

Cohen BL, Gawthrop A, Cavalier-Smith T (1998) Molecular phylogeny of brachiopods and phoronids based on nuclear-encoded small subunit ribosomal RNA gene sequences. Philos Trans R Soc Lond B Biol Sci 353: 2039–2061.

Conway Morris S (1993) The fossil record and the early evolution of the Metazoa. Nature 361: 219–225.

Conway Morris S (1998). Diversity in ancient ecosystems. Philos Trans R Soc Lond B Biol Sci 353: 327–345.

Conway Morris S, Peel J (1995) Articulated halkieriids from the lower Cambrian of North Greenland, and their role in early protostome evolution. Philos Trans R Soc Lond B Biol Sci 347: 305–358.

Davidson E, Peterson K, Cameron R (1995) Origin of bilaterian body plans: Evolution of developmental regulatory mechanisms. Science 270, 1319–1325.

De Robertis EM (1997) Evolutionary biology: The ancestry of segmentation. Nature 387: 25–26.

Doolittle RF (1994) Convergent evolution: The need to be explicit. Trends Biochem Sci 19: 15–18.

Doolittle WF (1999) Lateral genomics. Trends Genet 15: M5–M8.

Erwin DH (1991) Metazoan phylogeny and the Cambrian radiation. Trends Res Ecol Evoln 6: 131–134.

Erwin DH (1993) The origin of metazoan development: A palaeobiological perspective. Biol J Linn Soc 50: 255–274.

Erwin D, Valentine J, Jablonski D (1997) The origin of animal body plans. Am Sci 85: 126–137.

Foley R (1993) Striking parallels in early hominid evolution. Trends Res Ecol Evol 8: 196–197.

Fortey RA, Briggs DEG, Wills MA (1996) Cambrian explosion, clades and morphology. Biol J Linn Soc 57: 13–33.

Friday, AE (1994) Adaptation and phylogenetic inference. In: Phylogenetics and Ecology (Eggleton P, Vane-Wright RI, eds), 207–217. London: Academic Press.

Fryer G (1997) In defence of arthropod polyphyly. In: Arthropod Relationships (Fortey RA, Thomas RH, eds), 23–33. London: Chapman and Hall.

Garcia-Fernandez J, Holland PWH (1994) Archetypal organization of the amphioxus Hox gene cluster. Nature 370: 563–566.

Gehring WJ (1996) The master control gene for morphogenesis and evolution of the eye. Genes and Cells 1: 11–15.

Halanych KM (1996) Convergence in the feeding apparatuses of lophophorates and pterobranch hemichordates revealed by 18S rDNA: An interpretation. Biol Bull 190: 1–5.

Hall BK (1992) Evolutionary Developmental Biology. London: Chapman and Hall.

Hall BK (1994) Introduction. In: Homology, the Hierarchical Basis of Comparative Biology (Hall BK, ed), 1–19. New York: Academic Press.

Haszprunar G, Salvini-Plawen L von, Rieger RM (1995) Larval planktotrophy—a primitive trait in the Bilateria? Acta Zoologica 76: 141–154.

Hodin J (2000) Plasticity and constraints in development and evolution. J Exp Zool 288: 1–20.

Holland LZ, Kene M, Williams NA, Holland ND (1997) Sequence and embryonic expression of the amphioxus *engrailed* gene (*AmphiEn*): The metameric pattern of transcription resembles that of its segment-polarity homolog in *Drosophila*. Development 124: 1723–1732.

Holland PWH (1992) Homeobox genes in vertebrate evolution. BioEssays 14: 267–273.

Holland PWH, Garcia-Fernandez J, Williams NA, Sidow A (1994) Gene duplications and the origins of vertebrate development. In: The Evolution of Developmental Mechanisms (Akam M, Holland PWH, Ingham P, Wray G, eds), 125–133. Cambridge: Company of Biologists.

Horder TJ. (1994) Partial truths: A review of the use of concepts in the evolutionary sciences. In: Models in Phylogeny Reconstruction (Scotland RW, Siebert DJ, Williams DM, eds), 65–91. Oxford: Clarendon Press.

Jägersten G (1972) Evolution of the Metazoan Life Cycle. London: Academic Press.

Kimmel CB (1996) Was Urbilateria segmented? Trends Genet 12: 320–331.

Kobayashi M, Takahashi M, Wada H, Satoh N (1993) Molecular phylogeny inferred from sequences of small subunit ribosomal RNA supports the monophyly of the Metazoa. Zool Sci 10: 827–833.

Knoll AH, Carroll SB (1999) Early animal evolution: Emerging views from comparative biology and geology. Science 284: 2129–2137.

Lake JA (1990) Origin of the Metazoa. Proc Natl Acad Sci USA 87: 763–766.

Lawrence PA (1992) The Making of a Fly: The Genetics of Animal Design. Oxford: Blackwell Science.

Levinton, J (1988) Genetics, Palaeontology, and Macroevolution. Cambridge: Cambridge University Press.

Li W-H, Graur D (1991) Fundamentals of Molecular Evolution. Sunderland Mass: Sinauer.

Løvtrup S (1974) Epigenetics. London: Wiley.

Lowe CJ, Wray GA (1997) Radical alterations in the roles of homeobox genes during echinoderm evolution. Nature 389: 718–721.

Maley LE and Marshall CR (1998) The coming of age of molecular systematics. Science 279: 505–506.

Manak JR, Scott MP (1994) A class act: Conservation of homeodomain protein functions. In: The Evolution of Developmental Mechanisms (Akam M, Holland PWH, Ingham P, Wray G, eds), 61–71. Cambridge: Company of Biologists.

Manton SM (1977) The Arthropods: Habits, Functional Morphology, and Evolution. Oxford: Oxford University Press.

Maynard Smith J, Burian R, Kauffman S, Alberch P, Campbell J, Goodwin B, Lande R, Raup D, Wolpert L. (1985) Developmental constraints and evolution. Q Rev Biol 60: 265–287.

Mayr E (1969) Principles of Systematic Zoology. New York: McGraw-Hill.

Mayr E (1994) Cladistics and convergence. Trends Res Ecol Evol 9: 149–150.

McEdward LR, Janies DA (1997) Relationships among development, ecology and morphology in the evolution of echinoderm larvae and life cycles. Biol J Linn Soc 60: 381–400.

McHugh D, Rouse GW (1998) Life history evolution of marine invertebrates: New views from phylogenetic systematics. Trends Res Ecol Evol 13: 182–186.

McKitrick MC (1994) On homology and the ontological relationship of parts. Syst Biol 43: 1–10.

McShea DW (1996) Complexity and homoplasy. In: Homoplasy: The Recurrence of Similarity in Evolution, (Sanderson MJ, Hufford L, eds), 207–226. San Diego, Calif.: Academic Press.

Minelli A (1993). Biological Systematics: The State of the Art. London: Chapman and Hall.

Minelli A (1998) Molecules, developmental modules, and phenotypes: A combinatorial approach to homology. Mol Phylogenet Evol 9: 340–347.

Minelli A (2000) Holomeric vs meromeric segmentation: a tale of centipedes, leeches and rhombomeres. Evol Dev 2: 35–48.

Minelli A, Schram FR (1994) Owen revisited: A reappraisal of morphology in evolutionary biology. Bijdragen tot de Dierkunde 64: 65–74.

Moore J, Willmer PG (1997) Convergent evolution in invertebrates. Biol Rev 72: 1–60.

Newman SA (1993) Is segmentation generic? BioEssays 15: 277–283.

Newman SA (1994) Generic physical mechanisms of tissue morphogenesis: A common basis for development and evolution. J Evol Biol 7: 467–488.

Newman SA, Müller GB (2000) Epigenetic mechanisms of character origination. J Exp Zool 288: 304–317.

Nielsen C (1977) The relationships of Entoprocta, Ectoprocta and Phoronida. Am Zool 17: 149–150.

Nielsen C (1985) Animal phylogeny in the light of the trochaea theory. Biol J Linn Soc 25: 243–299.

Nielsen C (1995) Animal Evolution. Oxford: Oxford University Press.

Olive PJW (1985) Covariability of reproductive traits in marine invertebrates, implications for the phylogeny of the lower invertebrates. In: The Origins and Relationships of Lower Invertebrates (Conway Morris S, George JD, Gibson R, Platt HM, eds), 28–41. Oxford: Clarendon Press.

Patterson C (1982) Morphological characters and homology. In: Problems of Phylogenetic Reconstruction (Joysey KA, Friday AE, eds), 21–74. London: Academic Press.

Peterson KJ, Davidson EH (2000) Regulatory evolution and the origin of the bilaterians. Proc Natl Acad Sci USA, 97: 4430–4433.

Philippe H, Chenuil A, Adoutte A (1995) Can the Cambrian explosion be inferred through molecular phylogeny? In: The Evolution of Developmental Mechanisms (Akam M, Holland PWH, Ingham P, Wray G, eds), 15–25. Cambridge: Company of Biologists.

Popkov DV (1993) Polytrochal hypothesis of origin and evolution of trochophora type larvae. Zool Zh 72: 5–17.

Quiring R, Walldorf U, Kloter U, Gehring WJ (1994) Homology of the eyeless gene of *Drosophila* to the *Small eye* gene in mice and *Aniridia* in humans. Science 265: 785–789.

Raff RA (1996) The Shape of Life. Chicago: University of Chicago Press.

Raff RA, Kaufmann TC (1983) Embryos, Genes, and Evolution. New York: MacMillan.

Raff RA, Marshall CR, Turbeville JM (1994) Using DNA sequences to unravel the Cambrian radiation of the animal phyla. Annu Rev Ecol Syst 25: 351–375.

Remane A (1963) The enterocoelic origin of the coelom. In: The Lower Metazoa (Dougherty EC, ed), 78–90. Berkeley: University of California Press.

Riedl R (1978) Order in Living Organisms. New York: Wiley.

Rosa R de, Grenier JK, Andreeva T, Cook CE, Adoutte A, Akam M, Carroll SB, Balavoine G (1999) Hox genes in brachiopods and priapulids and protostome evolution. Nature 399: 772–776.

Roth VL (1988) The biological basis of homology. In: Ontogeny and Systematics (Humphries CJ, ed), 1–26. London: British Museum.

Rowell AJ (1982) The monophyletic origin of the Brachiopoda. Lethaia 15: 299–307.

Salvini-Plawen L von, Mayr E (1977) On the evolution of photoreceptors and eyes. In: Evolutionary Biology (Hecht MK, Steere WC, Wallace B, eds), 207–263. New York: Plenum Press.

Sanderson MJ, Donoghue MJ (1989) Patterns of variation in levels of homoplasy. Evolution 43: 1781–1795.

Sanderson MJ, Hufford L (eds) (1996) Homoplasy: The Recurrence of Similarity in Evolution. San Diego, Calif.: Academic Press.

Schaeffer B (1987) Deuterostome monophyly and phylogeny. In: Evolutionary Biology, vol 21 (Hecht MK, Wallace B, Prance GT, eds), 179–235. New York: Plenum Press.

Schlichting CD, Pigliucci M (1998) Phenotypic Evolution: A Reaction Norm Perspective. Sunderland Mass: Sinauer.

Shubin N, Tabin C, Carroll S (1997) Fossils, genes and the evolution of animal limbs. Nature 388: 639–648.

Slack JMW, Holland PWH, Graham CF (1993) The zootype and the phylotypic stage. Nature 361: 490–492.

Storch V (1979) Contributions of comparative ultrastructural research to problems of invertebrate evolution. Am Zool 19: 637–645.

Strathmann RR (1988) Larvae, phylogeny, and von Baer's law. In: Echinoderm Phylogeny and Evolutionary Biology (Paul CRC, Smith AB, eds), 53–68. Oxford: Clarendon Press.

Strathmann RR (1993). Hypotheses on the origins of marine larvae. Annu Rev Ecol Syst 24: 89–117.

Tautz D (1992) Redundancies, development and the flow of information. BioEssays 14: 263–266.

Thomson KS (1988) Marginalia: Ontogeny and phylogeny recapitulated. Am Sci 76: 273–275.

Valentine JW (1975) Adaptive strategy and the origin of grades and ground-plans. Am Zool 15: 391–404.

Van Valen L (1982) Homologies and causes. J Morphol 173: 305–312.

Wagner GP, Misof BY (1993) How can a character be developmentally constrained despite variation in developmental pathways? J Evol Biol 4: 449–455.

Wainright PO, Hinkle G, Sogin ML (1993) Monophyletic origins of the Metazoa: An evolutionary link with fungi. Science 260: 340–342.

Wake DB (1991) Homoplasy: The result of natural selection or evidence of design limitations? Am Nat 138: 543–567.

Wheeler WC, Cartwright P, Hayashi CY (1993) Arthropod phylogeny: A combined approach. Cladistics 9: 1–39.

Williamson DI (1992) Larvae and Evolution: Towards a New Zoology. London: Chapman & Hall.

Willmer PG (1990) Invertebrate Relationships: Patterns in Animal Evolution. Cambridge: Cambridge University Press.

Willmer PG, Holland PWH (1991) Modern approaches to metazoan relationships. J Zool 224: 689–694.

Wolpert L (1990) The evolution of development. Biol J Linn Soc 39: 109–124.

Wolpert L (1994) The evolutionary origin of development: Cycles, patterning, privilege, and continuity. In: The Evolution of Developmental Mechanisms (Akam M, Holland PWH, Ingham P, Wray G, eds), 79–84. Cambridge: Company of Biologists.

Wray GA (1992) The evolution of larval morphology during the post-Palaeozoic radiation of echinoids. Paleobiology 18: 258–287.

Wright AD (1979) Brachiopod radiation. In: The Origin of Major Invertebrate Groups (House MR, ed), 235–252. London: Academic Press.

4 Homology: The Evolution of Morphological Organization

Gerd B. Müller

These mighty Leviathan skeletons, skulls, tusks, jaws, ribs, and vertebrae are all characterized by partial resemblances to the existing breeds of sea-monsters, but at the same time bear on the other hand similar affinities to the annihilated antechronical Leviathans, their incalculable seniors.
—Herman Melville, 1851

The evolution of organismal form consists of a continuing production and ordering of anatomical parts: the resulting arrangement of parts is nonrandom and lineage specific. The organization of morphological order is thus a central feature of organismal evolution, whose explanation requires a theory of morphological organization. Such a theory will have to account for (1) the generation of initial parts; (2) the fixation of such parts in lineage-specific combinations; (3) the modification of parts; (4) the loss of parts; (5) the reappearance of lost parts; and (6) the addition of new parts. Eventually, it will have to specify proximate and ultimate causes for each of these events as well.

Only a few of the processes listed above are addressed by the canonical neo-Darwinian theory, which is chiefly concerned with gene frequencies in populations and with the factors responsible for their variation and fixation. Although, at the phenotypic level, it deals with the modification of existing parts, the theory is intended to explain neither the origin of parts, nor morphological organization, nor innovation. In the neo-Darwinian world the motive factor for morphological change is natural selection, which can account for the modification and loss of parts. But selection has no innovative capacity: it eliminates or maintains what exists. The generative and the ordering aspects of morphological evolution are thus absent from evolutionary theory.

The inability of evolutionary theory to account for phenotypic organization has been recognized by numerous authors, with regard to both biochemical and morphological evolution (e.g., Kauffman, 1993; Fontana, Wagner, and Buss, 1994; Müller and Newman, 1999; Newman and Müller, 2000). In expanding on the question of morphological evolution, I argue that comparative anatomy has always dealt with organization implicitly: the essence of its vast body of knowledge is embodied in what has been called the "concept" of homology. I propose that homology is not merely a concept or a conceptual tool, as it is often understood, but rather the manifestation of morphological organization processes. It thus represents a major, unsolved problem in evolutionary biology. Homology assessments, by illustrating the dynamics of parts in morphological evolution, provide the descriptive basis for a theory of morphological organization.

After the recent publication of two comprehensive volumes on homology (Hall, 1994; Bock and Cardew, 1999) and a host of papers emphasizing its developmental aspects (e.g., Hall, 1995; Bolker and Raff, 1996; Minelli, 1997; Laubichler, 2000; Gilbert and

Bolker, 2001), not much new can be added to the classical debate. Indeed, Wake (1999) has expressed doubt that any original thought can be generated on the subject. But the persisting difference of opinions shows that perceptions of the problem are still diverse; no consensual model has emerged that would explain the evolution of homology. I propose that viewing homology as organization represents a step forward. After discussing why we need to consider homology despite the persistent problems associated with its conceptualization, I review the dimensions of homology in morphological evolution and present a model for a causal explanation. I conclude by proposing an organizational homology concept.

An Elusive Concept of Similarity?

The term *homology* has a number of connotations, not all of them positive. Negative connotations include "idealistic," "typological"; some have gone so far as to call it a term "ripe for burning" (see Tautz, 1998). In contrast, a large group of biologists, paleontologists foremost among them, use the term in their daily work, seemingly without major doubts or emotional problems. Whence this disagreement? And is such an "elusive concept of similarity" a problem that we need to consider at all in modern discussions of organismal form? I argue that homology is neither elusive nor a concept, that it is not about similarity, and that none of the other objections raised against it is well founded. Difficulties associated with the term *homology* fall into the following five areas: (1) contrasting definitions; (2) semantic haze; (3) the character problem; (4) cross-level justification; and (5) the search for a locus.

Contrasting Definitions

It is true that the conceptual roots of homology lie in the idealistic morphology of the eighteenth century and even earlier (Spemann, 1915). But after Owen's first, precise, and closely comparative definition (1843, 379): "the same organ in different animals under every variety of form and function," Darwin (1859, 456) introduced a historical explanation:

On this same view of descent with modification, all the great facts in Morphology become intelligible, whether we look to the same pattern displayed in the homologous organs, to whatever purpose applied, of the different species of a class, or to the homologous parts constructed on the same pattern in each individual animal and plant.

From that time on, homology had a strictly scientific meaning, albeit with a host of changing definitions and uses. Certain periods favored different emphases, so that the general understanding of homology can be seen as having gone through several phases (see table 4.1). Although there is nothing unscientific about the evolution of a scientific term, and *homology* is probably no more elusive a term than *gene,* this multitude of contrasting definitions has

Table 4.1
Conceptualizations of homology

Type	Content	Representatives (examples)
Idealistic	Implicit use of homology without definition	Belon, Camper, Cuvier, Goethe, Linnaeus
Ahistorical	Explicit, comparative definition of homology as sameness	Owen
Historical	Homology as indicator of common descent	Darwin, Gegenbauer, Haeckel
Methodological	Homology used for systematic and taxonomical categorization	Henning, Remane
Explanatory-monocausal	Emphasis on causes rooted in specific levels of organization	Holland, Roth, van Valen, Wagner
Explanatory-systemic	Emphasis on interconnected causes of multiple levels of organization	Müller, Riedl, Striedter

Note: Represented here are prevailing notions, rather than a strictly historical sequence of definitions. Although most historians would probably call Owen's conceptualization "idealistic," I prefer to distinguish his conscious application of the homology concept from previous, implicit usages.

been a source of considerable confusion for the present-day (molecular) biologist, and has seemed to create an unscientific, "holistic" aura around the term. But different definitions are not in themselves proof of a problem with the phenomenon they attempt to define. Far from mutually exclusive, the different definitions of homology usually serve different purposes. That being the case, as long as their respective purposes are clearly stated, they pose no real problem. Most important, *definitions* of homology must be distinguished from *criteria* for its identification and from *explanations* of its causal origins (Panchen, 1994).

Semantic Haze

Two habits associated with speaking and writing about homology have helped to surround the term in a semantic haze. One is the frequent use of the combined term *homology concept*. Although there is nothing intrinsically wrong with this, that linkage seems to indicate to some that homology is merely "a concept" and does not refer to a biological reality. But just as "evolution" is not simply "a theory" but refers to biological facts *about* which different theories exist, so "homology" is not simply "a concept" but refers to biological facts *about* which several concepts have been proposed. Although these concepts differ depending on the aspect of homology they emphasize, they all refer to the conserved, lineage-specific combinations of structural parts resulting from morphological evolution.

The second habit of semantic laxity is the frequent equation of homology with "similarity" or "resemblance." Again there is nothing intrinsically wrong with this: initial observations of morphological similarity lead to the detection of homology. But once classified as such, homology is properly a statement about *sameness,* not about similarity. *Homology* thus denotes the *identity* of parts in the structural composition of different organisms, whereas, in speaking of homoplasy, convergence, and parallelism, it is *analogy* that denotes the similarity of parts. The usage of *synapomorphy,* the cladistic term for homology (Nelson, 1994), supports this point: a "shared derived character" means that—the *same* character is shared, however difficult the ascertaining of a specific case may be. Unfortunately even highly authoritative accounts, which specifically emphasize the point that Owen's original definition meant identity, can conclude that homology should be treated as similarity (Panchen, 1994), although this notion is explicitly rejected in other chapters of the same volume (Hall, 1994).

The Character Problem

Originally, *homology* referred uniquely to macroscopic elements of morphological design, "organs" in Owen's definition (1843). Later the usage was extended to "characters," which made it possible to include developmental, histological, and molecular traits, and even behavioral or functional ones. Clearly, this extension of the homology concept into nonmorphological domains was a source of multiple confusions, but appropriate definitions for each domain can help. The true problem lies in the uncertainty of what should be considered a "character" in the various domains (for a detailed discussion, see Wagner, 2001).

Already in the morphological realm, a major difficulty arises from the indiscriminate use of the term *character* in the sense of individual morphological traits, such as a certain bone, or muscle, or brain nucleus, as opposed to "character states," such as size and number of elements, or their biometric shape and proportions. The latter are essentially quantitative traits, whereas homology is foremost a qualitative property. Owen (1843) made this distinction very clear by defining *homology* as independent of "every variety of form and function." Quantitative traits, such as biometric shape, even when acquired by the most sophisticated morphometrics, cannot serve as homologous characters (Bookstein, 1994). Character states must therefore be excluded from homology assessments, even though they can be useful in cladistic analyses, where size of an element, or even its absence, may serve as a shared, derived (taxonomical) character—a synapomorphy. Moreover, a taxonomical character is not necessarily the same as a morphological character. All homologues are synapomorphics, but not vice versa. Absence of an anatomical element, for instance, can be a taxonomical synapomorphy but not a homology because homologues are positive anatomical parts. Hence *synapomorphy* and *homology* should not be considered as synonymous.

Another difficulty arises in describing larger character assemblies as "homologues." Thus it makes no sense to speak of complex assemblies such as vertebrate limbs or heads as "homologous" because this would mean that their subelements were arranged identically in all forms, which is almost never the case. One should therefore reserve "homologue" for indentically arranged individual elements of body construction.

Cross-Level Justification

First defined for and applied to morphological characters, the homology principle has been successfully extended to other levels of organismal organization, such as development, behavior, and genetics. Because, however, homology can only exist for entities that belong to the same level of organization, whether structure, behavior, or genes, major problems arise when we attempt to ascertain homology at one level by comparing characters or processes that belong to a different level (figure 4.1). Thus, for example, even though a certain behavior is shown to be homologous for all members of a clade, this does not mean that the behavior is carried out via the same muscles and the same nervous circuitry, or that it is controlled by the same set of genes (Müller, 2001). The same is true for structural characters and their developmental makeup. Homology of a character in different species does not

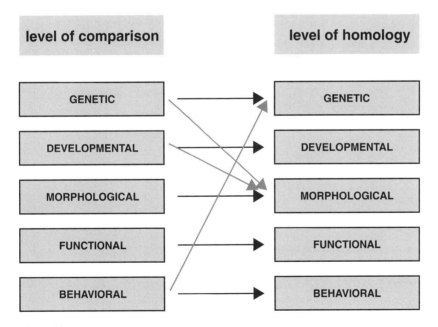

Figure 4.1
Legitimate and illegitimate usages of the homology concept. Homology can only be legitimately inferred from comparisons at the same level of organization (black arrows). Inference or explanation across levels (gray arrows) can lead to substantially invalid conclusions of homology.

mean that this character is generated by identical developmental processes, is controlled by the same genes, or has an identical protein composition (see below).

It is important to note that the reverse of this argument is also true. The existence of homology at lower levels of organization does not automatically generate homology at higher levels. Thus, although the detection of homology can be greatly assisted by studying other levels, and even the explanation for the origin of a certain homologue can come from different levels of analysis, character identity itself cannot be securely determined by these means. Homology can only be ascertained from phylogenetic comparison of characters within the same level of organization.

Search for a Locus

The understanding of homology is often hampered by the attempt to identify a specific causal agent or to rely on a specific class of data, which is held to contain the key information (Lauder, 1994). After seeking such loci of determination for homology in structure and connectivity (see Rieppel, 1988; Shubin and Alberch, 1986; see also Rasskin-Gutman, chapter 17, this volume), later researchers sought them primarily in development and genetics (see below). Every one of these loci, however, was eventually shown not to contain unequivocal evidence for homology (summarized in Lauder, 1994), which has reinforced the growing tendency to reject homology as a serious scientific topic.

Development was particularly important in the search for mechanistic causes of homology. Already the idea of recapitulation contained a mechanistic notion of character origination. Although recapitulation no longer plays a role in modern discussions of homology, there has been a revival of the notion that development contains the key to homology; indeed, several recent concepts of homology rest on developmental definitions (e.g., van Valen, 1982; Roth, 1984; Wagner, 1989a,b). At the same time, it has become abundantly clear that it is not particular mechanisms of development that are responsible for the maintenance of homology in every case. Thus it was shown that the genetic control, molecular makeup, cell populations, inductive interactions, and ontogenetic trajectories could all be modified by evolution, while the resulting homologues are maintained (Wagner and Misof, 1993; Hall, 1994; Bolker and Raff, 1996). Hence neither the developmental mechanisms nor the developmental origins of a particular trait, though they can tell us much about phylogenetic commonalities of generative processes, alone represent sufficient causes or unique loci of determination for homology.

Genetics, of course, is presently the most fashionable candidate for the true locus of homology, despite the fact that numerous treatments of this notion, beginning particularly with that of de Beer (1971), have revealed its insufficiency (Hall 1995; Bolker and Raff, 1996; Minelli, 1997). Homologues should best be identified, some have proposed, by shared expression patterns of homologous developmental genes (e.g., Hickman, Roberts, and Hickman, 1988; Holland, Holland, and Holland, 1996). The idea leads to astonishing consequences. The exciting discovery, for instance, that homologues of the mammalian *Pax-6* gene are expressed in the early eye morphogenetic pathway of many vertebrate and invertebrate species, has led some to conclude that vertebrate and cephalopod eyes—a classical example for nonhomology—may not have evolved by convergence after all (Quiring et al., 1994). But this finding can be interpreted quite differently. Whereas the genetic basis of the embryonic initiation of eye formation may be homologous, the resulting anatomical structures are most emphatically not. The famous cephalopod eye, although constructed almost identically and initiated by *Pax-6* orthologues, is *not* homologous to the vertebrate eye: its anatomical structures evolved completely independently from vertebrate eyes; vertebrates and cephalopods do not share a common ancestor with anatomically differentiated eyes.

Thus some of the most notoriously conserved (homologous) developmental control genes, e.g., of the homeobox kind, exhibit nonhomologous expression domains in comparative maps of vertebrate and invertebrate embryos (Duboule, 1994), and vice versa—homologous structures can be characterized by nonhomologous genes. This is exemplified, for instance, by the differences in the "master regulatory genes" controlling the sex determination pathways in different dipteran insects (Meise et al., 1998; Saccone et al., 1998). Several other forms of dissociations between genes and structures can be distinguished (Wray, 1999), including radical changes in the developmental roles of control genes (Lowe and Wray, 1997) and extensive genetic "rewiring" of developmental circuits (Salazar-Ciudad, Newman, and Solé 2001; Salazar-Ciudad, Solé, and Newman, 2001; Szathmary, 2001).

From these incongruences between genetic background and phenotype (see also Gilbert, Bissell et al., and Larsen, chapters 6, 7, and 8, this volume), a set of possible relations between genes and morphological homology can be extracted. Only one of the six possibilities represents a correspondence between genetic and morphological homology (figure 4.2). Hence the detection of homologous genes, although immensely useful in combination with careful morphological and phylogentic analyses, does not provide an infallible guide to the locus of homology (see also de Beer, 1971; Dickinson, 1995; Bolker and Raff, 1996; Abouheif, 1997).

As we enter a phase of postgenomic explanation, emphasizing the organismal level of evolution, homology has attracted greatly renewed interest, as documented by numerous recent publications (e.g., Hall, 1994; Bock and Cardew, 1999). The reason for this is, I would argue, that there is no other handle on the organization problem. Homology is the manifestation of an ordering principle in morphological evolution. It poses the concrete question: How do we explain the establishment, conservation, and organization of individualized constructional elements in the evolution of organismal forms? This legitimate and

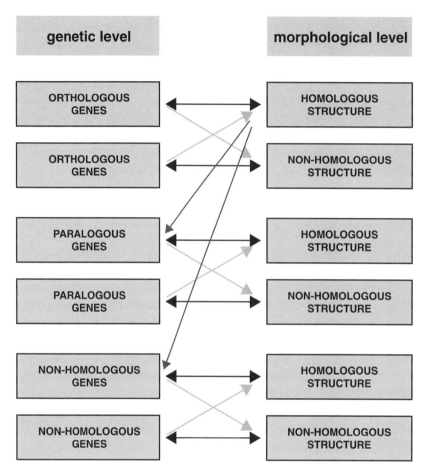

Figure 4.2
Associations between genes and structure (double-pointed arrows). Evolutionary shifts of genetic control are possible (gray arrows), as are shifts in the association of a given homologous structure (upper right) between homologous (orthologous) genes, paralogous genes, and nonhomologous genes (thin arrows).

strangely neglected scientific question lies at the center of the phenomenon of organismal evolution; it is not satisfactorily answered by current evolutionary theories. All attempts to brush the question aside, to dismiss *homology* as an uncertain, or confusing term, to brand the homology principle as "idealized" (Tautz, 1998) should not blind us to the clearly observed and established phenomena it denotes. *Morphological homology* is a manifestation of structural organization that maintains identical building elements despite variation in their molecular, developmental, and genetic makeup. The recognition of this biological fact is crucial for an understanding of the origin and diversification of organismal form.

Dimensions of Homology

The continuous production and fixation of parts has resulted in the thirty-seven or so presently known extant body plans (fixed sets of homologues) and even more existed in the past. All minor clades are modifications of those major body plans, each characterized by distinct, hierarchical combinations of homologues, altogether represented in several million species. Based on the recognition of sameness of anatomical construction, the early systematists reconstructed the natural system of organisms, which was essentially a naming and classification exercise using natural body plan units, namely, homologues. Although many mistakes were made, especially with regard to the smaller clades, this systematic grouping of homologues led to a system of phylogentic relationships that would later be largely corroborated by molecular data. Homologues represent the units of morphological evolution: all parts that define a clade are homologues. Evolution along a phylogenetic lineage proceeds by creating an increasing order of hierarchically nested sets of homologous building elements. The "natural system" of systematics thus reflects the homology order.

Only few attempts have been made to quantify the extent of homology order in nature. Riedl (1978) provides a rough approximation of the possible orders of magnitude by multiplying the sum of single homologues that make up an anatomical system by the number of species in which it is represented. Multiplying, for instance, 4.4×10^3 homologues for the vertebral column by 10^5 vertebrate species, or 2×10^3 homologues for the insect nervous system by 10^6 species, he arrives at magnitudes near 10^8 and 10^9 for the homology represented in those systems. The vertebrate nervous system is thought to roughly approximate 7×10^4 single homologues in 10^5 species. Other attempts to calculate the magnitude of homology for a certain system use different principles (Zarapkin, 1943; Olson and Miller, 1958).

Although it is difficult to judge the exactness and meaning of such numbers, these exercises show that, in principle, morphological homology is a quantifiable phenomenon. Clearly, an exact calculation is hampered by a number of problems, such as how to deal with serial homology or with the weighing of characters. But using proper mathematical tools, it should be possible to arrive at reasonable estimates of the homology order realized in morphological evolution. Homologues might even be treated as agents in dynamical systems, which could greatly improve our understanding of the evolutionary roles of morphological organization (Fontana and Buss, 1994).

Origins of Homology: A Three-Stage Model

Few attempts have been made to approach homologues as real entities of organismal design (but see Riedl 1978, Wagner, 1989a,b; Minelli and Peruffo, 1991), and even in those efforts their existence is usually taken as given. But how do individualized elements arise that become the homologues of anatomical construction? And what are the mechanisms for their integration and eventual fixation in a body plan? Müller and Newman (1999) proposed that the evolution of homology consists of three distinguishable stages—(1) generative, (2) integrative, and (3) autonomized—each governed by distinctive mechanisms and properties.

Modes of Generation: Pre-Mendelian and Mendelian

A *morphological novelty* can be defined as "a structure that is neither homologous to any structure in the ancestral species, nor homonomous to any other structure of the same organism" (Müller and Wagner, 1991, 243; see also Wagner and Chiu, chapter 15, this volume). This excludes simple continuous variation as a sufficient mechanism for the origination of morphological homologues, and hence requires explanations that go beyond the standard Darwinian model, which applies to variation. Moreover, the mechanisms that led to the primordial building elements of the first metazoans seem to have been recognizably distinct from those that cause innovation in established, heritable body plans. Therefore, in this first stage of the evolution of homology, it may be useful to distinguish between a pre-Mendelian and a Mendelian mode (Newman and Müller, 2000).

The first mode is called "pre-Mendelian" because the generic properties of early metazoan cell assemblies, not yet routinized by heritable systems of genetic control, must have had an immediate influence on the generation of initial structures. Cell adhesiveness was likely a key property in the processes of multicellular interaction from which the generation of a limited number of primary building structures resulted. Subtle modifications of adhesivity, together with alterations of cell proliferation and cell size, can account for virtually all basic structures of simple metazoans, such as invagination, compartmentalization, multilayering, and segmentation. These tissue behaviors were shown to result generically from the physical properties of cells and condensed, chemically active tissue masses (Newman and Comper, 1990; Newman, 1994).

In this pre-Mendelian scenario, the physical properties of primitive, multicellular assemblages would have been the primary agents for the generation of first morphological elements that eventually became homologues of more elaborate, metazoan body plans. These basic elements are thought not to have arisen as a consequence of established "programs" for development but rather to have "fallen into place" spontaneously, like proteins assuming tertiary structures generic to their molecular and thermodynamic properties. Newman and

Müller (2000) proposed that such generic structures provided "morphogenetic templates" for the biochemical sophistication of cell interaction and became only secondarily co-opted by the genome. This ensured heritability and resulted in the more deterministic relationship between genetic circuitry and form generation observed in evolved organisms.

Once basic metazoan body assemblies had formed, new morphological detail was added in what can be called the "Mendelian phase" of homology evolution, which would be characterized by the existence of genetically stabilized and heritable developmental systems. But such systems still depended on contextual information during the process of development, usually referred to as "epigenesis." Müller (1990) and Müller and Wagner (1991) suggested that these epigenetic properties were instrumental not only in genetic change but in the initiation of morphological novelty and additions to the designs that emerged from the first phase—again followed by genetic stabilization. Organisms in the later phases of evolution exhibit the familiar Mendelian alterations of phenotype brought on by mutation and allelic variation. But the particular forms that the phenotypes take on will continue to depend on the generic and other material properties of tissues and are thus still subject to epigenetic determination.

This concept of epigenetic character initiation in the Mendelian phase of evolution does not contradict classical neo-Darwinian mechanisms. It does not deny that evolutionary change is based on heritable, genetic variation, causing small phenotypic variation. Natural selection may, for example, act on parameters such as the onset, offset, or rates of developmental processes, influencing cell cycle properties, cell proliferation or differentiation, relative size and position of morphogenetic primordia, and so on. It can result only in gradual modifications of existing structures, however, but does not produce independent, new elements, that is new homologues, which can arise only when the gradual, evolutionary modifications of developmental parameters reach critical thresholds of one or several processes (Müller, 1990; Streicher and Müller, 1992; Raff, 1999). Developmental thresholds may consist of the number of cells required for a blastema, the distance between inductive tissues, the concentration of a morphogen controlling pattern formation, and many other properties. At such threshold points, the phenotypic outcome of a parameter change depends on the developmental reaction norm of the affected system, which may result in the loss of a structure or the appearance of a new one.

Thus, in the Mendelian phase, incipient morphological novelty often arises as a by-product of the evolutionary modification of developmental systems. This concept represents the core of the "side effect hypothesis" for the origin of new characters in morphological evolution (Müller, 1990). It means that the origin of new homologues in a given body plan is contingent on the epigenetic structure of the particular developmental system acquired by an evolving lineage. Developmental individualization (Wagner, 1989a,b) and constructional individualization then follow.

Modes of Integration

The generation of new, structural elements does not in itself establish homology. Homology only arises when a new element is locked into the constructional body plan and is maintained in at least one derived species. This process can be called "integration"; it represents the second stage of the evolution of homology and is likely to take place through different mechanisms, acting among and between different levels of organization—genetic, developmental, phenotypic, and functional.

Modes of genetic integration have been discussed outside the context of homology (Waddington, 1957, 1962): they apply equally to phenotypic traits that represent nascent, anatomical homologues. One mode of genetic integration could occur through the co-optation of redundant developmental control genes that become available in evolution, for example, as a result of gene duplication, via a mechanism Roth (1988) calls "genetic piracy." Many cases are known in which orthologous or paralogous regulatory genes acquire new associations and new developmental roles over the course of evolution (Wray, 1999). The evolving genome may thus gain control of the epigenetic conditions responsible for the initiation of new building elements, as described above, and the formative processes may increasingly stabilize and even become "overdetermined" (Newman, 1994). Because the process involves growing numbers of structures and pathways, it results in an ever-closer mapping between genotype and phenotype. Such transitions can be represented as a switch from emergent to hierarchical gene networks (Salazar-Ciudad, Newman, and Solé, 2001; Salazar-Ciudad, Solé, and Newman, 2001). Other modes of genetic integration (see Kaneko and Nanjundiah, chapters 12 and 14, this volume) reflect increasing agreement on viewing genetic consolidation as secondary to phenotypic innovation (e.g., Budd, 1999; Newman and Müller, 2000; Endress, 2001).

But not all integration necessarily corresponds to early, genetic fixation. Many phenotypic characters can be experimentally suppressed by changing the epigenetic conditions of their formation (Hall, 1983; Gilbert, 2001). Obviously, the developmental systems of cell and tissue interactions represent in themselves a form of integration. Because, as noted, an important mode for the origin of new parts is through epigenetic by-products, a first requirement for integration is merely the maintenance of the epigenetic conditions under which they originated. The fixation of these conditions may arise from stabilizing selection, and may faithfully result in the generation of the novel character in each generation, without direct genetic fixation of the character itself. In such cases, epigenetic integration actually precedes genetic integration and can be maintained for extensive periods of time, stabilizing incipient homologues long before their genetic hardwiring (Johnston and Gottlieb, 1990).

Evolving structure-function interrelationships at the phenotypic, structural level also assists in the integration of new characters (Olson and Miller, 1958; Wake and Roth, 1989;

Galis, 1996). The hierarchical position of a new character in the body plan of an organism, and the number of structural and functional interdependencies in which it is involved, gains increasing importance as more design and functional differentiation are added. This development again has consequences at the genetic level, with selection favoring the genetic linkage of functionally coupled characters (Wagner, 1984; Bürger, 1986).

Thus genetic, developmental, and constructional mechanisms of morphological organization become increasingly interwoven and result in the progressive integration of character complexes. Epigenetic mechanisms of development are central in this process of organization of body design because they provide templates for both phenotypic and genetic integration. These integrative processes "lock in" the new characters that arise as a consequence of conditional mechanisms, generating the heritable building units of the phenotype called "homologues." In principle, the number of genetic, developmental, structural, and functional interdependencies accumulated by a homologue can be quantified, allowing us to define the "strength" of fixation of a homologue and of the probability of it becoming "undone" during subsequent steps of evolution, a notion implicit in the concepts of "burden" (Riedl, 1978), "developmental constraint" (Maynard Smith et al., 1985), "generative entrenchment" (Wimsatt, 1986), "epigenetic traps" (Wagner, 1989a), and "evolutionary ratchet" (Arnold, 1989).

Modes of Autonomization

The third and puzzling stage in the evolution of homology is the source of much of the present confusion surrounding the term. Homology, in its most evolved form, is characterized by an increasing independence from the underlying developmental, molecular, and genetic processes that led to the first appearance (generation) and subsequent fixation (integration) of basic, individual homologues. This increasing independence of the structural design from the generative and variational mechanisms has been called "autonomization" (Müller and Newman, 1999).

The empirical evidence supporting autonomization continues to accumulate in molecular studies of development. These results indicate that the same phenotypic endpoint can be reached via different developmental pathways, and that indeed such pathways have changed importantly in the case of structures that are clearly homologous (Hall, 1995). It is seen that evolutionary modification can affect both early and late developmental events that lead to the formation of a homologue (see above; Steinberg and Striedter, chapters 9 and 16, this volume), and yet that element is maintained at the structural level. Experimental studies also demonstrate that genetic and morphological variation can be poorly correlated (Atchley, Newman, and Cowley, 1988). These observations of comparative developmental biology are paralleled by the incongruences between rates of genetic and morphological divergence (Meyer et al., 1990; Sturmbauer and Meyer, 1992; Bruna, Fisher, and Case, 1996). And it is well known that the correlation between genome and morphological complexity is slight (Miklos and Rubin, 1996).

Autonomization of homologues is a consequence of a decoupling between genetic and morphological evolution. Homologues, the design units of the phenotype, transcend their underlying molecular, epigenetic, and genetic constituents and assume an independent, organizational role in morphological evolution. With their integration into the body plan of a phylogenetic lineage, homologues take on constructional identity and come to act as accretion points of further structural organization. They can be understood as "attractors" of morphological design (Striedter, 1998; Müller and Newman, 1999), around which more design is added, and will be conserved even in the face of changing adaptive conditions. Therefore, at this autonomized stage, homologues are more influential for the further path of morphological evolution than the primary generic conditions underlying their origin and the biochemical circuitry that controls their developmental formation. Homology is the product of an increasing autonomization of a design principle from its mechanistic underpinnings (Müller and Newman, 1999).

The Organizational Homology Concept

If, as here proposed, homology results from organizing processes that integrate and fixate generic and conditionally generated building elements into a stable body plan, then neither genetic nor developmentally based definitions alone can capture the essence of this phenomenon. A more encompassing concept that represents the integration between different levels of organization is needed, one I propose to call the "organizational homology concept." This concept is based on seven premises (four established by earlier authors, and three proposed in the current chapter):

1. Homologues are constant elements of organismal construction; they are independent of changes in form and function (Owen, 1843);

2. Homology signifies identity, not similarity (Owen, 1843);

3. Homologues are fixated by hierarchically interconnected interdependencies ("burden"; Riedl, 1978);

4. Homologues are developmentally individualized building units (Wagner, 1989a,b);

5. Homology denotes constancy of constructional organization despite changes in underlying generative mechanisms;

6. Homologues act as organizers of the phenotype; and

7. Homologues act as organizers of the evolving molecular and genetic circuitry.

Premise 5 refers to a constitutive characteristic of homologues: they are more permanent than the generative mechanisms that establish them. Homologues remain constant over long, phylogenetic periods despite significant changes in the molecular, genetic, and developmental mechanisms that execute their realization. As a consequence, we must not assume that the mechanisms controlling the development of specific homologues in extant organisms are the same ones that acted at their origination, let alone "caused" these structures in evolution. The premise also indicates that the position of homologues in the organizational hierarchy of the phenotype is more important than the pathways of their construction.

Premise 6 refers to the independent, organizing role that homologues assume in morphological evolution. As more and more constructional detail is added, homologues come to act as organizers of the phenotype. Following an order-on-order principle, earlier homologues serve as accretion points for new elements that become parts of evolving body plans. At this stage, homologues are autonomized: they become constructionally independent building elements, which can be understood as attractors in the epigenetic and phenotypic landscapes of evolutionary lineages.

It was argued that the homologues that arise from generic properties of cell masses, and later from conditional interactions between cells and tissues, provided "morphogenetic templates" for the biochemical sophistication of cell interaction (for a discussion of the relation of homology to homoplasy in this framework, see Newman and Müller, 2000). Although their co-optation by the evolving genome results in the genetic "programs" of development, the structure of these genetic control systems is based on the morphogenetic templates already present, as premise 7 indicates. Hence the close mapping between genotype and morphological phenotype observed in evolved organisms should be interpreted as a consequence of evolution and not as its cause.

The organizational homology concept I am proposing is a biological homology concept (Roth, 1984; Wagner, 1989b): it refers to the biological mechanisms that underlie the origination of homology, rather than to the genealogical and taxonomic aspects emphasized by historical homology concepts. On the other hand, the concept is less restrictive than current definitions of biological homology, which are predominantly based on the notion of developmental constraints (but see Wagner and Chiu, chapter 15, this volume, for a modified definition). The organizational homology concept includes developmental constraints, but regards the constructional rules of the morphological phenotype as equally important. It gives priority to the active role of organizing processes rather than passive limitations. It is thus based on the following, preliminary definition: "Homologues are autonomized elements of the morphological phenotype that are maintained in evolution due to their organizational roles in heritable, genetic, developmental, and structural assemblies."

This definition permits us to formulate new strategies of empirical and theoretical research into the nature of the homology problem in morphological evolution. It emphasizes that a central task must be the elucidation of the organizational rules that govern the relation between genetic, epigenetic, and constructional determinants in the three-dimensional and four-dimensional processes of development.

A full appreciation and analysis of these processes will require the use of new computational tools in developmental research to accurately represent the relationship between gene activation, cell behavior, and morphogenesis (Jernvall, 2000; Streicher et al., 2000; Streicher and Müller, 2001). Data obtained by these and other means can be used to arrive at formal models of the organizing mechanisms relating genotype and phenotype (Salazar-Ciudad, Newman, and Solé, 2001; Salazar-Ciudad, Solé, and Newman, 2001; Szathmary, 2001; see also Britten, Bissell et al., and Nijhout, chapters 5, 7, and 10, this volume), with the prospect of making explicit and testable predictions about the evolution of homology. Other computational approaches can be based on the morphological homologues themselves (Rasskin-Gutman, chapter 17, this volume). These new uses of homology will eventually make it possible to integrate rules of developmental and phenotypic organization into the framework of evolutionary theory.

References

Abouheif E (1997) Developmental genetics and homology: A hierarchical approach. Trends Ecol Evoln 12: 405–408.

Alberch P (1982) Developmental constraints in evolutionary processes. In: Evolution and Development (Bonner JT, ed), 313–332. Berlin: Springer.

Arnold SJ (1989) How do complex organisms evolve? In: Complex Organismal Functions: Integration and Evolution in Vertebrates (Wake DB, Roth G, eds), 403–433. New York: Wiley.

Atchley WR, Newman SA, Cowley DE (1988) Genetic divergence in mandible form in relation to molecular divergence in inbred mouse strains. Genetics 120: 239–253.

Bock GR, Cardew G (eds) (1999) Homology. Chichester, England: Wiley.

Bolker JA, Raff RA (1996) Developmental genetics and traditional homology. BioEssays 18: 489–494.

Bookstein FL (1994) Can biometrical shape be a homologous character? In: Homology (Hall BK, ed), 197–227. San Diego, Calif.: Academic Press.

Bruna EM, Fisher RN, Case TJ (1996) Morphological and genetic evolution appear decoupled in Pacific skinks. Proc R Soc Lond B Biol Sci 263: 681–688.

Budd GE (1999) Does evolution in body patterning genes drive morphological change—or vice versa? BioEssays 21: 326–333.

Bürger R (1986) Constraints for the evolution of functionally coupled characters: A nonlinear analysis of a phenotypic model. Evolution 40: 182–193.

Darwin C (1859) On the Origin of Species by Means of Natural Selection. London: John Murray.

de Beer SG (ed) (1971) Homology, an unsolved problem. Oxford Biology Readers, vol. 11. London: Oxford University Press.

Dickinson WJ (1995) Molecules and morphology: Where's the homology? Trends Genet 11: 119–121.

Duboule D (ed) (1994) Guidebook to the Homeobox Genes. Oxford: Oxford University Press.

Endress PK (2001) Origins of flower morphology. In: The Character Concept in Evolutionary Biology (Wagner GP, ed) 493–510. San Diego; Calif.: Academic Press.

Fontana W, Buss LW (1994) The arrival of the fittest: Toward a theory of biological organization. Bull Math Biol 56: 1–64.

Fontana W, Wagner G, Buss LW (1994) Beyond digital naturalism. Artif Life 1: 211–227.

Galis F (1996) The application of functional morphology to evolutionary studies. TREE 11: 124–129.

Gilbert SF (2001) Ecological developmental biology: Developmental biology meets the real world. Devel Biol doi:10.1006/dbio.2001.0210 Available online at www.idealibrary.com.

Gilbert SF, Bolker JA (2001) Homologies of process and modular elements of embryonic construction. J Exp Zool (Mol Dev Evoln) 291: 1–12.

Hall BK (1983) Epigenetic control in development and evolution. In: Development and Evolution (Goodwin BC, Holder N, Wylie CG, eds), 353–379. Cambridge: Cambridge University Press.

Hall BK (1995) Homology and embryonic development. Evol Biol 28: 1–37.

Hall BK (ed) (1994) Homology. San Diego, Calif.: Academic Press.

Hickman CP, Roberts LS, Hickman FM (1988) Integrated Principles of Zoology. Saint Louis: Mosby.

Holland LZ, Holland PW, Holland ND (1996) Revealing homologies between body parts of distantly related animals by *in situ* hybridization to developmental genes: Amphioxus versus vertebrates. In: Molecular Zoology (Ferraris JD, Palumbi SR, eds), 267–295. New York: Wiley-Liss.

Jernvall J (2000) Evolutionary modification of development in mammalian teeth: Quantifying gene expression patterns and topography. Proc Natl Acad Sci USA 97: 14444–14448.

Johnston TD, Gottlieb G (1990) Neophenogenesis: A developmental theory of phenotypic evolution. J Theor Biol 147: 471–495.

Kauffman SA (1993) The Origins of Order. New York: Oxford University Press.

Laubichler MD (2000) Homology in development and development of the homology concept. Am Zool 40(5): 777–788.

Lauder GV (1994) Homology, form, and function. In: Homology (Hall BK, ed), 151–196. San Diego, Calif.: Academic Press.

Lowe CJ, Wray GA (1997) Radical alterations in the roles of homeobox genes during echinoderm evolution. Nature 389: 718–721.

Maynard Smith J, Burian R, Kauffman S, Alberch P, Campbell J, et al. (1985) Developmental constraints and evolution. Q Rev Biol 60: 265–287.

Meise M, Hilfikerkleiner D, Dubendorfer A, Brunner C, Nothiger R, et al. (1998) *Sex-lethal*, the master sex-determining gene in *Drosophila*, is not sex-specifically regulated in *Musca domestica*. Development 125: 1487–1494.

Melville H (1851) Moby-Dick. London: Richard Bentley.

Meyer A, Kocher TD, Basasibwaki P, Wilson AC (1990) Monophyletic origin of Lake Victoria cichlid fishes suggested by mitochondrial DNA sequences. Nature 347: 550–553.

Miklos G, Rubin H (1996) The role of the Genome Project in determining gene function: Insights from model organisms. Cell 86: 521–529.

Minelli A (1997) Molecules, developmental modules, and phenotypes: A combinatorial approach to homology. Mol Phylogenet Evoln 9: 340–347.

Minelli A, Peruffo B (1991) Developmental pathways, homology, and homonomy in metameric animals. J Evol Biol 2: 429–445.

Müller GB (1990) Developmental mechanisms at the origin of morphological novelty: A side-effect hypothesis. In: Evolutionary Innovations (Nitecki MH, ed), 99–130. Chicago: University of Chicago Press.

Müller GB (2001) Homologie und Analogie: Die vergleichende Grundlage von Morphologie und Ethologie. In: Konrad Lorenz und seine verhaltensbiologischen Konzepte aus heutiger Sicht (Kotrschal K, Müller GB, Winkler H, eds), 148–157. Fürth: Filander.

Müller GB, Newman SA (1999) Generation, integration, autonomy: Three steps in the evolution of homology. In: Homology (Bock GR, Cardew G, eds), 65–73. Chichester, England: Wiley.

Müller GB, Wagner GP (1991) Novelty in evolution: Restructuring the concept. Annu Rev Ecol Syst 22: 229–256.

Müller GB, Wagner GP (1996) Homology, Hox genes, and developmental integration. Am Zool 36: 4–13.

Nelson G (1994) Homology and systematics. In: Homology (Hall BK, ed), 101–149. San Diego, Calif.: Academic Press.

Newman SA (1994) Generic physical mechanisms of tissue morphogenesis: A common basis for development and evolution. J Evol Biol 7: 467–488.

Newman SA (1995) Interplay of genetics and physical processes of tissue morphogenesis in development and evolution: The biological fifth dimension. In: Interplay of Genetic and Physical Processes in the Development of Biological Form (Beysens D, Forgacs G, Gaill F, eds), 3–12. Singapore: World Scientific.

Newman SA, Comper WD (1990) "Generic" physical mechanisms of morphogenesis and pattern formation. Development 110: 1–18.

Newman SA, Müller GB (2000) Epigenetic mechanisms of character origination. J Exp Zool (Mol Dev Evol) 288: 304–317.

Olson E, Miller R (1958) Morphological Integration. Chicago: Chicago University Press.

Owen R (1843) Lectures on the comparative anatomy and physiology of the invertebrate animals, delivered at the Royal College of Surgeons, in 1843. London: Longman, Brown, Green, and Longmans.

Panchen AL (1994) Richard Owen and the concept of homology. In: Homology (Hall BK, ed), 21–62. San Diego, Calif.: Academic Press.

Quiring R, Walldorf U, Kloter U, Gehring WJ (1994) Homology of the *eyeless* gene of *Drosophila* to the *small eye* gene in mice and *aniridia* in humans. Science 265: 785–789.

Raff R (1999) Larval homologies and radical evolutionary changes in early development. In: Homology (Bock GR, Cardew G, eds), 110–121. Chichester, England: Wiley.

Riedl R (1978) Order in Living Organisms. Chichester, England: Wiley.

Rieppel OC (1988) Fundamentals of Comparative Biology. Basel: Birkhäuser.

Roth VL (1984) On homology. Biol J Linn Soc 22: 13–29.

Roth VL (1988) The biological basis of homology. In: Ontogeny and Systematics (Humphries CJ, ed), 1–26. New York: Columbia University Press.

Saccone G, Peluso I, Artiaco D, Giordano E, Bopp D, et al. (1998) The *Ceratitis capitata* homologue of the *Drosophila* sex-determining gene *sex-lethal* is structurally conserved, but not sex-specifically regulated. Development 125: 1495–1500.

Salazar-Ciudad I, Newman SA, Solé R (2001) Phenotypic and dynamical transitions in model genetic networks: 1. Emergence of patterns and genotype-phenotype relationships. Evol Dev 3: 84–94.

Salazar-Ciudad I, Solé R, Newman SA (2001) Phenotypic and dynamical transitions in model genetic networks: 2. Application to the evolution of segmentation mechanisms. Evol Dev 3: 95–103.

Shubin NH, Alberch P (1986) A morphogenetic approach to the origin and basic organization of the tetrapod limb. In: Evolutionary Biology (Hecht MK, Wallace B, Prance GT, eds), 319–387. New York: Plenum Press.

Spemann H (1915) Zur Geschichte und Kritik des Begriffs der Homologie. In: Allgemeine Biologie (Chun C, Johannsen W, eds), 63–86. Leipzig: Teubner.

Streicher J, Donat MA, Strauss B, Spörle R, Schughart K, et al. (2000) Computer-based three-dimensional visualization of developmental gene expression. Nat Genet 25: 147–152.

Streicher J, Müller GB (1992) Natural and experimental reduction of the avian fibula: Developmental thresholds and evolutionary constraint. J Morphol 214: 269–285.

Streicher J, Müller GB (2001) 3D modeling of gene expression patterns. Trends Biotechnol 19: 145–148.

Striedter GF (1998) Stepping into the same river twice: Homologues as recurring attractors in epigenetic landscapes. Brain Behav Evol 52: 218–231.

Sturmbauer C, Meyer A (1992) Genetic divergence, speciation, and morphological stasis in a lineage of African cichlid fishes. Nature 358: 578–581.

Szathmary E (2001) Developmental circuits rewired. Nature 411: 143–145.

Tautz D (1998) Evolutionary biology: Debatable homologies. Nature 395:17.

van Valen L (1982) Homology and causes. J Morphol 173: 305–312.

Waddington CH (1957) The Strategy of the Genes. London: Allen and Unwin.

Waddington CH (1962) New Patterns in Genetics and Development. New York: Columbia University Press.

Wagner GP (1984) Coevolution of functionally constrained characters: Prerequisites for adaptive versatility. BioSystems 17: 51–55.

Wagner GP (1989a) The origin of morphological characters and the biological basis of homology. Evolution 43: 1157–1171.

Wagner GP (1989b) The biological homology concept. Annu Rev Ecol Syst 20: 51–69.

Wagner GP (ed) (2001) The Character Concept in Evolutionary Biology. San Diego, Calif.: Academic Press.

Wagner GP, Misof BY (1993) How can a character be developmentally constrained despite variation in developmental pathways. J Evol Biol 6: 449–455.

Wake (1999) Homoplasy, homology, and the problem of "sameness" in evolutionary biology. In: Homology (Bock GR, Cardew G, eds), 24–33. Chichester, England: Wiley.

Wake DB, Roth G (eds) (1989) Complex Organismal Functions: Integration and Evolution in Vertebrates. New York: Wiley.

Wimsatt WC (1986) Developmental constraints, generative entrenchment, and the innate-acquired distinction. In: Integrating Scientific Disciplines (Bechtel W, ed), 185–208. Dordrecht: Nijhoff.

Wray (1999) Evolutionary dissociations between homologous genes and homologous structures. In: Homology (Bock GR, Cardew G, eds), 189–203. Chichester, England: Wiley.

Zarapkin S (1943) Die Hand des Menschen und der Menschenaffen: Eine biometrische Divergenzanalyse. Z menschl Vererb- und Konstitutionslehre 27: 390–414.

III RELATIONSHIPS BETWEEN GENES AND FORM

In this part, several approaches are presented to the organizing principles that relate genes to the construction of biologicalから. Gene structure and activity are the best understood aspect of the developmental process, but do they suffice to provide a full account of it? Roy Britten answers in the affirmative, stating that "only details determine"—to understand the construction of an organism, we need primarily characterize the interactions among genes and their products, refined by natural selection (however daunting this task may be). At the other philosophical pole, Ellen Larsen contends that "trying to understand the coordination of development from a knowledge of molecules and genes is akin to putting Humpty-Dumpty together again." The range of phenomena set out by Scott Gilbert and by Mina Bissell and colleagues reminds us that the "details that determine" can be considered to occur at multiple structural and dynamical levels and that the determinants of developmental gene expression are often conditionalities that may extend beyond the organism itself.

Roy Britten (chapter 5) points out that there are no individual genes known to encode large amounts of developmental information. Genes simply specify RNA molecules and proteins. Their products can enter into macromolecular assemblies, act to throw molecular switches at the transcriptional or posttranscriptional levels, but not much else. Leaving open the possibility that "details" did not determine the evolutionary route to modern organisms as thoroughly as they determine their current structures (a theme that is taken up in other chapters), Britten lists nine propositions that developmental systems need to fulfill in order to integrate the staggering molecular details that determine developmental episodes. The system he proposes would work through self-assembly and automatic coordination of local activities, but without "overarching or global control mechanisms." The need for high-powered digital computation to establish concrete models is clear, and Britten initiates this important task with a novel formal framework based on a numbering system for specifying cell states, analogous to that used by Kurt Gödel to explore the logical structure of mathematics.

Scott Gilbert (chapter 6) seeks to extend the locus of developmental determination beyond the organism's perimeter. Taking as his starting point the history of embryology, he notes that explanatory models have become progressively more reductionistic over the past century and a half. As the field moved from an original grounding in comparative anatomy to a more mechanistic concern with underlying physiological processes, and then on to the contemporary concentration on gene action and presumed "genetic programs" of development, the centrality of the organism as a whole and the often great extent to which its structural features are determined by its natural context were eclipsed. Gilbert contends that the narrowed focus necessary to acquire the data relevant to these frameworks of causation led to the use of experimental systems that lack the plasticity and conditionality that characterize much of development, what he calls the "reactive genome." By extension of the embryological concepts of primary and secondary induction-intraembryonic interactions that establish body plan and organ form, he proposes the concept of "tertiary induction" to denote the interactions by which environmental factors influence the phenotype on the developing organism. With a few well-chosen examples, Gilbert persuasively establishes the generality of evolved capacities to incorporate external cues into developmental repertoires.

Mina Bissell, Saira Mian, Derek Radisky, and Eva Turley (chapter 7) show that the cellular properties involved in organogenesis are more extensive than the adhesive interactions which form the sole basis of tissue organization in many accounts and which drive sorting-out behavior of different cell types (see Steinberg, chapter 9, this volume). They provide evidence, moreover, that even cellular phenotype is plastic and dependent on context. Thus a full picture of the development of form must take into account not only how preestablished, differentiated cells build three-dimensional structures, but how those structures act back on the cellular components and alter their differentiated states. In this view (as in Scott Gilbert's) the very "details that determine" are subject in a reciprocal fashion to what is being determined. Reviewing their earlier work on the relationship of lumenal organization of mammary epithelium to the extracellular environment (including the discovery that even mammary tumor cells can be coaxed to assemble into normal glandular structures given a normal microenvironment), Bissell and colleagues explore the question, also raised by Roy Britten, of what computational framework would be appropriate for capturing the unique properties of this class of complex systems. They consider the theory of "highly optimized tolerance" in which evolved design features ("details") permit interconnected systems to gain a measure of robustness against uncertainties in one domain by becoming more sensitive in others, and the phenomenon of "stochastic resonance" in which external fluctuation ("noise") elicits organization in a complex system, which may then be inherited if a means of recording, such as genetic change, is available.

By explicitly considering evolution in her formulation of the problem of organismal form, Ellen Larsen (chapter 8) homes in on a paradox in any purely reductionist attempt to understand the relationship between genes and form. If the majority of molecules regulating development (with some important exceptions) are common to all metazoan taxa, how can we account for the extensive disparities in morphological phenotype in contemporary organisms? Larsen presents a framework in which "cells and their properties form both a material and conceptual link between genotype and phenotype." This framework permits integration of the genetic "details determine" perspective of Britten and the self-assembly phenomena reviewed by Steinberg. In particular, Larsen's classification of genes into "worker genes," which provide material for cell behaviors (e.g., cytoskeletal and membrane proteins) and "bureaucrat genes," whose products control the activity of other genes, permits one to see how the determinants of organ self-assembly can be fine-tuned over the course of evolution by changes in regulatory coupling strengths rather than gene composition. Finally, Larsen convincingly argues that studies of what forms arose in evolution need to be combined with studies of what forms are possible. She presents an experimental approach for exploring the range of possible morphologies that can be achieved by a developmental system via small changes in cell behavior.

5 Only Details Determine

Roy J. Britten

There are no known genes that individually encode large amounts of information specifying the structure or patterns of development of an organism. Although we may simply not know how to appreciate or find such genes, I assume they do not exist. Logically then, many individual genes and molecules (details) participate and interact in the process of development, as a result of a long process of natural selection. These details determine everything. The system works through self-assembly and automatic coordination of their activities. The details include, among many others, specific binding of individual macromolecules to form the structure of the cell; the many details required for the differentiated state of cells and for cell structures such as filopods; transcription factors and their binding sites; many details needed for expression and modification of transcription factors and many other molecules; signaling molecules and their receptors; intercellular adhesion sites and receptors; control molecules capable of switching on chains of control processes leading to complex structures; and macromolecules and their receptors active in guiding neurons or cell movements in development.

I propose a numbering system with a set of entries specific to each cell state and with an individual number assigned to each of the details and to some important summaries such as cell differentiated state. These details and summaries include measures of expression of genes; rate of synthesis of gene products including transcription factors and signaling molecules; their modification; description of concentration and location (intra- or extracellular, nuclear or gradient); description of cell structure and differentiation; the timing and lineage of cells; and significant extracellular structures and states. Because we may be able to demonstrate that biology depends just on the details without knowing about all of them and the way they work, consider what form useful and enjoyable knowledge might take once many of the details were recorded in an immense computer library.

I appreciate the invitation to write this chapter for the Vienna Series in Theoretical Biology for many reasons, chief among which is that I was forced to think about the way life works and have further explored a concept that can be labeled "only details determine." Also, Vienna brought Kurt Gödel to mind, which led me to propose a numbering system for all of the details specific to each state. Though it is not mentioned in papers I have seen, I do not believe that the concept that "only details determine" is new, but actually underlies much research. The abstract for a paper entitled "Biological Computation" (Brenner, 1998, 106) states: "Genes can only specify the properties of the proteins they code for and any integrative properties of the system must be 'computed' by their interactions." No application was considered for the specification of eukaryotic form. I do not believe that the implications of the "details determine" (DD) thesis have been examined as far as can be done with present knowledge.

A Propositional Model

In an attempt (Britten, 1998) to show that living systems could successfully develop based on the details without any overarching or global control mechanisms, I constructed a verbal model that started with ten propositions. Nine of the original propositions and comments are listed here (proposition 9 has been omitted), with deletions and modifications, and with some added comments placed in square brackets.

My model makes use of a simplified terminology (for a good source of appropriate references and customary terminology, see, for example, Davidson, 1994). The macromolecules that make up the structure of the cells and extracellular parts of organisms are termed *units* if they can be or are bound into a structure, regardless of additional biochemical or other capabilities. The unconventional usage *assembly of units* means "growth from egg to adult." *Local,* when describing cell-to-cell relationships, refers primarily to adhesion to and signaling between adjacent cells; *signaling* is used comprehensively to cover all possible processes whereby adjacent and more distant cells and tissues can affect the pattern of gene expression in a nucleus. *Adhesion* has its usual meaning for specific binding between cells, assuming that all cells have specific adhesion molecules on their surface that locate them by binding to matching sites on other cells, although these are not always fixed, and cells can migrate in controlled fashion. *Organ* has its usual meaning. The timing of cell stages and their lineage, to the extent that it is definable, is central to development. Only through development can all of the details be expressed and utilized.

1. *The control of development is by means of local interactions.* Local means "binding between macromolecules and interactions between adjacent or nearby cells" and does not exclude morphogen or hormone interactions. It does exclude potential global control mechanisms containing much information specifying form because only mechanistically understood kinds of control are acceptable in the model and overarching regulatory principles are not. A few long-range diffusing molecules have important roles, for example, in sex determination. There exist important gradients such as *bicoid* in *Drosophila* eggs that act over large distances and contain information important to development. Though not precisely local, such gradients do not contain global information specifying the form of the embryo. As an example of other large-scale processes, the environment of the embryo may supply minerals, nutrition, and hormones—and may, under abnormal circumstances, damage the embryo or upset control processes—but it does not specify the form of the embryo. The extent of plant growth, leaf size, and other aspects respond to conditions, largely as a choice between alternatives, each of which is controlled by local cell and tissue interactions. Because there is no evidence for global control processes specifying form (nor have any complete theoretical models of such control been devised), this model is restricted to

local control except for certain well-known long-range processes such as hormone action and control factor gradients, each not including large amounts of information that specify form. [At this time, I would replace "local interaction," which is not easily defined, with "detailed interaction."]

2. *A macromolecule near a specific site will bind by mass action.* [As macromolecules are synthesized within a cell, they are modified to join a structure. They bind specifically to the preexisting partial structure in the correct, genetically determined, locations and orientations. To reach their binding sites is a complex process that likely involves the Golgi apparatus and vesicles, and may involve the organized transport of vesicles. Diffusion and mass action may be restricted to the final binding process. When bound, each unit and adjacent units then expose one or more binding sites where the next macromolecules will specifically bind. The succeeding extended series of binding events establishes each cell, with its specific adhesion and signaling receptors.]

3. *Starting with a precursor cell, all cells are assembled automatically by specifically binding new macromolecules.* Although there is little direct evidence for this logically required proposition, no alternatives are known. During cleavage, the units present in the precursor cell become available for the daughter cells, which often differ in detail from each other, and are bound in place. According to the model, all of these controlled events are due to the specific genes expressed as required in each of the cells in all of the lineages and set by transcription control factors and other regulatory mechanisms. ["Assembled automatically" is taken to mean that supplying the elements and cofactors that go into a structure causes them to be assembled without any overall guidance.]

4. *At the surface of cells are specific adhesion sites that determine how all cells bind to each other.* This proposition derives from a wide range of studies (Edelman, Cunningham, and Thiery, 1990; Brümmendorf and Rathjen, 1994; Jones et al., 1997; Miller and McClay, 1997a,b).

5. *Both the molecular and cellular binding processes are specific, and an organ will assemble automatically when the parts (macromolecules, extracellular structures and cells) appear as specified by nuclear control factors.* Automatic assembly with cells as building blocks is how the model works. Because the adhesion sites present on the cells are determined by the state of the control factors, no other information is required for assembly. The process is not mass action: many cell locations are the result of their duplication in place; other cells move in controlled fashion. For this and other reasons, a dissociated organ will not usually self-assemble again. Plants may lack specific adhesion and cell motion, but the assembly during growth is automatic in any case.

6. *The set of nuclear control factors in each cell is a combination of inherited factors from precursor cells and factors derived by signaling from other cells.* This proposition,

though obvious, introduces a powerful concept. Adjacent cell signaling will establish most patterns, but diffusible ligands or factors are also important. In early cleavage, the control factors are mostly maternal. Later, signaling control becomes more important as cells become differentiated.

7. *The macromolecules that determine specific binding, cell adhesion, and signaling are produced as specified by the nuclear control factors, and, in a grand feedback, the cell adhesion and signaling systems determine the nuclear factor patterns that control the expression of these macromolecules.* The pattern of transcription is established by the transcription factors that are determined by cell-cell interactions, signaling, and diffusible factors. The factor pattern is changed from outside the cells and in turn affects the intercellular binding sites. The expression of the units, their binding, and the set of transcription control factors operate as a linked set, different of course in each part of the developing embryo. Because all the elements in a feedback loop are subject to experimental manipulation, a macromolecule might seem to be in charge, when in fact no single part of the feedback loop is in control (Callaerts, Halder, and Gehring, 1997; Desplan, 1997).

Propositions 1 to 6 are general descriptions of the processes that underlie development. Proposition 7 refers to the most important control processes including those receiving the greatest attention from researchers at the heart of development. Propositions 8 and 10, based on a plethora of specific examples where the factors and signaling systems have been observed and the stages of morphogenesis followed, deal with the formation of the precursors of organs and ultimately the organs themselves. Model animal systems differ widely: "In some embryos, specification depends on intercellular interaction during cleavage, while in others this cannot be so since specification occurs while nuclei are syncytial; some rely on invariant cell lineages, while others develop from populations of migratory cells of no fixed lineage; some generate autonomously specified founder cells, while others have none; and so forth" (Davidson, 1994, 604). My proposed model, by contrast is intended to apply to all, although autonomous specification and syncytial cases need comment, and some rephrasing is required. In insects, the syncytial nuclei are formed in place and the specific cell-to-cell adhesion is delayed until cellularization. The nuclei do influence each other's factor pattern; before cellularization, many specific interactions are formed, in *Drosophila,* for example, that lead to the "stripes" establishing the precursors of segments. The anteroposterior and dorsoventral axes are established also.

8. *The embryonic precursor cells of organs known as "precursor groups" (pgroups) are linked by specific adhesion and signaling relationships.* [The pgroup is a radical way of expressing what may be an ordinary concept in the minds of developmental biologists. The original pgroup discussion including proposition 9 is omitted here because it involves additional hypotheses that are not necessary.] The ability of natural selection to establish all

of the series of specific relationships of signaling and adhesion that carry a cell lineage through the many duplications and steps to a functional location in an adult organ has a great fascination. This is simultaneously performed for many lineages leading to perhaps more than 10^{12} ultimate cells, the relationships among which maintain their function in the adult. That is the essence of the model.

10. *Organs are held together by cell adhesion in functional relationships.* Thus the form and function of the organism are specified entirely by local control mechanisms that establish the organs. The form of the organs is decided by the complex cell-to-cell relationships and the adhesion molecules that bind the cells. The organs are formed in association with each other; the adhesion characteristics of the appropriate parts bind them to each other, establishing the form of the organism. Some organs will contribute more to the form than others because they may be larger or external. In a sense, the division of an organism into organs is arbitrary and the pgroups are considered to apply to the actual requirements for assembly where much is to be learned. For indirect development, the organs may become parts of the larva or pupa, while a subset of lineages form the pgroups of the imaginal disks or rudiment and become parts of the adult.

Gödel in Vienna

Some seventy years ago a 24-year-old from Vienna changed forever the face of mathematics (Gödel, 1931). Kurt Gödel used a numbering system to identify statements about each mathematical theorem and showed that any consistent mathematical theory that includes the natural numbers is incomplete and in addition cannot contain a proof of its own consistency. Mathematically defining completeness and consistency would give a firm logical foundation to mathematics, or so leaders such as David Hilbert and Bertrand Russell expected. Their theoretical great expectations were disappointed. Of course biology lacks the strong logical structure of mathematics, but analogies between the fields are clear. An example of failed great expectations in biology is that we could find a tree of relationship among all organisms based on their descent, including the relationship of molecular sequences. As more information on the genomes of microorganisms has become available, however, the prospect of finding such a tree has become ever more remote, due in part to the horizontal transfer of many genes and to differences in the rate of DNA evolution (e.g., Pennisi, 1999). Although other great expectations, that we can understand embryonic development by recognizing the control systems or "master genes," have yet to be tested, I will point out some rough spots on the road to such understanding.

The cell state numbering system I am proposing, though analogous to Gödel's numbering of mathematical statements, thus far lacks a clear logical structure. Ultimately, this numbering system could record all of the significant details of living systems; for now, it is quite abstract and will require an immense amount of labor to complete.

Neural Guidance and a Balance of Molecules and Receptors

Recent work has shown that the quantitative level of certain control elements in specific neural connections of the brain is important (Bashaw and Goodman, 1999; Seeger and Beattie, 1999; Serafini, 1999) including members of the Ephrin, Netrin, Semaphorin, and Slit protein families. The quantitative levels actually control whether particular control molecules cause attraction or repulsion. The importance of quantitative balance of control elements is not novel or specific to neural connections. It is crucial that a modest number of different control elements can in combination specify a large number specific relationships or, in the case of neurons, connections. In Britten, 1998, 9374, I wrote that "it is very unlikely that 10^{10} different and effective adhesion and signaling relationships exist." I believe that was an error; as I demonstrate below, 10^{14} possible choices required for neural connections can be decided by the level of expression of a few genes, described abstractly for this demonstration as a few numbers. In an unrealistic extreme case, 1 byte can represent 256 possible states; 2 bytes, $256^2 = 65,536$ states. For n bytes, the number is $256^n = 10^{2.408n}$. Thus even as few as 6 bytes could represent more than 10^{14} states and establish the choice between that many outcomes. Of course, it does not work like that: the number of significantly distinct concentrations for a control element might be low, perhaps only 2 or 3. If it were only 2 (i.e., a switch being on or off) and $2^n = 10^{0.301n}$, then 10^{14} states would require 47 bits or switches, which is still a very reasonable number. In the "state numbering system," we may still be required to set aside a byte for each state to allow for those cases where small quantitative changes in concentration are significant. It is unlikely that 47 different control elements are required for neural connections in the brain. It will be interesting to see just what kind of economy natural selection has managed for the set of control elements, differentiated cell states, and regional restrictions that make the choices and establish the brain neuronal connections. Another aspect of this numerical analysis is that it assures us that the small number of genes (10^5 maximum) can support all of the complexity of a cell or organism structure with a number of states or decisions well in excess of 10^{12} or 10^{14}. A "state numbering system" could work and give insights into how this is accomplished.

Shared Functions

Many parts of living systems are developed or maintained by sets of parts that carry out very similar roles. Examples include parallel processing in the brain and brain neurons with parallel functions; the binding sites in the 5′ control regions of genes, many of which bind the same transactivator or repressor protein; apparently parallel signaling pathways; perhaps clusters of Hox genes; ribosomal cistrons; histone gene clusters; and other duplicated genes.

In most of these examples, a group of similar effectors carry out almost identical roles. We may be able to add to this list the purported proteins involved in self-assembly of cells and the adhesion molecules involved in self-assembly of organs or developmental structures from cells. A central question is, how are the many macromolecules with nearly identical functions maintained in evolution? If they were exactly redundant, then the loss of one of them would make no difference to the survival or reproduction of the animal and some copies would drift in sequence, most likely becoming nonfunctional. This would not occur if each part contributed, so that the loss of any one represented a selective disadvantage, with the role of each part being, say, slightly different, or with each having a role in different circumstances. Another possibility is that, though identical, each part contributes quantitatively and the total has selective advantage over any reduced number. With large populations of animals, the selective advantage of the total over a smaller number of parts could be very small and still give enough selective advantage to maintain the complete set.

Shared function is perhaps important in the self-assembly of cells. A model of this process would require many specific binding sites and many proteins in specific locations to form all of the structures, junctions, and adhesion sites. Although the genetic defects associated with these processes are not evident in the record, there are nevertheless so many of them that they might in principle dominate a list of defects. One explanation for this is that because, in many cases, there are multiple genes for families of closely related proteins that share these functions, the genetic defects are not recognizable in ordinary screening. Indeed, they would be important only in long periods of evolution, where small differences in viability would be sufficient to maintain the whole family. Shared function of similar elements might be important to the balance of regulatory elements that permit choice between large numbers of connections, and the subtle balance of similar elements might be very difficult to analyze.

Consideration of a Cell and Cells

Although development depends on the state and structure of the many cells that participate in what may be described as a "contingent fashion," we know little of how the cells themselves form during development. In many cases, certain control molecules are necessary for a particular differentiated state, but the downstream processes are not known. The formation of the differentiated cell might require a process of controlled synthesis, modification, transport, and assembly of perhaps 10^{12} molecules to form the supporting structure, membrane, and specialized parts. In the most extreme case there are many hundreds of different brain neurons. I bring up cells at this point in the argument because as far as I know no one has or could suggest a master gene with the information content that could specify the role 10^{12} cells in the structure. The logic is clear: timed control of synthesis, transport and modification of molecules, and self-assembly of the cell. A lot of details formed by a lot of details, and timing is crucial.

Of course every cell is daughter to another cell, and much may be inherited by the daughter cell. That may include an initial state made up of levels and balances among the control systems as well as some physical structure. Something of the sort must be present to be modified for the succeeding differentiated state. It is not a magic element of cellness or life that is passed on to daughter cells, but simply a state (possibly defined by a large number). This process occurs for every daughter cell, and later is presumably switched by signaling systems to the new state and structure. During cleavage in the absence of growth, as in sea urchin embryos, much of the process involves rearranging molecules, with minimal new synthesis. Many of the external features of this process have been well studied, for example, which cells are totipotent and which are effectively differentiated, and some of the signals have been identified. Nevertheless, there is still an enormous amount to be learned about just how the 10^{12} or more molecules get into place each time in the formation of what may end up being 10^{12} cells. (The exponent 12 is somewhat arbitrary, standing for what is indubitably a large number.) However the controls are set, the result must be self-assembly involving what may multiply to 10^{23} molecules in the organism. There are many details to get more or less right every time so that the next step may function because the details are in place. In summary, we cannot start with DNA and grow a cell because there must be an adequate initial state of a cell with a vast multitude of details under control, and we simply do not know enough to set up such a state.

Determination

Moreno and Morata (1999, 873) nicely capture the current phraseology and attitude of genetic determination:

Segment identity along the anteroposterior body axis of *Drosophila* is determined by the genes of the Hox complex (McGinnis and Krumlauf, 1992) but there are two regions that develop normally in the absence of Hox gene activity: the anterior head region and the most posterior body segment which includes the anal structures of larvae and adults.... We propose that *cad* [caudal] is the Hox gene that determines the development of the fly's most posterior segment.

Moreno and Morata show specific local expression of *cad* and that it is "required for normal development of analia structures." They examine the other control genes that are required in the process downstream of *cad*. They also show that ectopic expression leads to ectopic formation of anal structures in the head region and several other but not all locations.

Clearly "determined by" means specifically "switched on by" and does not mean that the information for the anal structure is carried by the *cad* gene or gene product. Indeed, even the *cad* gene's switching ability depends on the environment. The gene also has a maternal embryonic function that establishes the anteroposterior body axis of *Drosophila* (Macdonald and Struhl, 1986), although I do not feel that the grand implications of this function are widely different from, for example, those of the *Pax-6* (*ey*) gene, which determines the *Drosophila* eye. Several control genes can participate, and their presence with the *ey* gene can cause the ectopic production of many more fragments of eye structures. This example is striking: although the chicken analogue of the *ey* gene will replace the *Drosophila ey* gene, *Drosophila* eyes are formed, rather than chicken eyes. This neatly demonstrates that no information beyond switching is required to replace the *ey* gene function (Callaerts, Halder, and Gehring, 1997; Chen et al., 1997; Halder, Callaerts and Gehring, 1995; Pignoni et al., 1997).

Applying Numbers as a Descriptive Method: Three Reasons

The first and simplest reason to identify each stage, cell, and cell part with a Gödel-like number is to count them. The second and primary reason for numbering is to define a "state number set" that lists all of the significant elements representing each stage, cell, or cell part, namely, transcription control factor concentrations; other controls, signals, receptors, and receptor linkages to the nucleus; cell adhesion elements; and state of differentiation of the cell, timing, and lineage. Included may be descriptions of organs or organ parts and relationships to other parts of an organ, for example, brain neural connections. The extent of the list is awe-inspiring; it could hardly be built or dealt with it without computer help.

The third reason for numbering is to assemble a preliminary, shorter form of the just-described list for testing. I visualize a computer model that will progress from building differentiated cells and forming functional organs with them, to forming zygotes and fertilizing them to produce a population, which would then be asked to survive and evolve in some niche. Using such a method, the "details determine" concept might be shown to be a workable description of life.

Details Determine—Living with the Concept

Although I am logically forced to the position that details determine, I am daunted by the staggering profusion and complexity of details and linkages between some of the details, for example, networks of regulatory molecules. A cell might have 10^{12} macromolecules, each binding or fitting its neighbors, but we do not know how many distinct macromolecules, nor how many genes determine them, nor how the molecules specifically bind. Nor do we know which genes form the many differentiated cells, some with microvilli and filopods, and including of course the neurons in their hundreds of specialized varieties. Nor which elements must be specified for the cells to specifically bind together to form organs, among which is the brain with all its neurons and specific junctions. Clearly, no individual could list or deal with all these details, although computer programs will likely succeed. By producing sublists of values that were significant for particular issues (suppressing all zero or constant standard values) and by carrying such sublists through the stages of development, computer programs might enlighten us on its processes.

How will we learn to live with "details determine"? Mostly, we will do what we have always done. The level of integration is a matter of taste. Using a good model system, we might, for example, find an interesting part of development and focus on it, trying to identify the important regulatory molecules. On the other hand, knowing what the overall system is like may help, making us focus more on relationship to the whole, perhaps helping us evaluate the part of the state number that is different with time or place, and thus bringing us closer to an overall image or at least to a glimpse of it.

What Difference Does "Details Determine" Make in the Interpretation of Biological Knowledge?

It seems to me unlikely that "Details Determine" (DD) will be replaced by a large set of regulatory genes that include much information and that determine pattern formation. Researchers have long explored the magnitude of phenotypic effects of mutations in evolution (see, for example, Fisher, 1930; Kimura, 1983; Marshall, Orr, and Patel, 1999). As far as I know, no one has shown that mutations with large phenotypic effects occur in genes with large information content where that content plays a determining role. Even candidate genes such as the highly conserved Hox genes can be considered as switches whose major effects occur as a result of many downstream processes (Weatherbee et al., 1998).

Without order and summary principles, our world is too complex to parse and remember. How then to judge and understand—indeed, to make sense of—the immense biological knowledge with all of its values behind the details of DD? That a major control gene has little information makes it no less interesting. The DD point of view focuses on how such a major switch works. What are the differentiated cells? What other control elements and what downstream genes are involved? Although these questions are much the same as they were without DD, the formal way of presenting the answers is likely to differ, particularly if the numbering system turns out to be practical.

Phenomena encapsulated in principles such as organizers, morphogens, selector genes, and induction are valuable and, I believe, indispensable ways of understanding. The number and strength of regulatory relationships from the DD point of view will ultimately allow us to describe these phenomena at the molecular level; indeed, we have already begun to do so, as in the case of the gastrula organizer (Nieto, 1999).

The question arises as to what form useful and enjoyable knowledge will take once the details are recorded in an immense computer library. There might be summaries that are less "functional descriptions of mechanism" and more "needed ways of understanding and remembering." For those of us trained in molecular approaches, a satisfactory goal might be a DD "state number set" of very large size with technical quantitative descriptions of examples of typical interactions for reference in our computers. Presumably, there would be many ways of viewing and comparing regions and deriving grand summaries. For many others, such a set and descriptions would lack the essence of which knowledge is made.

Nagel (1998) states that there are "two varieties of antireductionism: epistemological and ontological." The first holds that, "given our finite mental capacities, we would not be able to grasp the ultimate physical explanation even if we knew the laws governing their ultimate constituents." The second holds that "certain higher order phenomena cannot even in principle be fully explained by physics, but require additional principles that are not entailed by the laws governing the basic constituents." The issue is not antireductionism, however. It is a direct question of whether genes with large information content exist. I am excited by the prospect of proving the grand principle that biology depends just on the details, that living systems function entirely on the basis of many detailed interactions, for which genes with little information carry all that is needed. Although it would be virtually impossible to prove that genes with large amounts of information encoding shape or function were absent, strong suggestive evidence could come from modeling to show that such a system could work.

References

Bashaw GJ, Goodman CS (1999) Chimeric axon guidance receptors: The cytoplasmic domains of Slit and Netrin receptors specify attraction versus repulsion. Cell 97: 917–926.

Brenner S (1998) Biological computation. In: Novartis Foundation Symposium 213: The Limits of Reductionism in Biology (Bock, GR, Goode, JA, eds), 106–116. Chichester, England: Wiley.

Britten RJ (1998) Underlying assumptions of developmental models. Proc Natl Acad Sci USA 95: 9372–9377.

Brümmendorf T, Rathjen FG (1994) Protein Profile. Vol 1, no 9. London: Academic Press.

Callaerts P, Halder G, Gehring WJ (1997) *Pax-6* in development and evolution. Annu Rev Neurosci 20: 483–532.

Chen R, Amoui M, Zhang Z, Mardon G (1997) Dachshund and eyes absent proteins form a complex and function synergistically to induce ectopic eye development in *Drosophila*. Cell 91: 893–903.

Davidson EH (1994) Molecular biology of embryonic development: How far have we come in the last ten years? BioEssays 16: 603–615.

Desplan C (1997) Eye development: Governed by a dictator or a junta? Cell 91: 861–864.

Edelman GM, Cunningham BA, Thiery JP (eds) (1990) Morphoregulatory Molecules. New York: Wiley.

Fisher RA (1930) The Genetical Theory of Natural Selection. Oxford: Oxford University Press.

Gödel K (1931) Über formal unentscheidbare Sätze der *Principia Mathematica* und verwandter Systeme, part 1. Monatshefte für Mathematik und Physik 38: 173–198.

Halder G, Callaerts P, Gehring WJ (1995) Induction of ectopic eyes by targeted expression of the *eyeless* gene in *Drosophila*. Science 267: 1788–1792.

Jones FS, Kioussi C, Copertino DW, Kallunki P, Holst BD, Edelman GM (1997) *Barx2*, a new homeobox gene of the *Bar* class, is expressed in neural and craniofacial structures during development. Proc Natl Acad Sci USA 94: 2632–2637.

Kimura M (1983) The Neutral Theory of Molecular Evolution. Cambridge: Cambridge University Press.

Macdonald PM, Struhl G (1986) A molecular gradient in early *Drosophila* embryos and its role in specifying the body pattern. Nature 324: 537–545.

Marshall CR, Orr HA, Patel NH (1999) Morphological innovation and developmental genetics. Proc Natl Acad Sci USA 96: 9995–9996.

McGinnis W, Krumlauf R (1992) Homeobox genes and axial patterning. Cell 68: 283–302.

Miller JR, McClay DR (1997a) Changes in the pattern of adherens junction-associated beta-catenin accompany morphogenesis in the sea urchin embryo. Dev Biol 192: 310–322.

Miller JR, McClay DR (1997b) Characterization of the role of cadherin in regulating cell adhesion during sea urchin development. Dev Biol 192: 323–339.

Moreno E, Morata G (1999) *Caudal* is the Hox gene that specifies the most posterior *Drosophila* segment. Nature 400: 873–877.

Nagel T (1998) Reductionism and antireductionism. In: Novartis Foundation Symposium 213: The Limits of Reductionism in Biology (Bock, GR, Goode, JA, eds), 3–14. Chichester, England: Wiley.

Nieto MA (1999) Reorganizing the organizer 75 years on. Cell 98: 417–425.

Pennisi E (1999) Is it time to uproot the tree of life? Science 284: 1305–1307.

Pignoni F, Hu B, Zavitz KH, Xiao J, Garrity PA, Zipursky SL (1997) The eye-specification proteins *so* and *eya* form a complex and regulate multiple steps in *Drosophila* eye development. Cell 91: 881–891.

Seeger MA, Beattie, CE (1999) Attraction versus repulsion: Modular receptors make the difference in axon guidance. Cell 97: 821–824.

Serafini T (1999) Finding a partner in a crowd: Neuronal diversity and synaptogenesis. Cell 98: 133–136.

Weatherbee SD, Halder G, Kim J, Hudson A, Carroll S (1998) Ultrabithorax regulates genes at several levels of the wing-patterning hierarchy to shape the development of the *Drosophila* haltere. Genes Dev 12: 1474–1482.

6 The Reactive Genome

Scott F. Gilbert

The final production of a phenotype is regulated by differential gene expression. However, the regulators of gene expression need not all reside within the embryo. Environmental factors such as temperature, photoperiod, diet, population density, or the presence of predators can produce specific phenotypes, presumably by altering gene expression patterns. Here "signals from above" interact with internal signals to produce the particular phenotype. This chapter looks at some of the historical trends that mediated the removal of environmental considerations from embryology. It provides evidence that internal factors, alone, cannot give a complete explanation of development. In many instances, significant developmental phenomena are given their specificity by the particular circumstances of the environment. Whereas it is usually thought that genes provide the instructional specificity for development (why the moth is brown or white, why the head is a particular shape, why a certain bone develops in a particular place) and that the environment is merely permissive, there are many cases wherein the genome is permissive and environmental instructions elicit a particular phenotype (see also Nanjundiah, chapter 14, this volume).

The Organism Out of Context: Embryology from Evolution to Physiology

Embryology is a science that links egg to adult. It is a science of becoming. Developmental biology, the anagenetic descendent of embryology, retains the fundamental questions of embryology and has added others. Developmental biology studies not only embryos, but also such diverse developmental phenomena as regeneration, metamorphosis, and the formation of blood cells and lymphocytes in the adult. Because there are no sets of techniques, levels of organization, or types of organism that limit the field of inquiry, the techniques and contexts for studying development have come largely from other disciplines.

In the mid to late nineteenth century, the context for studying development came from evolutionary biology, and the methods were, therefore, those of the evolutionary biologist, that is, of comparative anatomy. Embryology was to assist paleontologists and evolutionary biologists reconstruct the phylogeny of life. Homologies derived from anatomical studies and cell lineage analyses were the fruits of these endeavors.

With the twentieth century, the science of developmental mechanics (*Entwicklungsmechanik*) or physiological embryology emerged. This approach sought the physiological mechanisms by which the egg became an adult. Comparative anatomy was to be superseded by mechanistic physiology. Just as physiology was considered the "new" biology, so physiological embryology would be the "new" embryology. The anatomical tradition and its evolutionary context were considered old-fashioned. Experimentation was the watchword of the day. As Nyhart (1995) notes, after Roux's programmatic statement of

Entwicklungsmechanik, developmental mechanics and experiment became synonymous with modernity among a group of younger enthusiasts. This did not go unchallenged. Nyhart has documented the indignation and resentment felt by many of the developmental anatomists over this change in attitude. First, the evolutionary context that they had taken for granted was not being used, and the questions they had invested their life pursuing were not thought worth following. Second, the discipline was being brought indoors. Many of the older embryonic anatomists worried that students trained in the new experimental embryology would no longer be aware of the developing organism in its natural context. Third, the questions that were asked were radically different questions and they called out for different techniques. Rather than carefully observing embryos, these physiological embryologists were going to manipulate their embryo parts in controlled experiments.

In some cases, the new physiological embryology attempted to keep some of its environmental roots, and some of the first experiments done on embryos (such as those performed by Curt Herbst and by members of the Institute for Experimental Biology—the Vivarium—in Vienna) sought to change the environmental parameters, such as temperature or ionic conditions, in which embryos developed. The mainstream of physiological embryology, however, was the experimental program to determine the relationships between the embryonic parts by means of defects, cell and tissue isolation, rearrangement, and transplantation. The dominant scientists in this program included Hans Driesch, Hans Spemann, Sven Hörstadius, Ross G. Harrison, T. H. Morgan, Theodor Boveri, and their students. Embryology moved from the seashore (where the anatomical tradition still was strong) into the laboratory; developmental biology became a laboratory discipline, wherein scientists would manipulate the embryo, one variable at a time.

It was in this physiological context that the "model system" approach emerged in embryology. In the anatomical tradition, the choice of organism was dominated by the particular question and by the availability of the organism. We learned much about the anatomy of those organisms living close to the scientists, and we learned much about organisms (such as penguins and gorillas) whose anatomy might divulge information about phylogeny. In many cases, the seasonal availability of embryos dictated which organism was to be studied. However, with the advent of physiology, the variety found in nature had to be diminished so that the populations being tested would be as identical as possible, a necessary precondition for controlled experiments and for comparing conclusions between different laboratories. Thus animals came to be bred in the laboratory.

As Bolker (1995) has pointed out, though, very few organisms are capable of developing in the laboratory. Such organisms must be selected for the inability of their development to be influenced by specific environmental cues. Sea urchins, flatworms, and frogs, for example, found favor because they could develop readily in seawater or pond water in an aerated beaker. They did not even need to be fed during early development. Similarly, the chick has always been a model system (in both the anatomical and physiological contexts) because its egg has within it all the environment it needs (just add temperature) and because domestication has made the incubation time uniform and rapid. Thus both the influence of environment and environmental sources of phenotypic diversity were progressively eliminated under the physiological context of embryology.

The Genome Out of Context: Embryology from Physiology to Genetics

Starting in the 1930s, genetics, whose epistemology, methodology, and source of questions are based on the gene, began to provide a third context for developmental biology. The embryologist-turned-geneticist T. H. Morgan explicitly redefined embryological problems in terms of genetic ones. Moreover, according to Morgan (1924, 728), the cytoplasm was unimportant: "It is clear that what the cytoplasm contributes to development is almost entirely under the influence of the genes carried by the chromosomes." His student H. J. Muller, in his aptly entitled 1926 book *The Gene as the Basis of Life,* similarly concluded that the cytoplasm of the cell was inconsequential and that "the primary secrets common to all life lie further back, in the gene material itself" (Keller, 1995, 8).

Morgan's student and protégé Alfred Sturtevant (1932, 304) claimed that all of development could be explained by gene action in his talk to the 1932 International Congress of Genetics: "One of the central problems of biology is differentiation—how does an egg develop into a complex many-celled organism? That is, of course, the traditional problem of embryology. But it also appears in genetics in the form of the question: How do genes produce their effects."

Notice that Sturtevant (whose thought experiment on how snail shells coiled was one of the early triumphs of the genetic approach over physiological embryology) has simplified embryology into the question of determination. The other major embryological questions—morphogenesis, growth, evolution, and reproduction—are no longer mentioned. Evolutionary explanation would be taken over by the geneticists; the other portions of embryology will be assumed to be epiphenomena of differentiation, just as differentiation was assumed to be an epiphenomenon of gene expression (see Gilbert, 1988, 1998). This assumption was made explicit when molecular biology entered into embryology. In 1948, Sol Spiegelman could argue that cell differentiation was synonymous with differential protein synthesis and could be studied more readily in *Escherichia coli* or yeast than in metazoan embryos. Embryogenesis could be modeled by differential gene expression in unicellular microbes.

The questions of genetics differed from the questions of physiology in that the tissue and organism levels of explanation were derived from the genomic level of analysis. The methodology was certainly different in that it involved the isolating, analyzing, and mapping of discrete genetic mutations rather than performing experiments on embryos.

Genetics also challenged the dominant ontology of embryology. Physiological embryology was characterized by an interactive holism. A cell became what it became because of its position in the embryo. Whereas the effects of the environment may have been discounted, the fate of a particular cell was often found to be controlled by its interactions with other cells. As Hans Spemann noted in 1943: "We are standing and walking with parts of our body which could have been used for thinking had they developed in another part of the embryo" (Horder and Weindlung, 1986, 219). Genetics also brought with it a reductionist ontology (Roll-Hansen, 1978; Allen, 1985). The genes were responsible for the phenotype. Cells, tissues, organs, and the organism were epiphenomena of genes.

This reductionist and unidirectional ontology was reinforced by the program of molecular biology, which saw all life as manifestation of DNA (see Tauber and Sarkar, 1992). As Lewis Wolpert (1991, 77) claimed, "Ex omnia DNA." The central dogma—DNA makes RNA, and RNA makes proteins; DNA is thus primary—was also supported by the new sociobiology, which sought the origins of all behaviors in the genome. We see the confluence of these traditions in Dawkins (1986, 111; see also Gilbert, 2000a) reflecting on the willow seeds falling outside his window:

It is raining DNA outside. . . . The cotton wool is made mostly of cellulose, and it dwarfs the capsule that contains DNA, the genetic information. The DNA content must be a small proportion of the total so why did I say that it was raining DNA rather than raining cellulose? The answer is that it is the DNA that matters. . . . The whole performance—cotton wool, catkins, tree and all—is in aid of one thing and one thing only, the spreading of DNA around the countryside. . . . It is raining instructions out there; it's raining programs; it's raining tree-growing fluff-spreading algorithms. This is not metaphor, it is plain truth. It couldn't be plainer if it were raining floppy discs.

Although it is easy to separate the genetic tradition of methodology, ontology, and context from the physiological tradition, Richard Lewontin (1991) has emphasized an important continuity. Both traditions have gotten rid of the environment. The physiological tradition ignored the habitat in which the organism developed; the genetic tradition ignored the cytoplasmic and organismal environment in which the genes acted. "When those who react against the utter reductionism of molecular biology call for a return to consideration of the 'whole organism,' they forget that the whole organism was the first step in the victory of reductionism over a completely holistic view of nature" (Lewontin, 1991, ix–xix).

Integration of Developmental Contexts

Thus, across the three contexts in which development has been placed, there has been a progressive reductionism such that, under the physiological model, the importance of the environment has been diminished in favor of the whole organism, and under the genetic model, the importance of the whole organism has been diminished in favor of the genes.

But certain embryologists have had problems with both the reductionism of the genetic and physiological ontologies and the disappearance of environmental interactions in their epistemologies (see Gilbert, 1998; Gilbert and Sarkar, 2000).

Hertwig, Berrill, and Waddington were among the embryologists who had such doubts. Waddington, equally at home in genetics, evolutionary biology, and embryology, not only expressed these doubts eloquently, but also put forth solutions to them. Waddington's view of the genome was that it was both active and reactive. Unlike most researchers—both today and then—Waddington did not think of genes solely in terms of gene activity. Rather, he saw the genes in a dialectic of acting and being acted upon. In fact, in *Principles of Embryology* (Waddington, 1956) he called his chapter on developmental genetics "The Activation of the Genes by the Cytoplasm." He listed four examples "of the activation, by different types of cytoplasm, of different specifically corresponding genes": mosaic eggs, induction, chromosome puffs, and *Paramecium* G-antigens (Waddington, 1956, 348). He viewed the nucleus and the cytoplasm as being in a continual reciprocal dialogue. The *epigenotype* is the term Waddington (1939; see also Gilbert, 1991, 2000b) used to capture the idea of the interactions between the genes, gene products, and the environment that led from genotype to phenotype. Today we might think of the epigenotype as the networks of transcription factors, paracrine factors, and environmental influences that allow the genotype to realize the phenotype. In the epigenotype, the gene is not an autonomous entity; it is part of a network of interacting components.

Waddington (1940, 1957) not only mentioned cases where the cytoplasm obviously told the genes what to do, he also kept alive the tradition of the environmental regulation of development. He recognized that some major phenotypes, such as sex in echiuroids and in some reptiles, was determined by temperature, and he felt that this environmental effect was of the same magnitude as a mutation in a homeotic gene. The environment could act on genes, as well. Although the tradition of the environment acting on genes has been largely on the periphery of developmental biology (see van der Weele, 1999), recent studies have not only provided more examples of environmental regulation; they have become integrated into developmental biology literature. This has largely been due to the new interest in life history strategies as an intersection between developmental biology and ecology (see Stearns, 1992; Gilbert, 1997; van der Weele, 1999) and to the new interest in endocrine disrupters and their potential to influence sexual development and morphogenesis (Colborn, Dumanovski, and Myers, 1996; Gilbert, 1997).

Tertiary Induction

Numerous species, especially *Homo sapiens,* possess developmental plasticity wherein the organism inherits the ability to express certain phenotypes in some situations and other phenotypes under another set of conditions. Some species have polyphenisms that are distinct (either-or) phenotypes that are elicited by the environment. Other species have a range of phenotypes where the response to the environment can be incremental. This continuous range of phenotypes is called the "reaction norm" (*Reaktionsnorm;* Woltereck, 1909; see also Stearns, de Jong, and Newman, 1991; Nanjundiah, chapter 14, this volume).[1] Although, as inherited potentials, reaction norms and polyphenisms can be selected, in all these instances, the genotype can govern only the range of phenotypes produced. The actual phenotype of the particular individual is elicited by the environment.

Therefore, the organism, its genome, and its environment should be seen as porous and interactive compartments (Lewontin, 1991; Gottlieb, 1992). I wish to emphasize this by using a particular embryological term to unite these areas in development—*tertiary induction*. *Primary induction* is that set of inductive events which constitutes the individuality of the organism. By means of primary embryonic induction (e.g., in vertebrates, the induction of the central nervous system by the derivatives of the dorsal blastopore lip), each egg forms a single embryo. *Secondary induction* is that set of inductive events by which the developing body parts interact to generate the organs of the body. In an extension of this concept, *tertiary induction* is that set of interactions by which environmental factors influence the phenotype of the developing organism (Gilbert, 2000c). As with primary and secondary inductions, there is no distinction in the mechanisms proposed. As the following examples will show, the genome is reactive as well as active, and organisms have evolved to let environmental factors play major roles in phenotype determination.

Holtzer (1968) distinguished between permissive and instructive interactions. In instructive interactions, a signal from the inducer initiates new patterns of gene expression in the responding cells. In permissive interactions, the responding tissue contains all the information required to express the genes; it needs only the permissive context in which to activate them. It is usually assumed that the developing organism's environment constitutes a necessary permissive set of factors, whereas its genome provides the specificity of the interaction. In instances of developmental plasticity, however, the genome is permissive and the environment is instructive.

Context-Dependent Sex Determination

The scientific evidence against genetic determinism has been building up ever since embryologist Oskar Hertwig (1894) used location-dependent sex determination in the echiuroid worm *Bonellia* against what he called the "preformationism" of August Weismann. More recently, temperature has been demonstrated to be the prime determinant of whether many embryos become male or female. For instance, embryos of the turtle *Emys* all become males if incubated below 25°C during the last third of their incubation, and all females if incubated above 30°C (Pieau et al., 1994; at intermediate temperatures, different percentages of both sexes are formed). In some species, such as the fish *Menidia,* temperature-dependent sex determination is found in the parts of their range where it is adaptive to be one sex or the other during particular portions of the breeding season (Charnov and Bull, 1977; Conover and Heims, 1987).

Life history strategies are also seen as being examples of tertiary induction. The sex of the blueheaded wrasse, *Thalassoma bifasciatum,* is determined by the social structure into which the larva enters: if it enters a reef where there are no males, it becomes male; if there are other males present, a female. Usually, there is one male for each dozen females. When that male dies, the largest female develops testes within twenty-four hours and, within two weeks, is making functional sperm and mating with the remaining females (Warner, 1993). Some invertebrates, such as the crustacean "pillbug" *Armadillidium vulgare* (Rigaud, Juchault, and Mocquard, 1991) and a wide range of insects (Werren and Windsor, 2000) can have their sex determined by bacterial infection. Thus a phenotype as significant as the organism's sex can be determined by the environment. The genome is permissive and can allow one or the other phenotypes to be elicited.

Seasonal Polyphenisms

Sex is not the only phenotypic trait influenced by the environment. We now know that what Linnaeus had classified as two different species of butterfly are two phenotypes of the same species, *Araschnia levana,* and that they are regulated by the temperature and photoperiod experienced by the late instar larvae. More daylight and higher temperatures cause higher amounts of ecdysone and produce the dark summer morph. Less daylight and lower temperatures produce the orange spring morph (see figure 6.1a–b; Nijhout, 1991; van der Weele, 1999). The seasonal polymorphism of the butterfly *Bicyclus anyana* is also caused by temperature, which, in some yet unknown way, effects the stability of *distal-less* gene expression in the wing imaginal discs (Brakefield et al., 1996.)

Nutritional Polyphenisms

In some cases, phenotype can depend upon what a developing organism eats.[2] In numerous species of Hymenoptera, the worker, soldier, and queen castes are determined by the levels of food fed the respective larvae. In the ants *Pheidole* and *Pheidologeton,* the protein-rich diet causes elevated juvenile hormone titers, and these titers allow more growth, lengthening the time before which metamorphosis will occur. The differences in size, structure, and even cuticular proteins are often quite significant. In the moth *Nemoria arizonaria,* a nutritional polyphenism provides adaptive coloration for two different sets of caterpillars. Caterpillars that hatch in the spring feed on oak catkins; they develop a rugose, beaded, yellow-brown morphology that enables them to hide among the catkins. By contrast, caterpillars that hatch in the summer eat the oak leaves; their morphology changes to resemble that of a new oak twig (Greene, 1989).

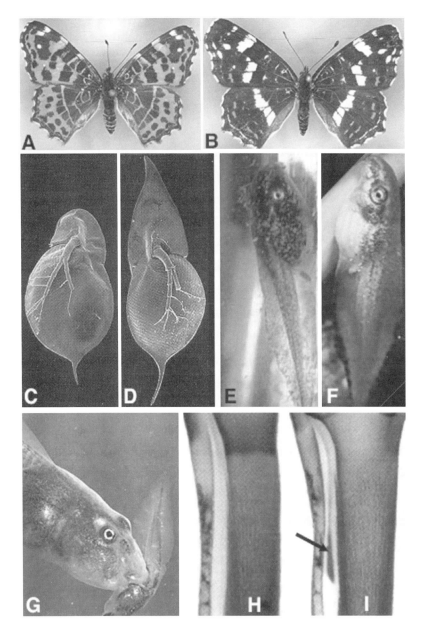

Figure 6.1
Instructive induction of morphological phenotypes by the environment. The spring (*A*) and summer heat-induced (*B*) morph of the European map butterfly, *Araschnia levana*. The uninduced (*C*) and *Chaoborus* kairomone–induced (*D*) morphs of *Daphnia cucullata*. The uninduced (*E*) and kairomone-induced (*F*) morphs of the tadpole of the gray tree frog *Hyla crycscelis*. *Scaphiopus* tadpoles, (*G*) with the uninduced morph in the jaws of the density-induced morph. Uninduced (*H*) and movement-induced (*I*) tissue in the embryonic chick hind limb. The arrow points to the movement-induced fibular crest, an important bone in bird evolution. (*A*, *B*: courtesy of H. F. Nijhout; *C*, *D*: courtesy of R. Tollrian; *E*, *F*: courtesy of J. van Buskirk; *G*: courtesy of T. Wiewandt; *H*, *I*: courtesy of G. B. Müller. From Gilbert, 2001.)

Predator-Induced Polyphenisms

Several vertebrates and invertebrates have evolved developmental responses to predators. By sensing a predator-secreted chemical in the environment, various species of rotifers, crustaceans, molluscs, fish, and reptiles develop differently (Tollrian and Harvell, 1999). In the parthenogenetic water flea *Daphnia cucullata,* the predator-induced defense is beneficial not only to itself, but to its offspring: when it encounters the predatory larvae of the fly *Chaeoborus,* its helmet grows to twice its normal size (figure 6.1c–d). This inhibits its being eaten by the fly larvae. This same helmet induction occurs if the *Daphnia* are exposed to extracts of water in which the fly larvae had been swimming. Agrawal, Laforsch, and Tollrian (1999) have shown that the offspring of such induced *Daphnia* will be born with this same altered head morphology.

Predator-induced polyphenism is abundant among amphibia, and tadpoles found in ponds or in the presence of other species may differ significantly from those tadpoles reared by themselves in aquaria. For instance, when newly hatched tadpoles of the wood frog *Rana sylvetica* are reared in tanks containing predatory larvae of the dragonfly *Anax* (confined in mesh cages so that they cannot kill the tadpoles), the tadpoles in the predator-filled tanks grow smaller than those in similar tanks without the caged predators. Moreover, their tail musculature deepens, allowing faster turning and swimming speeds to escape predator strikes (McCollum and Leimberger, 1997; van Buskirk and Relyea, 1998). In fact, what initially appeared to be a polyphenism may be a reaction norm that can assess the amount (and type) of predators. Adding ever greater numbers of predators to the tanks causes the tadpoles' tail fin and tail musculature to progressively deepen.

Tadpoles of related species are capable of producing different phenotypic changes, depending on the predator. The tadpole of the gray tree frog *Hyla cryoscelis* responds to soluble predator molecules both by changing its size and by developing a bright red tail coloration that deters predators (figure 6.1e–f; Relyea and Werner, 2000; McCollum and van Buskirk, 1996). The trade-off is that the noninduced tadpoles grow more slowly and survive better in predator-free environments (van Buskirk and Relyea, 1998; Relyea, 2001). Amphibian larvae have evolved to respond to other environmental cues, as well.

In addition to responding to cues from predators, *Rana* tadpoles also respond to cues from *competitors.* Wood frog and leopard frog tadpoles compete for the same food. The presence of the leopard frog tadpoles changes the responses of the wood frog tadpoles to predator-derived cues (Relyea, 2000). In some instances, the competitor- and predator-induced phenotypes go in opposite directions (with the former making shallower tails, for instance). In these cases, the competitor-induced phenotypes are more competitive (against other organisms competing for the same food source), but they suffer a higher predation.

Nowhere is predator-induced polyphenism more important than in mammals such as humans. Our major predators are microbes. We respond to them through an immune system based on clonal selection of lymphocytes that recognize specific predators and their products (see Gilbert, 2000c). Our immune system recognizes a particular microbe such as a cholera bacterium or a poliovirus by making lymphocytes, each expressing a different gene product on its cell surface. These genes for immunoglobulin and T-cell receptors form the receptor proteins of the lymphocytes. Each B-lymphocyte, for instance, makes one and only one type of antibody, and it places this antibody on the cell surface. One B-cell may be making an antibody to poliovirus, while its neighboring B-cell is making an antibody to diphtheria toxin. When a B-cell lymphocyte binds its foreign substance (the antigen), it begins a pathway that causes it to divide repeatedly and to differentiate into a cell that secretes the same antibody that originally bound the antigen. Moreover, some of the descendants of that stimulated B-cell remain in the body as sentinels against further infection by the same microorganism. Thus identical twins are not identical with respect to the cells of their respective immune systems. Their phenotypes (in this case, both the types of cells in their lymph nodes and their ability to respond against an infectious microorganism) have been altered by the environment.

Context-Dependent Development: Abiotic Conditions

The spadefoot toad *Scaphiopus* has a remarkable strategy for coping with a particularly harsh environment. The toads are called out from hibernation by the thunder that accompanies the first spring storm in the Sonoran desert. The toads breed in the temporary ponds caused by the rain, and the embryos develop quickly into larvae. After the larvae metamorphose, the young toads return to the desert, burrowing into the sand until the next year's storms bring them out.

The desert ponds are ephemeral pools that either dry up quickly or persist, depending on the initial depth and the frequency of the rainfall. One might envision only two alternative scenarios confronting a tadpole in such a pond: either (1) the pond persists until the tadpole has time to metamorphose, and it lives, or (2) the pond dries up before metamorphosis, and it dies. These toads (and several other amphibians), however, have evolved a third alternative. The time of metamorphosis is controlled by the pond. If the pond does not dry out, development continues at its normal rate, and the algae-eating tadpoles eventually develop into juvenile spadefoot toads. If, however, the pond begins to dry out and overcrowding occurs, some of the tadpoles embark on an alternative developmental pathway. They develop a wider mouth and more powerful jaw muscles that enable them to eat, among other things, other *Scaphiopus* tadpoles. These carnivorous tadpoles metamorphose quickly, albeit into a smaller version of the juvenile spadefoot toad.

The signal for this accelerated metamorphosis appears to be the change in water volume. *Scaphiopus* tadpoles are able to sense the removal of water from aquaria, and their acceleration of metamorphosis depended upon the rate at which the water was removed. A stress-induced corticotropin-releasing hormone signaling system appears to modulate this effect (Denver, Mirhadi, and Phillips, 1998; Denver, 1999). The two morphs can be obtained by feeding tadpoles the appropriate diets, and the rapid development of the cannibalistic tadpoles may be being driven by the thyroxin they acquire from their prey. The trade-off is that the toads generated by fast-metamorphosing tadpoles lack the fat reserves of those toads produced from the more slowly growing tadpoles, and their survival rate after metamorphosis is not as high as those toads developing from slower-growing larvae (Newman, 1989, 1992; Pfennig, 1992).

Vertebrates also respond significantly to abiotic conditions. In addition to stress-related muscle development, physical stress is needed to produce bones such as the mammalian patella and the avian fibular crest (figure 6.1h–i; Müller and Streicher, 1989; Wu, 1996). Corruccini (1984) and Varrela (1992) have speculated that the reason that nearly one-quarter of our population needs orthodontic appliances is that our lower jaw needs physical stress in order to grow. Such jaw anomalies (malocclusions wherein the teeth cannot fit properly in the jaw) are relatively new to European populations. Well-preserved skeletons from the fifteenth and sixteenth centuries show almost no malocclusion in the population (Mohlin, Sagna, and Thilander, 1978; Helm and Prysdö, 1979; Corrucini, 1984; Varrela, 1990). Corruccini and Varrela have hypothesized that the change in children's meals from a coarse diet to a mild-textured diet has resulted in decreased mastication and a decrease in jaw skeleton and muscle development. Increased chewing causes tension that stimulates mandible bone and muscle growth (Kiliardis, 1986; Weijs and Hillen, 1986). Placing young primates on a soft diet will cause malocclusions in their jaws, similar to those in humans (Corruccini 1984; Corruccini and Beecher 1982). Mechanical tension stress has been found to induce the expression of certain bone morphogenetic protein (BMP2, BMP4) genes in adult rats (Sato et al., 1999).

Bone density is also regulated by mechanical stress, and several genes for osteoblast and osteocyte functions are known to be regulated through physical load (Nomura and Takano-Yamamoto, 2000; Zaman et al., 2000). Astronauts experiencing weightlessness are at risk for such negative bone remodeling (losing about 1 percent of healthy bone mineral density per month in space), and studies on the space shuttles have shown that several genes, including the gene for the vitamin D receptor, are dramatically downregulated in microgravity (Hammond et al., 2000; Wassersug, 2000).

Coda

The environment is not merely a permissive factor in development. It can also be instructive. A particular environment can elicit different phenotypes from the same genotype. Development usually occurs in a rich environmental milieu, and most animals are sensitive to environmental cues. The environment may determine sexual phenotype; it may induce remarkable structural and chemical adaptations according to the season, or specific morphological changes that allow an individual to escape predation; it may induce caste determination in insects. The environment can also alter the structure of our neurons and the specificity of our immunocompetent cells. We can give a definite answer to the question posed by Wolpert (1994, 572): "Will the egg be computable? That is, given a total description of the fertilized egg—the total DNA sequence and the location of all proteins and RNA—could one predict how the embryo will develop?"

The answer has to be no. The phenotype depends to a significant degree on the environment, and this is a necessary condition for integrating the developing organism into its particular habitat. Development depends not only on signals "from below," but also on signals "from above." This means that reductionism cannot provide a complete explanation of development (Gilbert and Sarkar, 2000). Rather, a context-dependent organicism must integrate the signals from the genome, from the interactions between cells, and from the environment in which the organism develops.

Notes

1. In his history of the reaction norm concept, Sarkar (1999) has shown that Woltereck argued that what was inherited was a reaction norm and that thinking in these terms was better than viewing the inherited potential as a static genotype. Johannsen, who formulated the genotype concept (and its distinction from the phenotype) agreed that the reaction norm was almost equivalent to his genotype, but that the genotype was more a directive force, whereas the reaction norm was more an enabling agent. Although the reaction norm is an evolutionarily selectable trait and a product of gene selection (see Schlichting and Pigliucci, 1998; Nanjundiah, chapter 14, this volume), its variability should not be confused with the variability, suggested by other contributions in this volume (e.g., Müller, chapter 4, Newman, chapter 13) of traits that have not yet been stabilized by genes and their products.

2. One of the most interesting cases of the nutritional regulation of development involves the disease gulonolactone oxidase deficiency (hypoascorbemia; OMIM 240400). Homozygosity of a mutation in the gulonolactone oxidase gene on the short arm of chromosome 8 produces a syndrome that produces death in childhood due to connective tissue malfunction. Interestingly, this syndrome effects 100 percent of the human population. Gulonolactone oxidase is the final enzyme in the pathway leading to ascorbic acid, and we are all homozygous at this mutant locus (Nishikimi et al., 1994), which is why we need ascorbic acid (vitamin C) in our diet. Without this replacement therapy from the environment, we would all be dead. This example illustrates that it is impossible to parse environment and heredity into neat, nonintersecting categories. What is the effect of the environment on the human phenotype? 100 percent. The effects of genes and environment are interactive; they cannot be separated into component percentages. Our genotype programs us for an early death, and the fact that we are here is testimony to the power of the environment to circumvent our genetic heritage.

References

Agrawal AA, Laforsch C, Tollrian R (1999) Transgenerational induction of defenses in animals and plants. Nature 401: 60–63.

Allen GE (1985) Thomas Hunt Morgan: Materialism and experimentalism in the development of modern genetics. Trends Genet 1: 151–154.

Bolker JA (1995) Model systems in developmental biology. BioEssays 17: 451–455.

Brakefield PM et al. (1996) Development, plasticity, and evolution of butterfly eyespot patterns. Nature 384: 236–242.

Charnov EL, Bull JJ (1977) When is sex environmentally determined? Nature 266: 828–830.

Colborn T, Dumanoski D, Myers JP (1996) Our Stolen Future. New York: Dutton.

Conover DO, Heins SW (1987) Adaptive variation in environmental and genetic sex determination in a fish. Nature 326: 496–498.

Corruccini RS (1984) An epidemiologic transition in dental occlusion in world populations. Am J Orthod 86: 419–426.

Corruccini RS, Beecher CL (1982) Occlusal variation related to soft diet in a nonhuman primate. Science 218: 74–76.

Dawkins R (1986) The Blind Watchmaker. New York: Norton.

Denver RJ (1999) Evolution of the corticotropin-releasing hormone signaling system and its role in stress-induced phenotypic plasticity. Neuropeptides: Structure and Function in Biology and Behavior. Ann NY Acad Sci 897: 46–53.

Denver RJ, Mirhadi N, Phillips M (1998) Adaptive plasticity in amphibian metamorphosis: Response of *Scaphiopus hammondii* tadpoles to habitat desiccation. Ecology 79: 1859–1872.

Gilbert SF (1988) Cellular politics: Just, Goldschmidt, and the attempts to reconcile embryology and genetics. In: The American Development of Biology (Rainger R, Benson K, Maienschein J, eds), 311–346. Philadelphia: University of Pennsylvania Press.

Gilbert SF (1991) Induction and the origins of developmental genetics. In: A Conceptual History of Modern Embryology (Gilbert SF, ed), 181–206. New York: Plenum Press.

Gilbert SF (1998) Bearing crosses: The historiography of genetics and embryology. Am J Med Genet 76: 168–182.

Gilbert SF (2000a) Mainstreaming feminist critiques into the biology curriculum. In: Doing Science and Culture (Reid R, Traweek S, eds), 199–220. London: Routledge.

Gilbert SF (2000b) Diachronic biology meets evo-devo: CH Waddington's approach to evolutionary developmental biology. Am Zool 40: 729–737.

Gilbert SF (2000c) Developmental Biology (6th ed). Sunderland, Mass.: Sinauer.

Gilbert SF (2001) Ecological developmental biology: Developmental biology meets the real world. Dev Biol 233:1–12.

Gilbert SF, Faber M (1996) Looking at embryos: The visual and conceptual aesthetics of emerging form. In: The Elusive Synthesis: Aesthetics and Science (Tauber AI, ed), 125–151. Dordecht: Kluwer Academic.

Gilbert SF, Sarkar S (2000) Embracing complexity: Organicism for the twenty-first century. Dev Dyn 219: 1–9.

Gottlieb G (1992) Individual Development and Evolution. New York: Oxford University Press.

Greene E (1989) A diet-induced developmental polymorphism in a caterpillar. Science 243: 643–646.

Hammond TG, Benes E, O'Reilly KC, Wolf DA, Linnehan RM, Taher A, Kaysen JH, Allen PL, Goodwin TJ (2000) Mechanical culture conditions effect gene expression: Gravity-induced changes on the space shuttle. Physiol Genom 3: 163–173.

Helm S, Prysdö U (1979) Prevalence of malocclusion in medieval and modern Danes contrasted. Scand J Dent Res 87: 91–97.

Hertwig O (1894) The Biological Problem of To-day: Preformation or Epigenesis? (Mitchell PC, trans). New York: Macmillan.

Holtzer H (1968) Induction of chondrogenesis: A concept in terms of mechanisms. In: Epithelial-Mesenchymal Interactions (Fleischmajer R, Billingham R, eds), 152–164. Baltimore: William and Wilkins.

Horder TJ, Weindling PJ (1986) Hans Spemann and the Organiser. In: A History of Embryology (Horder TJ, Witkowski JA, Wylie CC, eds). Cambridge: Cambridge University Press.

Keller EF (1995) Refiguring Life: Metaphors of Twentieth-Century Biology. New York: Columbia University Press.

Kiliardis S, Engström C, Thilander B (1985) The relationship between masticatory function and craniofacial morphology. Eur J Orthod 7: 273–283.

Lewontin RC (1991) Foreword. In: Organism and the Origins of Self (Tauber AI, ed), ix–xix. Dordrecht: Kluwer Academic.

McCollum SA, Leimberger JD (1997) Predator-induced morphological changes in an amphibian: Predation by dragonflies affects tadpole shape and color. Oecologia 109: 615–621.

McCollum SA, van Buskirk J (1996) Costs and benefits of a predator induced polyphenism on the gray treefrog *Hyla chrysoscelis*. Evolution 50: 583–593.

Mohlin B, Sagna S, Thilander B (1978) The frequency of malocclusion and the craniofacial morphology in a medieval population in Southern Sweden. OSSA 5: 57–84.

Morgan TH (1924) Mendelian heredity in relation to cytology. In: General Cytology (Cowdry EV, ed), 691–734. Chicago: University of Chicago Press.

Müller GB, Steicher J (1989) Ontogeny of the syndesmosis tibiofibularis and the evolution of the bird hind limb: A caenogenetic feature triggers phenotypic novelty. Anat Embryol 179: 327–339.

Newman RA (1989) Developmental plasticity of *Scaphiopus couchii* tadpoles in an unpredictable environment. Ecology 70: 1775–1787.

Newman RA (1992) Adaptive plasticity in amphibian metamorphosis. BioScience 42: 671–678.

Nijhout HF (1991) The Development and Evolution of Butterfly Wing Patterns. Washington, D.C.: Smithsonian Institution Press.

Nishikimi M, Fukuyama R, Minoshima S, Shimizu N, Yagi K (1994) Cloning and chromosomal mapping of the human nonfunctional gene for L-gulono-gamma-lactone oxidase, the enzyme for L-ascorbic acid biosynthesis missing in man. J Biol Chem 269: 13685–13688.

Nomura S, Takano-Yamamoto T (2000) Molecular events caused by mechanical stress in bone. Matrix Biol 19: 91–96.

Nyhart LK (1995) Biology Takes Form: Animal Morphology and the German Universities, 1800–1900. Chicago: University of Chicago Press.

Pfennig DW (1992) Proximate and functional causes of polyphenism in an anuran tadpole. Funct Ecol 6: 167–174.

Pieau C, Girondot N, Richard-Mercier G, Desvages M, Dorizzi P, Zaborski P (1994) Temperature sensitivity of sexual differentiation of gonads in the European pond turtle. J Exp Zool 270: 86–93.

Relyea RA (2000) Trait-mediated indirect effects in larval anurans: Reversing competitive outcomes with the threat of predation. Ecology 81: 2278–2289.

Relyea RA (2001) The lasting effects of adaptive plasticity: Predator-induced tadpoles become long-legged frogs. Ecology. 82: 1947–1955.

Relyea RA, Werner EE (2000) Morphological plasticity of four larval anurans distributed along an environmental gradient. Copeia 2000: 178–190.

Rigaud T, Juchault P, Mocquard JP (1991) Experimental study of the sex ratio of broods in terrestrial crustacean *Armadillidium vulgare:* Possible implications in natural populations. J Evol Biol 19: 603–607.

Roll-Hansen N (1978) Drosophila genetics: A reductionist research program. J Hist Biol 11: 159–210.

Sarkar S (1999) From the *Reaktionsnorm* to the adaptive norm: The norm of reaction, 1909–1960. Biol Philos 14: 235–252.

Sato M, Ochi T, Nakase T, Hirota S, Kitamura Y, Nomura S, Yasui N (1999) Mechanical tension-stress induces expression of bone morphogenetic protein BMP-2 and BMP-4, but not BMP-6, BMP-7, and GDF-5 mRNA, during distraction osteogenesis. J Bone Miner Res 14: 1084–1095.

Schlichting CD, Pigliucci M (1998) Phenotypic Evolution: A Reaction Norm Perspective. Sunderland, Mass.: Sinauer.

Smith-Gill SJ (1983) Developmental plasticity: Developmental conversion versus phenotypic modulation. Am Zool 23: 47–55.

Spiegelman S (1948) Differentiation and the controlled production of unique enzymatic patterns. In: Growth in Relation to Differentiation and Morphogenesis (Danielli JF, Brown K, eds), 286–325. Cambridge: Cambridge University Press.

Stearns SC (1992) The Evolution of Life Histories. New York: Oxford University Press.

Stearns SC, de Jong G, Newman RA (1991) The effects of phenotypic plasticity on genetic correlations. Trends Ecol Evol 6: 122–126.

Sturtevant AH (1932) The use of mosaics in the study of the developmental effects of genes. In: Proceedings of the Sixth International Congress of Genetics, 304. New York: Macmillan.

Tauber AL, Sarkar S (1992) The human genome project: Has blind reductionism gone too far? Perspect Biol Med 35: 220–235.

Tollrian R, Harvell CD (1999) The Ecology and Evolution of Inducible Defenses. Princeton, N.J.: Princeton University Press.

van Buskirk J, Relyea RA (1998) Natural selection for phenotypic plasticity: Predator-induced morphological responses in tadpoles. Biol J Linn Soc 65: 301–328.

van der Weele C (1999) Images of Development: Environmental Causes in Ontogeny. Albany: State University of New York Press.

Varrela J (1990) Effects of attritive diet on craniofacial morphology: A cephalometric analysis of a Finnish skull sample. Eur J Orthod 12: 219–223.

Varrela J (1992) Dimensional variation of craniofacial structures in relation to changing masticatory-functional demands. Eur J Orthod 14: 31–36.

Waddington CH (1939) An Introduction to Modern Genetics. New York: Macmillan.

Waddington CH (1940) Organisers and Genes. Cambridge: Cambridge University Press.

Waddington CH (1956) Principles of Embryology. London: Allen and Unwin.

Waddington CH (1957) The genetic basis of the assimilated bithorax stock. J Genet 55: 240–245.

Warner RR (1993) Mating behavior and hermaphroditism in coral reef fishes. In: Exploring Animal Behavior (Sherman PW, Alcock J, eds), 188–196. Sunderland, Mass.: Sinauer.

Wassersug RJ (2000) Vertebrate biology in microgravity. Am Sci 89: 46–53.

Weijs WA, Hillen B (1986) Correlations between the cross-sectional area of the jaw muscles and craniofacial size and shape. Am J Phys Anthropol 70: 423–431.

Werren JH, Windsor DM (2000) *Wolbachia* infection frequencies in insects: Evidence of a global equilibrium? Proc R Soc Lond B Biol Sci 267: 1277–1285.

Wolpert L (1991) The Triumph of the Embryo. Oxford: Oxford University Press.

Wolpert L (1994) Do we understand development? Science 266: 571–572.

Woltereck R (1909) Weitere experimentelle Untersuchungen über Artveränderung, speziell über das Wesen quantitativer Artunderscheide bei Daphniden. Versuch Deutsch Zool Ges 1909: 110–172.

Wu KC (1996) Entwicklung, Stimulation und Paralyse der embryonalen Motorik. Wien Klin Wochenschr 108: 303–305.

Zaman G, Cheng MZ, Jessop HL, White R, Lanyon LE (2000) Mechanical strain activates estrogen response elements in bone cells. Bone 27: 233–239.

7 Tissue Specificity: Structural Cues Allow Diverse Phenotypes from a Constant Genotype

Mina J. Bissell, I. Saira Mian, Derek Radisky, and Eva Turley

Two decades ago, based on the literature and her laboratory experience, one of us (Bissell) concluded that,

> if there is one generalization that can be made from all the tissue and cell culture studies with regard to the differentiated state, it is this: Since most, if not all, functions are changed in culture, quantitatively and/or qualitatively, there is little or no *constitutive* regulation in higher organisms; i.e., the differentiated state of normal cells is unstable and the environment regulates gene expression. (Bissell, 1981, p. 27; emphasis added.)

This concept, more recently referred to as the "plasticity" of the differentiated state, has gained some credence as literature has accumulated that differentiation may not be as terminal or fixed as was once thought—witness the cloning of Dolly, mice, and cows from restricted stem cells derived from adult tissues, or even from single somatic cells.

There is ample evidence that all cells retain the ability to modulate most, if not all, of their functions; even enucleated red blood cells still regulate their behavior depending on the context and what they encounter. It may be that cells never completely lose an intrinsic ability to morph from one cell type to another, and that they maintain a stable phenotype by integrating cues from the extra- and intracellular milieu. Indeed, there is also ample evidence to support the notion that, for a cell to continue functioning properly in a tissue-specific way, it must receive continuous signals to prevent growth or apoptosis and to maintain an appropriate structure and differentiation state, which is to say, cells must be directed at all times to remember how to behave within an organ. If these active signals are withdrawn from a resting, differentiated cell, or if a wrong signal is given (as is often the case in cell culture), it will do one of three things: die, start growing, or function inappropriately. What, then, are the cues in vivo that cause a cell to continue functioning in a manner that is specific for its tissue?

This and related questions raise a larger question that directly bears on the theme of this volume. Are mutations really the cornerstone of evolution through natural selection, or could radical changes in the microenvironment, even without spontaneous genomic mutations, allow an organism to evolve into a different form? The central tenet behind this reasoning derives its strength from the obvious miracle of development: we all began as a single cell and all our diverse tissues and organs contain the same DNA sequence. Thus we need to know not just how cellular differentiation is derived, but also how it is maintained against a constant DNA background.

Three-Dimensional Microenvironments

To address these complex biological questions experimentally, researchers must develop tractable model systems that pare the subject in question down to its most essential components. In higher organisms, this strategy has focused on using monolayer cultures of homogeneous cell populations propagated in vitro. Although this approach has been very successful in elucidating many of the basic principles of cell survival and growth, it has generally ignored the fact that within an organism, no cell is an island: each exists in the context of a complex microenvironment. In response to this limitation, inherent to two-dimensional (2-D) monolayer culture systems, cell culture strategies began to be redefined in the context of three-dimensional (3-D) microenvironments. In the seventies, Ellsdale and Bard (1974) and later Michaelopolis and Pitot (1975) and Emerman and colleagues (1977) grew cells on gels of collagen I, that were then floated; the resulting 3-D structures regained some of their original functions. Thus, when grown as monolayers on rigid substrata such as tissue culture plastic or attached collagen I gels, luminal epithelial cells extracted from mouse or human mammary glands did not differentiate structurally or functionally. Growing mammary epithelial cells on and in gels of extracellular matrix (ECM) materials similar in composition to the basement membrane (BM) associated with mammary epithelial tissues in vivo obviated the need for flotation and led to the formation of normal cellular architecture and to gene expression profiles characteristic of differentiated cells (Barcellos-Hoff et al., 1989; Petersen et al., 1992) (figures 7.1 and 7.2).

How does a gelatinous basement membrane with essentially insoluble proteins communicate with the nucleus? We believe that maintenance of tissue specificity involves an intimate and profound communication between the microenvironment around the cells and the organization of the nucleus. This concept, put forward two decades ago for ECM, is known as "dynamic reciprocity" (Bissell, Hall, and Parry, 1982; figure 7.1). Many of the essential players in the depicted signaling events (not then identified, and indicated by question marks in the insets) have since been characterized. Indeed, the number of proteins and protein modifications known to be involved in cell-ECM interactions is immense, but how the signals are integrated to permit organ formation is still far from clear.

Molecular Cues from the Extracellular Environment

The molecular mechanisms behind signaling from the extracellular matrix molecules through their receptors (largely integrins, but other receptors are being identified as well) have been intensely studied and elucidated (for reviews, see Clarke and Brugge, 1995; Guan and Chen, 1996; Yamada, 1997; Schoenwaelder and Burridge, 1999; Giancotti and Ruoslahti, 1999). How the signals are transduced to the nucleus and then propagated to other cells and tissues is less obvious and is not well understood. We have known for some time that there are growth-factor- and hormone-response elements in the 5' regulatory region of many genes. The discovery of the first ECM-response element was made possible through the development of transfectable mammary epithelial cells that could respond to ECM by making milk proteins (Schmidhauser et al., 1990). A reporter gene was cloned behind 1,600 bp of the 5' sequence encoding the milk protein β-casein, and this construct was transfected into the functional mammary cell line. The reporter gene was from 50 to 150 times more active when cells were grown on ECM. Subsequent promoter deletion

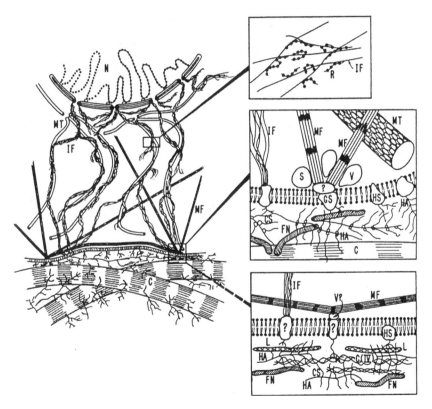

Figure 7.1
Dynamic reciprocity, the minimum required unit for tissue-specific functions. The postulated overall scheme for extracellular matrix–cell interactions. N, nucleus; MT, microtubules; IF, intermediate filaments; MF, microfilaments; C, collagen. (*Top inset*) Polyribosome attachment to cytoskeleton. R, ribosomes. (*Middle inset*) V, vinculin; S, *src* coded protein kinase; GS, Ganglioside (attaching fibronectin to membrane); FN, fibronectin; HA, hyaluronic acid; CS, chondroitin sulfate; HS, heparan sulfate. (*Bottom inset*) Possible attachment site to membranes in epithelial cells. L, laminin; C(IV), collagen type IV. (Reproduced with permission from Bissell et al., 1982.)

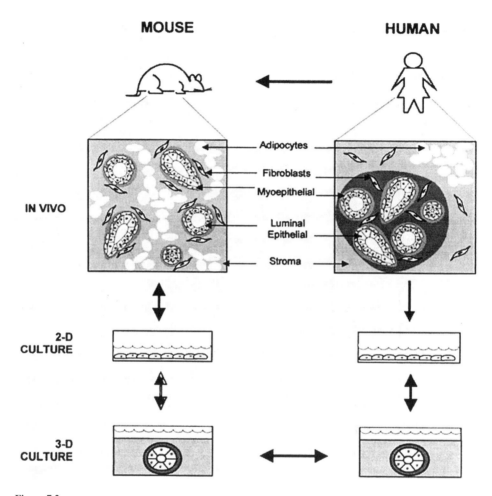

Figure 7.2
Consequences of culture in two versus three dimensions on tissue architecture. Modeling human breast function from studies of mouse mammary gland. As in humans, the mouse mammary tissue comprises multiple cell types, including luminal epithelial and myoepithelial cells, adipocytes, and stromal fibroblasts. Although mouse and human mammary tissue vary somewhat with respect to overall organization, the double-layered structure of the branching ducts and ductules is preserved in both organisms. In light of these fundamental similarities, it is not surprising that human and mouse epithelial cell types display similar behaviors in 3-D basement membrane (matrigel) cultures: both cell types undergo morphogenesis to form spherical structures that are similar to acini in vivo. (Reproduced with permission from Ronnov-Jessen, Petersen, and Bissell, 1996, and Schmiechel, Weaver, and Bissell, 1998, with minor modifications.)

analysis (Schmidhauser et al., 1992) identified a 160-nucleotide sequence that defined an ECM-response element in the regulatory sequences of this gene. Using site-specific mutagenesis, the response element was shown to be an enhancer and that C/EBPβ and STAT-5 transcription-factor-binding elements were essential for its activity (Myers et al., 1998). In this cell model, transcription factor binding was necessary, but not sufficient, to activate transcription in the absence of ECM. We now have found evidence that ECM signals can alter the histone acetylation/deacetylation status of chromatin, and that this change in chromatin structure is necessary to initiate the transcription of differentiation-specific genes (Myers et al., 1998; Boudreau and Bissell, 1998; Pujuguet et al., 2001).

Concurrent with the discovery of this ECM-response element, our laboratory and many others have investigated the ability of integrins, the largest and best studied class of ECM receptors, to activate signaling cascades (Streuli and Bissell, 1991; for reviews, see Clarke and Brugge, 1995; Yamada, 1997; Giancotti and Ruoslahti, 1999; Hynes and Zhao, 2000). Two important properties of these extracellular matrix receptors are essential to link the genome of a cell to its extracellular microenvironment, rendering the organism susceptible to evolutionary selection by epigenetic factors. The first is the ability of intracellular signaling molecules to modify the avidity of matrix receptors for their ligands, a property that in turn affects the subsequent intracellular signaling pathways that are activated. This two-way signaling across the membrane, referred to as "inside-outside" signaling, permits the cells to continuously interact with the extracellular microenvironment (Faull and Ginsberg, 1996; Brown and Hogg, 1996; Ruoslahti, 1997; Liu, Calderwood, and Ginsberg, 2000). The second important property is the ability of extracellular matrix receptors to functionally associate with growth factor receptors, thus linking the information conveyed by growth factors to the inside-outside paradigm (Hynes, 1992; Parsons and Parsons, 1997; Howe et al., 1998; Wang et al., 1998; Streuli and Edwards, 1998; Giancotti and Ruoslahti, 1999). In addition to this two-way flow of information mediated by extracellular matrix receptors, additional mechanisms permit extracellular matrix cues to be accessible to the genome. For instance, extracellular heparan sulfate traffics to the cell nucleus (Bhavanandan and Davidson, 1975; Hiscock, Yanagishita, and Hascall, 1994; Isihara, Fedarko, and Conrad, 1986; Liang et al., 1997) as a complex with high molecular weight forms of extracellular basic fibroblast growth factor (bFGF; Amalric et al., 1994; Maciag and Friesel, 1995; Nugent and Iozzo, 2000); together, they coordinate specific intracrine functions within the nucleus (Nugent and Iozzo, 2000); bestowing the ability, for example, to grow in serum-deprived conditions (Arese et al., 1999). Conversely, intracellular, phosphorylated forms of bFGF are secreted by a Golgi–endoplasmic reticulum–independent pathway and these forms are preferentially delivered to the nuclei of neighboring cells (Guillonneau et al., 1998). Similarly, extracellular hyaluronan can be transported from either the extracellular matrix or intracellular pools into the cell nucleus (Collis et al., 1998;

Evanko and Wight, 1999), where it likely associates with intracellular hyaluronan-binding proteins (hyaladherins; Toole, 1990; Sherman et al., 1994). Some of these proteins are known to associate with intracellular signaling molecules (Zhang et al., 1998) and to stabilize their conformation (Grammatikakis et al., 1999). Interestingly, this group of "intracellular" hyaluronan-binding proteins, can, like bFGF forms, also traffic outside of the cell into the extracellular matrix and onto the surface of neighboring cells, where, together with integrins, they regulate growth factor receptor signaling into the cell interior (Zhang et al., 1998; see also www.glycoforum.gr.jp/science/hyaluronan/HA11/HA11E.html). Trafficking hyaladherins and β-FGF resemble a growing group of proteins originally considered to function strictly as nuclear or cytosolic proteins; referred to as "messenger proteins" (Prochiantz, 2000), these include transcription factors such as *engrailed* that are exported out of a cell, taken up by neighboring cells, and transported back into the nuclei of neighboring cells. A viral mimic of this class of proteins is the HIV protein TAT. These mechanisms exist to ensure constant communication between the extracellular matrix and the cell nucleus, blurring the boundaries that once demarcated the cell and its microenvironment.

Experimental Evidence for the Role of an Intact Microenvironment

From our studies comparing simple monolayer cultures to cells maintained in a three-dimensional, basement membrane–containing environment, we know that when a cell is deprived of extracellular matrix signals, it loses its tissue-specific differentiation (figures 7.3, 7.5). When cells are maintained in an appropriate 3-D environment, extracellular matrix receptors are correctly engaged, and a cell is able to *coordinate* subtle combinations of signals to permit morphogenesis and differentiation to higher orders of organization (figures 7.2–7.5). This principle applies equally to cancer cells: Weaver and colleagues (1997) have manipulated the interactions between malignant mammary epithelial cells and their microenvironment to effect a reversion to a functionally normal phenotype (figure 7.6; for an overview, see Bissell et al., 1999). It is important to note that, in this example, the genome retained its malignancy, yet form and function were normalized (figure 7.7).

A web of functional connections among thousands of signaling pathways sustains the organization that is necessary for differentiation. Because pathways and cells are now interconnected in three dimensions, perturbation of any connection will be detected as a change throughout the tissues and organs. We argue that the computing power of a tissue is greater than the sum of its component cells in much the same way that the collective properties of an ant colony are greater than those of its member ants (figure 7.8). Researchers have concentrated on aspects of a tissue that come under the general rubric of "perception, cognition, and generation of action." While current experimental methods focus on the relatively

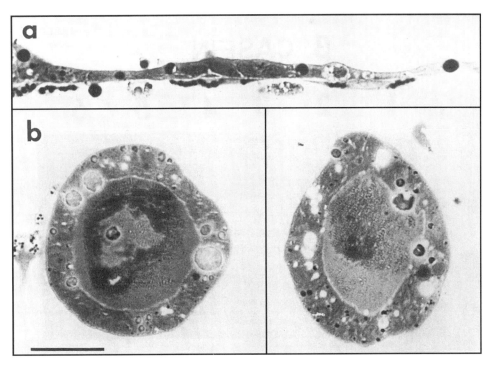

Figure 7.3
Electron micrograph of primary mouse mammary cells. Cross sections of primary mouse mammary epithelial cells on reconstituted basement membrane (matrigel). (*a*) Flattened cells on plastic. (*b*) Alveolar lumina formed by cells cultured on matrigel for 8 days in the presence of lactogenic hormones showing central lumen, minimal apical microvilli, and small lipid droplets, typical of cells in these cultures. Bar: 20 μm. (Reproduced with permission from Aggeler, Park, and Bissell, 1988.)

facile study of such aspects, robust techniques for quantitative and qualitative modeling of tissue evolution, reproduction, morphogenesis, and metabolism remain elusive.

Clearly, development of an organized extracellular matrix during evolution was an important step that enabled a collection of cells with individual features and characteristics to form a tissue capable of displaying aggregate behavior above and beyond those of its constituent cells. Knowing the essential attributes required to make this transition is akin to defining the minimal gene set necessary for cellular life. Thus we posit the need for a minimal tissue project whose goal is to elicit the necessary and sufficient features required to generate a functional tissue. The minimal genome project has estimated that 265–340 of the 517 genes of the bacterium *Mycoplasma genitalium* are essential for life (Hutchison et al., 1999). We propose that a minimal mammary gland tissue ecosystem includes luminal epithelial cells, myoepithelial cells, mesenchymal cells, lactogenic hormones, growth

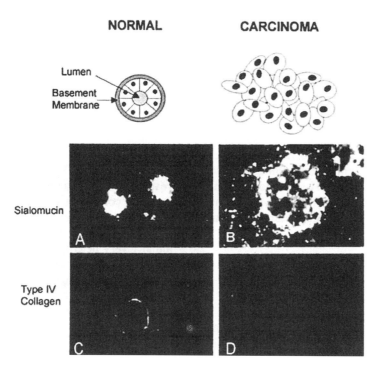

Figure 7.4
Three-dimensional basement membrane assay permits the expression of normal and malignant phenotypic traits by human breast cells. Primary cultures of normal breast epithelial cells (*A, C*) or breast carcinoma colonies were grown in 3-D matrigel for 7–10 days and were processed for immunofluorescence staining with antibodies directed against sialomucin (a marker for apical cell surfaces; *A, B*) or against type IV collagen (a marker for basement membrane; *C, D*). The staining here demonstrates that, whereas the normal breast epithelial cells grown in three dimensions are capable of forming organized spheres with central lumina and basally deposited basement membranes, their tumorigenic counterparts fail to undergo polarized morphogenesis and do not deposit endogenous basement membranelike material. (Reproduced with permission from Petersen et al., 1998, with minor modifications.)

factors, mesenchymal ECM, and basement membrane. Outstanding experimental challenges include determining the full repertoire of molecular and cellular components required to fabricate a minimal tissue ecosystem de novo.

Modeling the Role of Structural Cues

Current efforts to develop computational models of cells need to be accompanied by efforts to create computational models of tissues. The fundamental theoretical problems are how to create (1) mathematical formalisms with which to describe and define a tissue; (2) effi-

Tissue Specificity

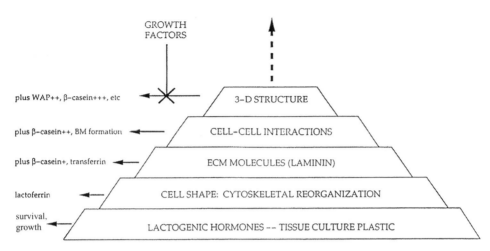

Figure 7.5
Hierarchy in regulation of mammary-specific gene expression in mouse epithelial cells. (Reproduced with permission from Bissell, 1997, with minor modifications.)

cient algorithms to estimate such models from incomplete, noisy, and heterogeneous data; and (3) techniques for generating nontrivial, accurate predictions at multiple levels of detail and abstraction. Candidate formalisms include stochastic process algebras (Hillston and Ribaudo, 1998) and graphical models (Jordan, 1998; but see also Britten and Rasskin-Gutman, chapters 5 and 17, this volume). The primary benefit of these approaches is their compositional nature: the components of a complex system and interactions between components can be modeled separately; the resultant models have clear structures, are easy to understand and can be constructed systematically by elaboration or refinement. Such techniques permit a library of reusable, hierarchical models to be developed and maintained. Because no single formalism will be adequate to represent all aspects of a tissue and no individual solution method will suffice to solve all models, an integrated approach will be necessary. With the completion of a draft human genome sequence in 2001, we are in a position to uncover some of the forces governing the interplay between the extra- and intracellular milieus that lead to formation and maintenance of a tissue.

A theory called "highly optimized tolerance" has been proposed to account for the tendency of interconnected systems to gain a measure of robustness against uncertainties in one area by becoming more sensitive in other areas (see the work of Doyle and colleagues at www.cds.caltech.edu/~doyle). If the unit of function in an organ (e.g., the mammary gland) possesses such a property, what are the common and designed-for uncertainties to which it is resilient? And what are the design flaws or rare events to which it is hypersensitive? Controlling and redesigning this highly optimized system so that transi-

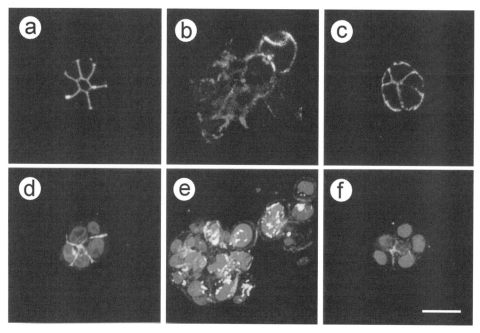

Figure 7.6
Reversion of mammary epithelial tumor cells. Treatment of T4-2 tumor cells with β1-inhibitory antibody leads to phenotypic reversion and acinar formation. Confocal immunofluorescence microscopy images of E-cadherin (FITC) and β-catenin (Texas red; dark gray) of phenotypically normal S1 cells (*a*), malignant T4-2 cells (*b*), and reverted T4-2 cells (*c*). In S1 (*a*) and T4-β1 reverted acini (*c*), E-cadherin and β-catenin were colocalized and superimposed at the cell junctions. In contrast, E-cadherin and β-catenin were often not colocalized in mock-treated T4 cells (*b*). (*d–f*) Confocal fluorescence microscopy images of F actin (FITC; light gray) and nuclei (propidium iodide; dark gray). Both the S1 (*d*) and the reverted T4-2 cells (*f*) showed acinar formation with basally localized nuclei (propidium iodide) and organized filamentous F-actin, whereas T4-2 mock-treated colonies had disorganized, hatched bundles of actin and pleiomorphic nuclei (*e*). Bar: 16 μm. (Reproduced with permission from Weaver et al., 1997, with minor modifications.)

tions to an aberrant state are minimized will require understanding its structure and behavior at many levels. One intriguing possibility is that the mammary gland exhibits the phenomenon known as "stochastic resonance" (SR), a mechanism whereby the presence of noise enhances the detection of weak signals; SR may be relevant to problems in sensory biology.

As a tissue evolves, it can adapt to, or learn from, its noisy environment. Whereas adaptation can be considered temporary with the system eventually resetting itself, learning involves a persistent and heritable change. Perhaps it makes sense to appropriate language from the machine learning community. Unsupervised learning, finding patterns or natural groups in data, might be the primary manner in which a collection of cells learns from

Tissue Specificity

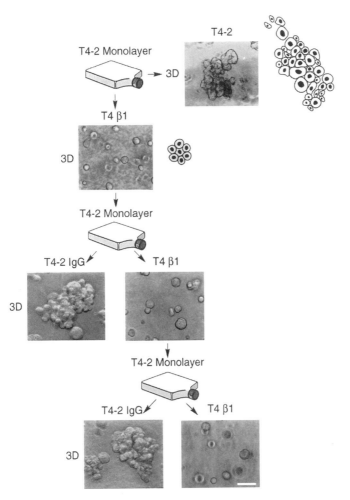

Figure 7.7
Evidence for phenotypic reversion rather than selection. Phase contrast micrographs of T4-2 (tumor) cells grown in matrigel and in the presence of β1 function blocking antibody (T4 β1), mock antibody (T4-2 IgG) or no antibodies (T4-2). Despite two rounds of treatment, these antibody reverted cells were able to resume their original tumorigenic phenotypes when cultured in the absence of antibody. (Reproduced with permission from Weaver et al., 1997, with minor modifications.)

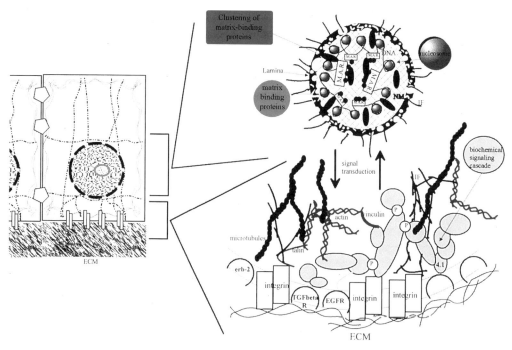

Figure 7.8
Central hypothesis: the bidirectional computing power of a tissue is greater than the sum of its component parts. Bidirectional flow of tissue-specific information is dependent on the nuclear and chromatin structures, the nature of membrane receptors, and environmental milieu.

its environment. Supervised learning necessitates the presence of a teacher to inform and guide data modeling and interpretation. For dynamic reciprocity, both the nucleus and ECM can play the role of teacher. Tissues might engage in what is known as "reinforcement learning." In the absence of a teacher, noisy feedback might serve to indicate how good an action was: different costs could be associated with alternative responses and this cost-benefit analysis of the stochastic environment might determine the actions implemented by the ECM or nucleus.

Cells inhabit an uncertain world. Signaling molecules that regulate intracellular and extracellular processes may be present in a few to a few hundred copies and display significant internal noise. Despite this, cells integrate and interpret myriad signals in a meaningful way, provided they are grown in natural tissue environments. Against a stable genome, cells differentiate into diverse phenotypes and associate into tissues, which in turn connect to form the entire organism. It is clear that understanding how these processes are regulated will require studies that utilize tissues, organs, and tissuelike model systems. We have taken the first steps along this pathway, but a long and interesting journey lies ahead.

Acknowledgments

We are grateful to Norene Jelliffe for expert administrative assistance. This work was supported by contract DE-AC03-76F00098 from the U.S. Department of Energy, Office of Biological and Environmental Research, and by National Institutes of Health grants CA64786 and CA57621 (to Mina Bissell), Canadian Institutes of Health Research grant MT9641 (to Eva Turley), and by Hollaender postdoctoral fellowship DE-AC03-SF00098 (to Derek Radisky).

References

Aggeler J, Park CS, Bissell MJ (1988) Regulation of milk protein and basement membrane gene expression: The influence of the extracellular matrix. J Dairy Sci 71: 2830–2842.

Amalric F, Bouche G, Bonnet H, Brethenou P, Roman AM, Truchet I, Quarto N (1994) Fibroblast growth factor-2 (FGF-2) in the nucleus: Translocation process and targets Biochem Pharmacol 47: 111–115.

Arese M, Chen Y, Florkiewicz RZ, Gualandris A, Shen B, Rifkin DB (1999) Nuclear activities of basic fibroblast growth factor: Potentiation of low-serum growth mediated by natural or chimeric nuclear localization signals. Mol Biol Cell 10: 1429–1444.

Barcellos-Hoff MH, Aggeler J, Ram TG, Bissell MJ (1989) Functional differentiation and alveolar morphogenesis of primary mammary cultures on reconstituted basement membrane. Development 105: 223–235.

Bhavanandan VP, Davidson EA (1975) Mucopolysaccharides associated with nuclei of cultured mammalian cells. Proc Natl Acad Sci USA 72: 2032–2036.

Bissell MJ (1981) The differentiated state of normal and malignant cells or how to define a "normal" cell in culture. Int Rev Cytol 70: 27–100.

Bissell MJ (1997) The central role of basement membrane in functional differentiation, apoptosis and cancer. In: Cell Death in Reproductive Physiology (Tilly JL, Strauss JF, Tenniswood M, eds), 125–140. Serono Symposia USA. New York: Springer.

Bissell MJ, Hall HG, Parry G (1982) How does the extracellular matrix direct gene expression? J Theor Biol 99: 31–68.

Bissell MJ, Weaver VM, Lelievre SA, Wang F, Petersen OW, Schmeichel KL (1999) Tissue structure, nuclear organization, and gene expression in normal and malignant breast. Cancer Res 59: 1757–1763.

Boudreau N, Bissell MJ (1998) Extracellular matrix signaling: Integration of form and function in normal and malignant cells. Curr Opin Cell Biol 10: 640–646.

Brown E, Hogg N (1996) Where the outside meets the inside: Integrins as activators and targets of signal transduction cascades. Immunol Lett 54: 189–193.

Clark EA, Brugge JS (1995) Integrins and signal transduction pathways: The road taken. Science 268: 233–239.

Collis L, Hall C, Lange L, Ziebell M, Prestwich R, Turley EA (1998) Rapid hyaluronan uptake is associated with enhanced motility: Implications for an intracellular mode of action. FEBS Lett 440: 444–449.

Ellsdale T, Bard J (1974) Cellular interactions in morphogenesis if epithelial mesenchymal systems. J Cell Biol 63: 343–349.

Emerman JT, Enami J, Pitelka DR, Nandi S (1977) Hormonal effects on intracellular and secreted casein in cultures of mouse mammary epithelial cells on floating collagen membranes. Proc Natl Acad Sci USA 74: 4466–4470.

Evanko SP, Wight TN (1999) Intracellular localization of hyaluronan in proliferating cells. J Histochem Cytochem 47: 1331–1342.

Faull RJ, Ginsberg MH (1996) Inside-out signaling through integrins. J Am Soc Nephrol 7: 1091–1097.

Giancotti FG, Ruoslahti E (1999) Integrin signaling. Science 285: 1028–1032.

Grammatikakis N, Lin JH, Grammatikakis A, Tsichlis PN, Cochran BH (1999) p50 (cdc37) acting in concert with Hsp90 is required for Raf-1 function. Mol Cell Biol 19: 1661–1672.

Guan JL, Chen HC (1996) Signal transduction in cell-matrix interactions. Int Rev Cytol 168: 81–121.

Guillonneau X, Regnier-Ricard F, Dupuis C, Courtois Y, Mascarelli F (1998) Paracrine effects of phosphorylated and excreted FGF1 by retinal pigmented epithelial cells. Growth Factors 15: 95–112.

Hillston J, Ribaudo M (1998) Stochastic process algebras: A new approach to performance modeling. In: Modeling and Simulation of Advanced Computer Systems (Bagchi K, Zobrist G, eds). New York: Gordon Breach.

Hiscock D, Yanagishita M, Hascall VC (1994) Nuclear localization of glycosaminoglycans in rat ovarian granulose cells. J Biol Chem 269: 4539–4546.

Howe A, Aplin AE, Alahari SK, Juliano RL (1998) Integrin signaling and cell growth control Curr Opin Cell Biol 10: 220–231.

Hutchison III CA, Peterson SN, Gill SR, Cline RT, White O, Fraser CM, Smith HO, Venter JC (1999) Global transposon mutagenesis and a minimal mycoplasma genome. Science 286: 2165–2169.

Hynes RO (1992) Integrins: Versatility, modulation, and signaling in cell adhesion. Cell 69: 11–25.

Hynes RO, Zhao Q (2000) The evolution of cell adhesion. J Cell Biol 150: 89–96.

Isihara M, Fedarko NS, Conrad HE (1986) Transport of heparan sulfate into the nuclei of hepatocytes. J Biol Chem 261: 13575–13580.

Jordan MI (ed) (1998) Learning in Graphical Models. Dordrecht: Kluwer Academic.

Liang Y, Haring M, Roughley PJ, Margolis RK, Margolis RU (1997) Glypican and biglycan in the nuclei of neurons and glioma cells: Presence of functional nuclear localization signals and dynamic changes in glypican during the cell cycle. J Cell Biol 139: 851–864.

Liu S, Calderwood DA, Ginsberg MH (2000) Integrin cytoplasmic domain-binding proteins. J Cell Sci 113: 3563–3571.

Maciag T, Friesel RE (1995) Molecular mechanisms of fibroblast growth factor-1 traffic, signaling, and release. Thromb Haemost 74: 411–424.

Michalopoulas G, Pitot HC (1975) Primary culture of parenchymal liver cells on collagen membranes: Morphological and biochemical observations. Exp Cell Res 94: 70–78.

Myers CA, Schmidhauser C, Mellentin-Michelotti J, Fragoso G, Roskelley CD, Casperson G, Mossi R, Pujuguet P, Hager G, Bissell MJ (1998) Characterization of BCE-1, a transcriptional enhancer regulated by prolactin and extracellular matrix and modulated by the state of histone acetylation. Mol Cell Biol 18: 2184–2195.

Nugent MA, Iozzo RV (2000) Fibroblast growth factor-2. Int J Biochem Cell Biol 32: 115–120.

Parsons JT, Parsons SJ (1997) Src family protein tyrosine kinases: Cooperating with growth factor and adhesion signaling pathways. Curr Opin Cell Biol 9: 187–192.

Petersen OW, Ronnov-Jessen L, Howlett AR, Bissell MJ (1992) Interaction with basement membrane serves to rapidly distinguish growth and differentiation pattern of normal and malignant human breast epithelial cells. Proc Natl Acad Sci USA 89: 9064–9068.

Petersen OW, Ronnov-Jessen L, Weaver VM, Bissell MJ (1998) Differentiation and cancer in the mammary gland: Shedding light on an old dichotomy. Adv Cancer Res 75: 135–161.

Prochiantz A (2000) Messenger proteins: Homeoproteins, TAT and others. Curr Opin Cell Biol 12: 400–406.

Pujuguet P, Radisky D, Levy D, Lacza C, Bissell MJ (2001) Trichostatin A inhibits beta-casein expression in mammalian epithelial cells J Cell Biochem 83: 660.

Ronnov-Jessen L, Petersen OW, Bissell MJ (1996) Cellular changes involved in conversion of normal to malignant breast: Importance of the stromal reaction. Physiol Rev 76: 69–125.

Ruoslahti E (1997) Integrins as signaling molecules and targets for tumor therapy. Kidney Int 51: 1413–1417.

Schmeichel KL, Weaver VM, Bissell MJ (1998) Structural cues from the tissue microenvironment are essential determinants of the human mammary epithelial cell phenotype. J Mammary Gland Biol Neoplasia 3: 201–213.

Schmidhauser C, Bissell MJ, Myers CA, Casperson GF (1990) Extracellular matrix and hormones transcriptionally regulate bovine beta-casein 5′ sequences in stably transfected mouse mammary cells. Proc Natl Acad Sci USA 87: 9118–9122.

Schmidhauser C, Casperson GF, Myers CA, Sanzo KT, Bolten S, Bissell MJ (1992) A novel transcriptional enhancer is involved in the prolactin- and extracellular matrix–dependent regulation of beta-casein gene expression. Mol Biol Cell 3: 699–709.

Schoenwaelder SM, Burridge K (1999) Bidirectional signaling between the cytoskeleton and integrins. Curr Opin Cell Biol 11: 274–286.

Sherman L, Sleeman J, Herrlich P, Ponta H (1994) Hyaluronate receptors: Key players in growth, differentiation, migration, and tumor progression. Curr Opin Cell Biol 6: 726–733.

Streuli CH, Bissell MJ (1991) Mammary epithelial cells, extracellular matrix, and gene expression. Cancer Treat Res 53: 365–381.

Streuli CH, Edwards GM (1998) Control of normal mammary epithelial phenotype by integrins. J Mammary Gland Biol Neoplasia 3: 151–163.

Toole BP (1990) Hyaluronan and its binding proteins, the hyaladherins. Curr Opin Cell Biol 2: 839–844.

Wang F, Weaver VM, Petersen OW, Larabell CA, Dedhar S, Briand P, Lupu R, Bissell MJ (1998) Reciprocal interactions between beta-1-integrin and epidermal growth factor receptor in three-dimensional basement membrane breast cultures: A different perspective in epithelial biology. Proc Natl Acad Sci USA 95: 14821–14826.

Weaver VM, Petersen OW, Wang F, Larabell CA, Brian P, Damsky C, Bissell MJ (1997) Reversion of the malignant phenotype of human breast cells in three-dimensional culture and in vivo by integrin blocking antibodies. J Cell Biol 137: 231–245.

Yamada KM (1997) Integrin signaling. Matrix Biol 16: 137–141.

Zhang S, Chang MC, Zylka D, Turley S, Harrison R, Turley EA (1998) The hyaluronan receptor RHAMM regulates extracellular-regulated kinase. J Biol Chem 273: 11342–11348.

8 Genes, Cell Behavior, and the Evolution of Form

Ellen Larsen

The "origination of organismic form" has two aspects, one evolutionary and one developmental. This chapter focuses on cells and the half a dozen cell behaviors responsible for multicellular forms in development, behaviors whose genetically based modifications may result in the evolution of form. It provides reasons for suspecting that relatively few changes in cell behavior may have dramatic effects on form without affecting the rest of development and suggests an experimental approach to exploring the range of achievable morphologies.

Darwin's maxim that evolution is the result of "descent with modification" is one of the most powerful statements in biology. It implies that different kinds of organisms look similar in some ways because of common ancestry, and look different because the shared heredity has been modified. Elsewhere in *The Origin of Species,* Darwin (1859) points out that modifications in those aspects of heredity concerned with development are the ones that may lead to new forms. Today, we might say that mutations in genes involved in development provide the material basis for the evolution of form.

Through mutational analysis and molecular biology, we have an overwhelming amount of knowledge about the molecules that play a role in development; with the genome sequencing of some organisms "complete," we actually have a catalogue of all the genes in organisms such as *Caenorhabditis elegans, Drosophila melanogaster,* and *Arabidopsis thaliana*. Although our better understanding of genetics has not resolved all of the issues relating genotype to phenotype, it has clarified ideas about evolution at the molecular level. For example, we can no longer maintain that new forms evolve merely by evolving new genes. Gene duplication and divergence are still a cornerstone of our understanding of evolutionary change, yet the overwhelming evidence is that the gene families used today are very ancient. Metazoans, at least, all use a remarkably similar set of genes not only for cell structure, function, and metabolism, but also for development. Twenty-five years ago, we accepted that enzymes in cell metabolism and structural molecules such as actin were highly conserved in evolution, but we had little idea whether this held for developmentally important molecules. Comparative studies of model organisms make it clear that most molecules regulating development are shared, with differences between taxa occurring in the number of members of a gene family and in how they are used. Another conserved aspect of development is that gene products involved in regulating development function in cascades, which are also highly conserved in evolution (see, for example, figure 8.1).

Because metazoans share so many genes and gene pathways, it is difficult to justify the idea that, if we knew enough about gene functions, we would be able to read the genome of an organism and determine its shape. From what we know about the genetic-molecular machinery of development, it is easier to explain similarities between taxa than their

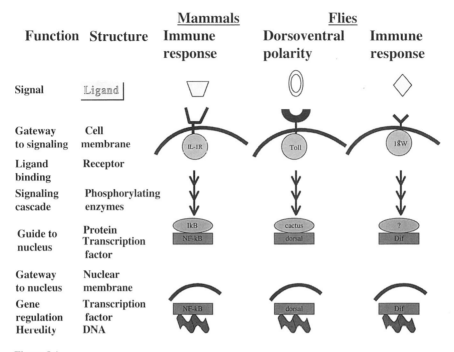

Figure 8.1
In signal transduction cascades, similar molecules play roles with different developmental outcomes. The generalized transduction of signals starts with a ligand-bound receptor, which sets off a cascade, of variable length, of intermediate enzymatic steps usually involving phosphorylation changes. At some point, this cascade impinges on the system for nuclear translocation of a transcription factor (as pictured here) or the transcription factor enters the nucleus unaided and binds to DNA receptors. In the cases illustrated here, the interleukin-1 pathway in mammals results in immune response activation, whereas putatively homologous pathways in flies result in dorsoventral patterning in embryos or immune responses in larval cells. The same basic pathway can mediate different end results by changing the binding properties of the cell membrane receptor or modifying target DNA so it can bind a new transcription factor. (Modified from Hultmark, 1994.)

apparent differences. We are faced with the problem that the genotype must be significant in evolution because it is the most important material passed from generation to generation, yet our current knowledge does not permit an intuitive leap from linear DNA to a three-dimensional organism. Thus we must look at the genotype-phenotype relationship in a fundamentally different way. This chapter shows how cells and their properties form both a material and conceptual link between genotype and phenotype.

The Cell as Pivot between the Gene and the Development of Form

Because genes, by themselves, cannot specify form and because metazoan forms are multicellular, it is natural to look to the cell as an intermediary between gene influence and multicellular development. When we consider what cells do in the course of morphogenesis, we find only six overlapping cell behaviors: (1) division, (2) growth, (3) death, (4) shape change, (5) matrix secretion, and (6) movement (Larsen and McLaughlin, 1987). Different taxa use these cell behaviors to different extents. Cell movement is less frequently used in arthropods than in vertebrates during morphogenesis, and not at all in plants, whose sister cells share a common cell wall. Nevertheless, plants make all the shapes basic to biological forms: rods, tubes, sheets, and spheres.

In addition to their role as providers of gene products for metabolism during the cell cycle, genes influence when and where cell behaviors occur. Thus they influence form through spatial and temporal coordination of cell assembly into multicellular structures. Genes that coordinate cell behaviors I call "bureaucrat genes"; these tell other genes when and where to turn on and off (Larsen, 1997). A brief look at the current literature in developmental genetics reveals that the number of such genes providing the signal transduction and transcription cascades central to developmental change is enormous. Although we are just beginning to see how these bureaucratic cascades effect changes in cell behavior, we know that mutations in the bureaucrats often lead to abnormalities in what cells do. For example, the famous fly mutation "Ultrabithorax" permits abnormal amounts of cell division in the anterior part of the presumptive haltere, turning this region into a winglike structure. Genes whose products actually provide the material for cell behavior modification I call "worker genes" (Larsen, 1997); these include genes for cytoskeleton and structural elements of membranes. The terms *bureaucrat* and *worker* are operationally defined: the same gene may perform both functions and thus be both a bureaucrat and a worker gene, such as β-catenin, which plays a structural role in cell adhesion in some contexts and also participates in the transcriptional control of other genes in others (Huber et al., 1996).

Clues as to the amount of genetic "effort" required to coordinate the production of form in taxa with different developmental mechanisms may be found by comparing ratios of bureaucrat to worker genes. The initial analysis of the *C. elegans* genome project suggests that this approach has potential. Of the approximately 19,100 protein coding genes found, the 20 most common protein motifs occurred at 2,910 loci. Of these, over 70 percent can be classified as "bureaucrats" for their putative signal transduction or transcription activation functions, whereas about 20 percent qualify as "workers" (*C. elegans* sequencing consortium, 1998). How many of these bureaucrats operate during development is as yet unknown. Because of the genome-sequencing projects currently under way or completed, comparative analyses should be increasingly feasible. We can look forward to comparing absolute numbers and relative numbers of bureaucrats and workers to see if these change in evolution and can be correlated to different kinds of coordination of cells and tissues in development.

Coordinating Development at the Tissue Level

In the hierarchy of metazoan structures, because cells are parts of tissues, it is natural that cell behaviors participate in and are regulated by tissue-level mechanisms. Although it is relatively easy to suggest how genes within a particular cell can influence that cell, once tissue-level behavior comes into play, the coordination of groups of cells poses a more difficult problem. The simplest tissue to consider is the monolayer epithelial sheet, one of the great metazoan inventions. Such an epithelial sheet consists of a single layer of cells, usually in a hexagonal array with a variety of junctions joining each cell to its neighbors. Typically, cells in an epithelium are polarized both with respect to the types of junctions from one end of the cell to the other and with respect to the location of subcellular organelles and cell surface constituents (Yeaman, Grindstaff, and Nelson, 1999). When an organism is multilayered, other tissues are usually derived from an epithelial sheet. For example, mesenchyme in vertebrate development is pinched off from an epithelial sheet. Some taxa are primarily composed of epithelia, arthropods being the most notable. Surprisingly, the cephalochordate amphioxus is also an epithelial animal, although related chordates are not.

How do cells in epithelia communicate? Some of the junctions allow the passage of small molecules, and some cells have receptors with ligands on neighboring cells that allow signaling between cells. In development, a great deal of attention has been focused on secreted molecules with a signaling range of a few cell diameters; a well-studied example is the decapentaplegic protein important in axis formation (Lecuit et al., 1996). Conceptually, the kind of communication seen in an epithelial tissue is two-dimensional; often, a given sheet of cells is not affected by other tissues in the organism. This tissue autonomy may account for the relationship between genes, cells, and tissues in epithelial modes of development. Experimental results suggest that tissue autonomy is associated with the cellular autonomy of important transcription factors. A gene is cell autonomous if it produces a mutant phenotype in a cell surrounded by genetically normal cells. This kind of test has established autonomy for such fly mutations as spineless aristapedia, in which the antenna is turned into a leglike structure. Surrounded by a sea of antennal cells, even a small patch of mutant cells will produce leg bristles, whereas larger patches will produce parts of legs (Postlethwait and Girton, 1974). Significantly, a single gene change in an epithelial sheet can have a dramatic effect, not only on cell differentiation but also on co-coordinating changes in morphology.

In contrast to arthropods, appendages in vertebrates arise as the result of an interaction between an overlying epithelium and mesenchymal tissue (see Wagner and Chiu, chapter 15, this volume). The latter is a mass of loosely cohering cells which migrate to the appropriate region under the epithelium and one way or two way chemically mediated interactions occur between the tissues. The morphogenesis of legs in vertebrates is an excellent example of a two-way interaction between tissues, where the mesenchyme determines the type of leg (fore or hind) and induces "legness" in the overlying epidermis. The epidermis feeds back on mesenchymal cell proliferation and is necessary for limb outgrowth (Fallon and Caplan, 1983). This three-dimensional strategy for patterning structures is widespread, not only in vertebrates, but in most metazoans, with exceptions such as the arthropods, amphioxus, and nematodes.

From the point of view of genetic regulation, one of the possible consequences of a three-dimensional strategy of development is that a single gene change affecting one of the interacting tissues will probably not transform the structure without complementary gene changes in its interacting partner. The requirement for two genetic changes (for example, in a receptor and its ligand) reduces the probability of large morphological change. I have suggested that the differences in relationship between genes and the epithelial sheet and tissue interaction modes of development may explain why, even though transformations of appendages from one type to another are fairly common in arthropods, one never sees people with feet where their hands should be (Larsen, 1992). To be sure, modification in structure can occur in the tissue interaction strategy; the evolution of tetrapod limbs attests to that. However, the changes appear to be variations on a theme rather than complicated transformational changes. For example, toe number can be experimentally altered by altering the number of cells in the appropriate presumptive part of the limb (Alberch and Gale, 1983). Presumably, polydactyly mutations act by increasing the number of cells in the presumptive digit regions so that the number but not the form of digits is altered.

Finally, following the strategy of lineage invariance, most organisms develop structures with a predictable cell number, division pattern, and final location of each cell. An example in fruit flies would be the formation of cells for sensory structures in the peripheral nervous system (Jan and Jan, 1993), each born in a particular place and in a particular order. There are few taxa, however, in which every cell division in the development of the organism can be mapped out and precisely what each cell will become is known. Among those taxa is the well-studied nematode *Caenorhabditis elegans,* a small organism with no more than 560 first-stage larval cells (Felix, 1999). Interestingly, the larger, marine nematode *Enoplus brevis* lacks this total lineage predictability (Voronov and Panchin, 1998). Thus it has been suggested that lineage invariance is a consequence of the phylogenetically derived strategy of being "paucicellular."

Can we ascribe selective advantages to particular tissue strategies under particular conditions? I have mentioned the possibility that lineage invariance may be a derived condition associated with having few cells. It may also be argued that tissue interaction strategies provide more developmental stability than autonomous sheet strategies because one gene change is likely to have less impact on structure. Alternatively, there may be no selective differences. At this point, the question is more interesting than the data available to sort out such speculations.

Molecules to Morphogenetic Fields: Can We Integrate These Aspects of Tissue Morphogenesis?

Trying to understand the coordination of development from a knowledge of molecules and genes is akin to putting Humpty-Dumpty together again. One of the problems is illustrated in figure 8.1 in which presumably homologous genes and cascades are associated with different developmental outcomes. Furthermore, the same molecule can be regulated by different genes in the same organism: *engrailed,* important for establishing segmental boundaries in flies, is regulated by *paired* in one group of cells in each parasegment and by *fushi tarazu,* four cells away (Manoukian and Krause, 1992); or by different genes in different organisms: *hedgehog,* which is important for setting up the anteroposterior axis, is controlled by *bicoid* in flies and *caudal* in the beetle *Tribolium* (Deardon and Akam, 1999). If the same genes or cascades produce different structures and different genes/cascades produce similar morphogenetic effects, how can we understand multicellular development? A reasonable conclusion is that the "meaning" of these molecular processes depends on the context in which they occur. The *morphogenetic field* might provide a generalized concept for context. This term refers to the regulative properties of a presumptive tissue. Consider presumptive leg tissue (which produces two legs when divided). In analogy to physical fields, the morphogenetic field of such tissue has a center of greater leg-forming ability, which diminishes with distance. Because morphogenetic fields behave according to rules and have dynamic properties, it is possible to model their behavior mathematically (Goodwin, 1997).

Although dynamic formulations may satisfy the needs of some developmental biologists, most cry out for the material basis of such abstractions. A bold attempt to provide this basis was put forward by Gilbert, Opitz, and Raff (1996), who proposed that morphogenetic fields be redefined in terms of gene cascades that appear to be crucial to the morphogenesis of the structures arising from the field. Although their proposal merits wide discussion, I have several reservations. First, it is not immediately obvious how these cascades can account for the approximately 50-cell requirement to have the properties of a multicellular field. Second, similar fields may have somewhat different players, which suggests that the cascades are not the primary expression of the morphogenetic field. Finally, regulative field phenomena such as regeneration can be set in motion by physically dividing a tissue in half. The tissue apparently "senses" the wound, and processes ensue that involve the ectopic expression of signal transduction cascades. This suggests to me that the morphogenetic field has emergent properties, independent of the particular molecular entities which carry out the behavior of the field.

Although I have no alternative proposal to explain the material basis of the field, I do have a way of connecting gene expression and cell behavior to morphogenetic fields. My suggestion is that we consider how gene products or their absence can alter cell behavior within fields and thereby evoke morphogenetic consequences. Two examples come to mind. Clark and Russell (1977) showed that a mutation, *suppressor of forked,* caused cell death in leg imaginal discs of flies, which was correlated with leg duplications. Presumably, the cell death instigated by the mutation functionally divided the disc as if it had been surgically cut, thus invoking a field response. More natural examples might be cases of classical individuation of fields. For example, in vertebrates there is a single eye field in the neural plate, which normally divides to produce two eyes; if this fails, a central, "cyclopic" eye results. One can imagine that altering the size of fields through cell division enhancement or reduction could also give rise to morphogenetic change (see Müller, chapter 4, this volume). This is consistent with previously described findings that experimentally changing the number of cells in developing amphibian limbs may change the number of digits (Alberch and Gale, 1983) and play a role in the development of (mutant) polydactyly phenotypes.

Evolution of Morphogenesis

If morphogenesis is viewed as the result of co-coordinating cell behaviors, the evolution of morphogenesis should involve changes in the coordination of such behaviors. But where should we look for changes in coordination? And can small coordinative changes lead to large morphological ones? Although signal transduction-transcription cascades would seem well suited to such behavior at the molecular level, as I have discussed, evidence suggests these cascades are ancient and widely used in metazoans, so that changing them seems not to be generally important in morphological evolution. What is more likely to be important is changing the receptors at the beginning of a cascade, the target genes at the end (which effect cell behavior changes), or both. Thus coupling and uncoupling cascades—or shuffling bureaucrat genes, as I have said previously (Larsen, 1997; Wagner and Altenberg, 1996)—should be a frequent hallmark of evolution.

Changing "who is coupled to whom" may occur at levels of organization above the molecular level. In fact, Simon (1973, p. 7) formalized this possibility in an essay on hierarchy theory and evolution:

One can show on quite simple and general grounds that the time required for a complex system, containing k elementary components, say, to evolve by processes of natural selection from those components is very much shorter if the system is itself comprised of one or more layers of stable component subsystems than if its elementary parts are its only stable components.

He went on to illustrate this with the parable of the two watchmakers assembling a 1,000-part watch. Each watchmaker is interrupted by a telephone call after assembling on average 150 elements of the watch, at which point any assemblies that are not stable fall apart completely. The first watchmaker makes stable subassemblies of 100 parts and then assembles these into the watch. The second watchmaker does not organize his work. On average, the first watchmaker will have finished a watch after eleven telephone calls, and on average, the second watchmaker will never succeed in finishing a watch. Simon also pointed out that, at a given level of a hierarchical system, components of the system are only partially coupled and so long as the initial function of a component is maintained, it is still possible for it to evolve. For example, so long as mitochondria supply a cell with sufficient ATP for its processes, its DNA is free to change.

Some of the factors I consider important for the evolution of form are summarized below:

1. Rates of morphogenetic evolution depend on changes in the coordination of developmental processes and more specifically the ease with which it is possible to change the timing and location of a small number of generic cell behaviors.

2. The hierarchical organization of biological systems facilitates relatively rapid evolutionary change because developmental systems are modular, modules are often partially coupled, and coupling and uncoupling modules can change the number of interacting components exponentially rather than linearly.

3. At the molecular level, surprisingly few genetic changes may lead to change in coupling between genes and their regulatory signal.

4. Gene changes affect morphogenesis by affecting cell behaviors, including the cell behaviors that modify morphogenetic fields.

5. At the tissue level, cell-autonomous genetic behavior in two-dimensional sheets may permit large and evolutionarily rapid changes in morphogenesis, compared to changes in interacting tissues that probably require complementary changes in both interacting tissues to achieve change.

6. Because there are alternative routes to achieve a morphology using different cell behaviors, the likelihood of having sufficient genetic variation for selection is increased.

Are There Developmental Constraints?

Evidence for the existence of *developmental constraints,* defined as developmental variations that bias the kinds of forms likely to evolve, comes chiefly from observing a paucity of organismic diversity. For example, despite genetic variation, there are relatively few body plans. We are now in a position to think about the way development is organized and the kinds of genetic variation which can modify that organization in order to assess the material reality of developmental constraints.

I suggest that there are properties of complex systems and specific aspects of morphogenesis that may bear on developmental constraints. Pertinent systems properties include the hierarchical nature of living organisms from molecules to individuals and the modular nature of hierarchies as well as the partial coupling of modules. Pertinent aspects of morphogenesis include its being the result of co-coordinating cell behaviors through changes in the coupling of regulatory cascades at intracellular, intercellular, and tissue levels of organization. Because experience demonstrates there is considerable genetic variation that can change cell behaviors, we should be able to imagine any morphology and conceive of several ways of organizing cells to make it. For example, isofemale fruit fly lines show significant differences between line variation for cell size and number (De Moed, De Jong, and Sharloo, 1997). Because of the flexibility of developmental coordination and the likely presence of genetic variation for cell behaviors, I envision no absolute developmental constraints on form.

There are, however, likely to be other kinds of constraints on the *generation* of form. Although I have not explicitly considered the mechanical properties of cells and tissues that contribute to morphogenesis (Newman and Comper, 1990; Newman, chapter 13, this volume), I assume that these properties can be influenced by genes. In addition to such physical constraints, because the minimal number of gene changes required to achieve a certain morphology will depend on the cellular route taken and the starting genome, there are likely to be constraints or biases in terms of the mechanisms used to evolve a new form. For example, I would predict that it would take more genetic change to make an ectopic fly limb using a tissue interaction strategy involving mesenchyme-epithelia interactions than it would be to create an ectopic vertebrate limb from vertebrate mesenchyme and epithelia. Why? Because we start with fly appendages made from a two-dimensional sheet (but see Percival-Smith et al., 1997) and would have to "create" mesenchymal cells and secreted molecules to elicit appendage growth in an overlying epithelium.

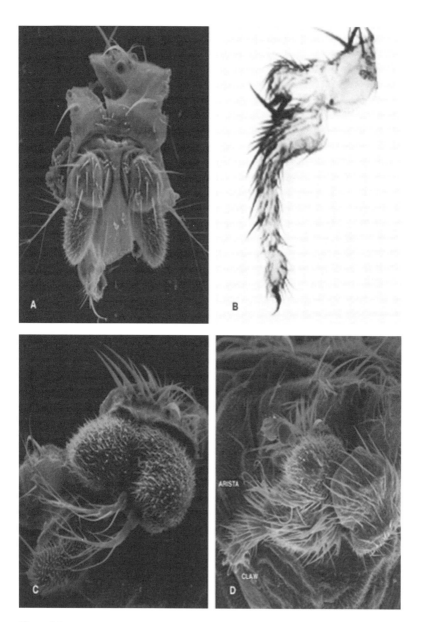

Figure 8.2
Steps in the plan to create a biramous appendage containing the distal regions of antenna and leg with a common base on fly heads. (*A*) Portion of a fly head with two antennae. (*B*) An antenna duplicated under the influence of the mutation, obake. Note the two branchlike aristae. (*C*) An antennal leg induced by Antennapedia protein in an antenna disc. (*D*) Combining *obake* and a mutation causing antenna-to-leg transformation, $Antp^{NS}$, a biramous appendage is formed with both claw and arista.

Designer Organisms: An Experimental Approach

The thrust of my arguments has been that relatively few changes in cell behavior may be responsible for large morphological changes without unraveling the rest of development. If this is so, we should be able to redesign structures in an organism with a few well-chosen mutations. In my first attempt, I set out to show that a uniramous fruit fly could be altered to produce a biramous appendage as is found in Crustacea (Williams and Müller, 1996). More specifically, I wanted to make an antenna-leg combination having the distal tips, arista, and claw, respectively, of each structure. The plan was to induce a duplication of the antenna morphogenetic field and then to transform one of the two fields into a leg field. Two mutations were used for this purpose, one gene (*obake*) to create a mirror duplicate of the antenna primordium and the other (*Antennapedia*) to transform one of the antennal primordia into a leg. Such biramous structures were achieved (figure 8.2), although at a lower rate than creating multiple legs on the fly heads (figure 8.3; Dworkin, Tanda, and Larsen, 2001). We are finding ways to create a higher proportion of the desired biramous structures.

Although the morphologies that we try to create are limited only by our imaginations, those designs which attempt to create in one taxon something found in another are likely to tell us the most about the minimum genetic change required to make a transition to the new

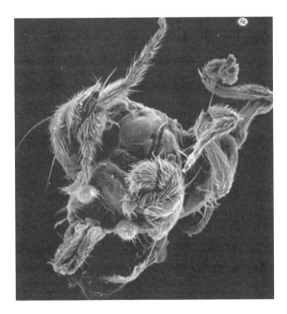

Figure 8.3
Portion of a fly head from an *obake, Antp^{NS}* fly with five leglike structures. More than one duplicated antenna was formed on one side of the head; all of these turned into legs.

morphology. Besides creating "new morphologies," we can also try to alter the processes that determine the number of repeated units for, say, stable differences in segment number.

I am not suggesting that "designer organism" studies will cast light on what actually happened in the evolution of a structure. Rather, these developmental studies can help us develop our intuition about the nature of constraints, to the extent they exist. They can also help us understand the nature of "genetic backgrounds" required to stabilize the development of these "new" structures, so that they can be produced more reliably. Understanding mechanisms for stabilizing the development of new structures may prove more elusive and more important evolutionarily than finding genes that participate in creating the forms. Combining studies of what forms arose in evolution with studies of what forms are possible will provide a more informed basis for understanding the nature of the origin of metazoan forms.

Acknowledgments

Many people have contributed to these ideas over the years, In particular, I would like to thank Mary Jane West-Eberhard, Hooley McLaughlin, Roy Pearson, Malcolm Telford, Sue Varmuza, and Ian Dworkin. Joel Atallah, Vidyanand Nanjundiah, and the volume editors also made helpful comments on earlier drafts of the manuscript. Funded by NSERC Canada.

References

Alberch P, Gale E (1983) Size dependence during the development of the amphibian foot, Colchicine-induced loss and reduction. JEEM 76: 177–197.

C. elegans Sequencing Consortium (1998) Genome sequence of the nematode *C. elegans*: A platform for investigating biology. Science 282: 2012–2018.

Clark W, Russell MA (1977) The correlation of lysosomal activity and adult phenotype in a cell-lethal mutant of *Drosophila*. Dev Biol 57: 160–173.

Darwin C (1859) The Origin of Species by Means of Natural Selection. London: John Murray.

Deardon P, Akam M (1999) Developmental evolution: Axial patterning in insects. Curr Biol 9: R591–R594.

De Moed, GH, De Jong G, Scharloo, W (1997) Environmental effects on body size variation in *Drosophila melanogaster* and its cellular basis, Genet Res 70: 35–43.

Dworkin I, Tanda S, Larsen E (2001) Are entrenched charactaers developmentally constrained? Creating biramous limbs in an insect. Evol Dev 3: 424–431.

Fallon JF, Fredrick JM, Carrington JL, Lanser ME, Simandl BK (1983) Studies on a limbless mutant in the chick embryo. In: Limb Development and Regeneration (Fallon JF, Caplan AI, eds), 33–43. New York: Liss.

Felix M-A (1999) Evolution of developmental mechanisms in nematodes. J Exp Zool (Mol Dev Evol) 285: 3–18.

Gilbert SF, Opitz JM, Raff RA (1996) Resynthesizing evolutionary and developmental biology. Dev Biol 173: 357–372.

Goodwin B (1997) Generic dynamics of morphogenesis. In: Physical Theory in Biology (Lumsden CJ, Brandts WA, Trainor LEH, eds), 187–207. Singapore: World Scientific.

Huber O, Korn R, McLaughlin J, Ohsugi M, Herrmann BG, Kemler R (1996) Nuclear localization of B-catenin by interaction with transcription factor LEF-1. Mech Dev 59: 3–10.

Hultmark D (1994) Ancient relationships. Nature 367: 116–117.

Jan YN, Jan LY (1993) The peripheral nervous system. In: The Development of *Drosophila melanogaster* (Bate M, Martinez Arias A, eds), 1207–1244. Plainview, N.Y.: Cold Spring Harbor Laboratory Press.

Larsen E (1992) Tissue strategies as developmental constraints: Implications for animal evolution. Trends Ecol Evol 7: 414–417.

Larsen E (1997) Evolution of development: The shuffling of ancient modules by ubiquitous bureaucracies. In: Physical Theory in Biology (Lumsden CJ, Brandts WA, Trainor LEH, eds), 431–441. Singapore: World Science.

Larsen E, Mc Laughlin H (1987) The morphogenetic alphabet: Lessons for simple-minded genes. BioEssays 7: 130–132.

Lecuit T, Brook WJ, Ng M, Calleja M, Sun H, Cohen SM (1996) Two distinct mechanisms for long-range patterning by decapentaplegic in the *Drosophila* wing. Nature 381: 387–393.

Manoukian AS, Krause HM (1992) Concentration-dependent activities of the even-skipped protein in *Drosophila* embryos. Genes Dev 6: 1740–1751.

Newman SA, Comper WD (1990) "Generic" physical mechanisms of morphogenesis and pattern formation. Development 110: 1–18.

Percival-Smith A, Weber J, Gilfoyle E, Wilson P (1997) Genetic characterization of the role of the two Hox proteins, Proboscipedia and Sex Combs Reduced, in determination of adult antennal, tarsal, maxillary palp and proboscis identities in *Drosophila melanogaster*. Development 124: 5049–5062.

Postlethwait JH, Girton JR (1974) Development in genetic mosaics of aristapedia, a homeotic mutant of *Drosophila melanogaster*. Genetics 76: 767–774.

Simon HA (1973) The organization of complex systems. In: Hierarchy Theory, (Pattee HH, ed), 2–27. New York: Braziller.

Voronov DE, Panchin YV (1998) Cell lineage in the marine nematode *Enoplus brevis*. Development 125: 143–150.

Wagner GP, Altenberg L (1996) Complex adaptations and the evolution of evolvability. Evolution 50: 967–976.

Williams TA, Müller GB (1996) Limb development in a primitive crustacean, *Triops longicaudatus:* Subdivision of the early limb bud gives rise to multibranched limbs. Dev Genes Evol 206:161–168.

Yeaman C, Grindstaff KK, Nelson WJ (1999) New perspectives on mechanisms involved in generating epithelial cell polarity. Physiol Rev 79: 73–98.

IV PHYSICAL DETERMINANTS OF MORPHOGENESIS

Unlike materials of the nonliving world, which are molded by purely physical forces, embryos and tissues seem to obey different rules: their forms appear to be expressions of intrinsic, highly complex, genetic "programs." Significantly, however, many forms and patterns assumed by embryos and their tissues and organs resemble outcomes dictated by the physics of nonliving materials. In some cases, this is because tissues are indeed subject to the same physical forces that mold and pattern nonliving materials. But in other cases, evolution seems to have produced genetic mechanisms that served to consolidate morphological outcomes originating in the action of physical processes on cell aggregates during earlier periods of metazoan history. In this second scenario, genetic circuitry acts to "overdetermine" the generation of forms and patterns, the originating causes of which may be barely discernible behind the molecular complexity of the modern-day developmental process.

The chapters of part IV present evidence that physical processes characteristic of nonliving, chemically active, condensed materials act, as well, within living embryos, where they are responsible for some of their more unusual, apparently goal-directed constructional properties. They generalize on such observations in order to extract principles by which the interplay of physical and genetic processes in tissues can be "deconstructed" to provide plausible and testable hypotheses for the evolution of biological form.

Malcolm Steinberg (chapter 9) discusses cell affinity, the defining characteristic of multicellular organisms, and demonstrates that variations in such affinities in a tissue mass, whether dictated by functional requirements, or even if incidentally present by virtue of experimental mixing of dissimilar tissue cells, leads to "self-organization" of the common cell mass into multilayered three-dimensional structures. Steinberg notes that "the specification of such a structure cannot be rationalized as the genetic fixation of a successful experiment of nature. Rather, it must reflect the expression of cell properties acquired for employment in a different context." He goes on to demonstrate that these properties are adhesive differentials, formally equivalent to the molecular interactions that cause two immiscible liquids to phase-separate when they are shaken up. This leads to the theoretical inference, borne out by numerous experiments, that regardless of the specific molecular bases of such cell adhesive differentials, they can drive morphogenetic rearrangements such that "many paths [lead] to a common goal."

Once multicellularity is established, the transmission of signals across the cell mass becomes both a physical inevitability and a generator of phenotypic or developmental variation. The simplest way that this can occur is by the production of gradients of secreted, diffusible molecules. Fred Nijhout (chapter 10) reviews the theoretical basis for the surprising richness of pattern-forming potential in spatially extended systems in which both diffusion and chemical reactivity are present. He shows that this capacity, originally analyzed by Turing in 1952, is an example of a virtually ubiquitous class of mechanism referred to as "local activation and lateral inhibition." Such pattern-forming processes are *generic* capacities of both living and nonliving "excitable media"—materials that store chemical or mechanical energy and respond actively when exposed to stimuli. Diffusion-like communication can be mediated by numerous, physically distinct signaling modalities, and activation and inhibition of biosynthesis can occur on multiple levels. Nijhout uses this insight to account for, and productively allay, misunderstandings that have arisen between "nuts and bolts" experimentalists who have been skeptical of pattern-forming models that appear indifferent to the cellular and molecular details by which activation and inhibition of cell behaviors arise, and theoreticians persuaded that generic generative principles can provide powerful insights into developmental mechanisms.

Another generic property of chemically excitable media is the ability to exhibit chemical oscillations—the periodic rise and fall of a component's concentration. Olivier Pourquié (chapter 11) reviews the prevalence of oscillatory phenomena in developing systems and provides a typology of such behaviors and their potential roles. The extent to which such rhythmic activities, which include calcium ion oscillations and the cell cycle itself, serve as developmental timekeepers (clocks) or are simply biochemical epiphenomena, is an open question. Pourquié discusses work in which he and his colleagues have determined that the presegmental mesoderm in vertebrate embryos is the site of a periodic dynamic wave of expression of the homolog of the *Drosophila* pair-rule gene *hairy*. New evidence suggests that this oscillation, which is intimately linked to the generation of somites from the segmental plate, may be a genuine developmental clock such as been predicted on theoretical grounds to underlie somitogenesis.

Because individual cells, much as multicellular tissues, are chemically excitable, they exhibit a wide range of dynamic behaviors which are reflected in characteristic steady-state, oscillatory, and even chaotically changing levels of their biochemical constituents. It is a standard expectation that identically prepared dynamical systems, such as cells with the same genotype, will exhibit the same average behaviors over time. But Kunihiko Kaneko (chapter 12) shows that this expectation is incorrect when interacting systems are considered. Using a mathematical model of identical, dynamically complex cells in metabolic communication with one another, he demonstrates that individual members of the cell community are forced into specialized dynamical states that persist as long as the interactions persist. Thus distinct cellular phenotypes arise *epigenetically* (by way of the physics of complex systems) in a genetically uniform population. Kaneko draws out implications of this "intra-inter dynamics" both for development, where it provides a means for initial cell diversification in otherwise uniform tissue masses, and for evolution, in which genetic change can stabilize the originally dynamically based and interaction-dependent phenotypic differences between cells or multicellular aggregates.

Given that modern-day multicellular organisms are produced by developmental mechanisms that are composites between interaction-dependent physical processes generic to semisolid excitable media and the more hierachically organized regulatory gene circuitry that is the subject of most contemporary work in developmental biology, the question arises as to the relation between these categories of process. Stuart Newman (chapter 13) proposes that the history of multicellular life has been characterized by the evolution not just of the organismal forms themselves but also of the nature of the processes by which they are generated. Taking the view that the evolutionary earliest multicellular aggregates were inevitably subject to an array of generic physical determinants that had not pertained to single-celled life (i.e., differential adhesion, reaction-diffusion coupling, chemical wavefronts, and intra-inter dynamics), he suggests that the result was an array of multilayered, hollow, segmented, and branched forms generated not by selection of incremental variations in form, but by the physics of these materials. Subsequent evolution of the genetically variable components of such systems led to organisms increasingly (though not completely) generated by programlike genetic routines, driving the transition from physics to modern development.

9 Cell Adhesive Interactions and Tissue Self-Organization

Malcolm Steinberg

At present, it would be futile to speculate further upon the possible subcellular factors that are engaged in cellular adhesiveness. It should be pointed out however that this principle is of universal significance in morphogenesis, and that, in connection with directed cell movements, it is deserving of more attention than it has received.
—Townes and Holtfreter, 1955

The Chordate Body Plan: Many Paths to a Common Goal

All members of the phylum Chordata possess, at some stage in their embryonic development, not only the notochord (from which "chordate" is derived), but also a trilaminar "basic chordate body plan," in which a layer of ectoderm is outermost. Beneath it, an ectodermally derived neural tube lies middorsally, flanked right and left by paired rows of blocklike mesodermal somites and more laterally and ventrally still by a sheet of mesoderm that will later give rise to kidneys, gonads, and other structures. Beneath the neural tube lies the mesodermal notochord, and beneath that, the innermost, tubular, endodermal gut. As is obvious from the many structural differences that distinguish man from turtle from pelican from sea squirt, the developmental pathways leading to the adult structure from the stage represented by the basic body plan have diverged greatly in the course of evolution.

Varying enormously in yolkiness and therefore in size, the zygotes of various chordates do not diverge continuously from a common starting point but actually converge upon a common stage along a broad range of developmental pathways. Whereas the cells of relatively yolk-free, isolecithal embryos of a "primitive" animal with direct development, such as the cephalochordate *Amphioxus* (now *Branchiostoma*), produce a spherical blastula and early gastrula, those of a highly yolky, telolecithal reptile or bird embryo, for example, are constrained to form a small disk lying outside the enormous yolk mass. This essentially two-dimensional embryo must rearrange its parts quite differently from its three-dimensional chordate ancestors. The evolution of placental mammals revolutionized the embryo's priorities, replacing the primacy of yolk accumulation with that of implantation, producing a reversion to an isolecithal zygote that cleaves rather like an untidy *Amphioxus* but still produces a blastodisk that gastrulates like that of its more immediate reptilian ancestors. Chordates with intermediate yolkiness, such as amphibians, develop along generally intermediate pathways.

Across the board, then, the chordate mesoderm takes its definitive position by a variety of procedures referred to by terms such as invagination, involution, delamination, ingression, and polyinvagination. In anurans, some or all of the prospective mesoderm may lie internally from the very beginning (Keller, 1986). The neural tube is depicted in most textbooks as being produced by the sequence: thickening of the neural plate, folding, fusion of the apposed folds and separation of the neural tube from the now-overlying epidermis. In *Amphioxus,* however, separation from the epidermis does not await contact between apposed neural folds. Rather, it occurs while the neural plate is still wide open, the free margins of the epidermis then converging over the still-open neural plate in pursestring fashion while the detached plate proceeds to roll up (Hatschek, 1881; Conklin, 1932). In teleost, ganoid, and cyclostomatous fishes, the neural plate never folds at all but thickens enormously and sinks inward as a solid "neural keel" which only later cavitates (Dean, 1896; Wilson, 1899; deSelys-Longchamps, 1910).

In higher vertebrates, anterior and posterior portions of the same individual's neural tube may develop by different morphogenetic pathways (Schoenwolf, 1991). Here, the neural plate forms in the textbook manner anteriorly. In frogs, however, more posterior regions are initially solid and cavitate secondarily, whereas in the zebrafish tail bud, cells of the prospective spinal cord actually sort out from an original admixture with future mesodermal cells (Kanki and Ho, 1996). Many other examples of pathway differences could be cited, including major differences in the manner in which the endoderm and the cavities of somites are produced. Yet, for all of those differences, these pathways all lead to the formation of the same basic body organization. One could scarcely imagine that in the continuous succession of ontogenies constituting the evolution of all of these closely related animals, the outcomes of these developmental processes could have remained fundamentally constant while the mechanisms underlying them were nevertheless changed.

This raises the question of what mechanisms are capable of specifying a morphological outcome without specifying a specific pathway leading to it, which brings us to a body of work describing self-assembly processes in metazoan systems.

Self-Organization in Animal Systems

This subject has its roots in Abraham Trembley's historic experiments on the reorganizational capacities of *Hydra* (described in his 1744 Memoirs; see Lenhoff and Lenhoff, 1986) and in H. V. Wilson's subsequent discovery (1907) that dissociated cells of marine sponges are able to reconstitute themselves into functional sponges.

Tissue and Cell Affinities

Some thirty years later, Johannes Holtfreter discovered that fragments of amphibian embryonic tissues placed in a physiological salt solution would round up, adhering and spreading over one another's surfaces when placed in mutual contact. These

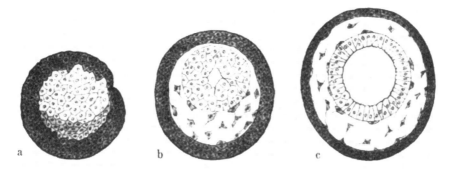

Figure 9.1
(*a*) Prospective amphibian gastrula endoderm plus mesoderm are wrapped in ectoderm. (*b*) Endoderm and mesoderm rearranged, the latter forming a mesenchyme that envelopes the endoderm and separates it from the outer ectoderm. (*c*) Endoderm forms an intestinal vesicle whose lumen arises by secondary cavitation. (From Holtfreter, 1939.)

rearrangements were not random. Rather, when tissues that normally cooperated to form a particular structure were combined, they rearranged to form a semblance of that structure. In one example, when fragments of gastrula ectoderm, mesoderm, and endoderm were combined, the ectoderm took up an external position and the endoderm the innermost position, with the mesoderm in between: their anatomically correct positions (figure 9.1). It seemed evident that this self-organizing behavior in vitro must result from "tendencies which are probably of great importance for normal development as well" (Holtfreter, 1939, p. 201). Holtfreter noted that some tissue pairs seem to associate preferentially, such as mesoderm with either ecto- or endoderm, but that others, such as ectoderm with endoderm, seem to avoid association. He therefore called these associative preferences "tissue affinities" (Holtfreter, 1939, p. 198), in reference to "the forces that are instrumental in these processes of attraction and repulsion." As to the nature of these forces, he wrote:

> the question of a chemotropic distant effect between cells has not even been touched upon. All the phenomena here described occurred while the various kinds of cells and tissues were in direct mutual contact. What was actually observed was an orderly union as well as non-unions and self-isolations. The events proceeded in an age- and tissue-specific manner, removed from the embryo as a whole, in a purely protective, indifferent medium and without the participation of a physically structured substrate. We, therefore, called them autonomous events and ascribed them to mutual cell-specific stimulation which we interpreted as an expression of affinities. Their chemical or physical nature was left undiscussed (Holtfreter, 1939, pp. 223–224). . . .

And again,

We are still far from grasping the physico-chemical processes involved in this "self-mobilization." (Holtfreter, 1939, p. 220)

The next step in Holtfreter's analysis of the capacities of embryonic tissues to self-organize appeared in a paper on the development of the amphibian pronephros:

New experiments have shown that the tendency of embryonic cells to organize themselves into a typical pattern of organs is manifested even in material that has undergone complete disintegration. By exposing amphibian explants to a pH of around 10.0, the tissues fall apart and form a suspension of free cells. This cell heap can be further disorganized by stirring it with a glass needle. When returned to normal pH conditions the cells reaggregate into firm bodies which continue differentiating. Instead of retaining their chaotic cell pattern, the aggregates become organized into well separated tissues and organs of a topographic pattern hardly less perfect than the one developed in a corresponding untreated explant. (Holtfreter, 1944a, pp. 235–236)

Up to this point, Holtfreter's experiments had dealt with the organizational behavior of combinations of apposed tissue fragments of various kinds. The technique of cell dissociation and reaggregation now allowed him to increase the resolution of these experiments from the level of tissues to the level of cells. A comprehensive series of cell-mixing experiments conducted with his student, P. L. Townes, demonstrated that the intermixed and reaggregated cells of dissociated embryonic tissues could sort themselves out to reconstitute their tissues of origin, arranged in a semblance of their normal organization (Townes and Holtfreter, 1955). For example, intermixed prospective epidermal and neural plate cells coaggregated to form a sphere within which the two kinds of cells sorted themselves out, the neural plate cells sinking into the interior—their normal position—to be replaced by epidermal cells emerging from the interior.

In some cases, Townes and Holtfreter compared the behavior of combined intact embryonic tissue fragments with that of their component cells after dissociation, mixing and reaggregation. An impressive example is illustrated in figure 9.2. When a fragment of neural plate including the prospective neural folds was apposed to a mass of endoderm, the plate folded normally and was partially enveloped by the endodermal mass. Some neural fold cells emigrated from the plate, as they normally do, to produce an ectomesenchyme surrounding the neural tube, whereas others formed a patch of overlying epidermis. The final structure is depicted as resembling a rudimentary late neurula. In a parallel experiment in which comparable tissue fragments were dissociated, intermixed, and reaggregated, not only did the cells reassociate selectively to reconstitute the original tissues but these reconstituted tissues were organized in their correct anatomical arrangements. The neural mass developed a lumen and was surrounded by ectomesenchyme, which was in turn apposed to the patch of epidermis, all lying on an endodermal mass. This embryo-like structure and the comparable structure self-assembled by the rearrangements of intact tissue

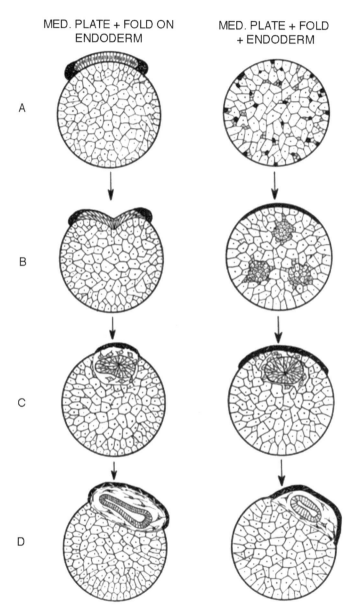

Figure 9.2
Left sequence (*A–D*): When medullary (neural) plate and neural fold are combined with endoderm, the neural plate portion invaginates partially and neural fold-derived epidermis caps a mass of neural crest-derived mesenchyme, which in turn enveloped the neural tissue. Right sequence (*A–D*): When dissociated cells of these same tissues were intermixed, they segregated and rearranged to form a similar structure, following an entirely different pathway. (From Townes and Holtfreter, 1955.)

fragments were indistinguishable, at least as depicted in these drawings. Thus the "tissue affinities" of 1939 became the "cell affinities" of 1955.

Selective Association and Selective Positioning of Embryonic Cells

It is important to note that in unscrambling and reorganizing themselves, these reconstituting embryonic tissues did two things simultaneously: (1) they chose to congregate with tissues of the same kind; and (2) they chose to settle into their "correct" position relative to the other tissues. Any explanation of "tissue affinities" must account for both of these behaviors and not only for a preference for association with "self." Holtfreter struggled with this realization and, after considering many possible mechanisms that might contribute to tissue affinities, ultimately proposed two separate mechanisms, acting in succession, to explain these two behaviors. These were directed cell movements, by which particular kinds of cells migrated either inward or outward within cell aggregates, followed by a "cell specificity of adhesion." He suggested that the cell movements might be directed by a gradient of substances lowering cell surface tension within cell aggregates, thereby inducing the polarized protrusion of pseudopodia (Holtfreter, 1944b). Or, as he would later write: "It seems necessary to assume the existence of a concentration gradient of some sort between inner and outer milieu of the aggregate towards which the different cell types react differently" (Townes and Holtfreter, 1955, p. 107).

Having postulated mutual attractions and repulsions between cells (Holtfreter, 1939), he proposed that not only the selectivity of cell-cell adhesions but also their relative intensities were important (Holtfreter, 1944b). He would later summarize both this plurality of competing explanations and the uncertainty with which they were regarded: "Morphogenetic movements and cellular adhesiveness obey quite different controlling factors . . . in morphogenesis, the forces controlling directed movements must overcome those of cell adhesion, . . . Unfortunately, no satisfactory answer can be given to the question of what makes the cells move either inward or outward under normal or experimental conditions" (Townes and Holtfreter, 1955, p. 110).

Programmed Assembly of Novel Structures

Thus far, we have focused on the ability of normal cell and tissue combinations—combinations that participate in the formation of a normal structure—to reorganize, after disarrangement, to form a semblance of that structure. Much insight into the mechanisms underlying this behavior has been gained from study of the exceptions to it—cases in which the normal structure either is not produced or does not even exist. For example, the three-layered structure depicted in figure 9.1 was produced only when the endoderm and mesoderm were entirely covered by ectoderm at the outset. If, on the other hand, ectoderm initially shared the surface with endoderm, the end result was as depicted in figure 9.3. In the final configuration,

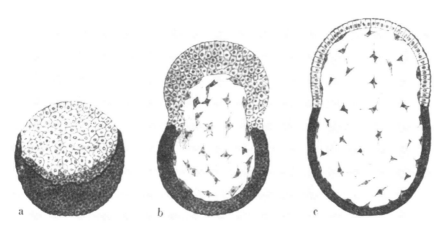

Figure 9.3
The same three tissues used in figure 9.1 were combined, with ectoderm and endoderm sharing the surface from the outset. (*a, b, c*) In this case, a mesenchyme-filled cyst was produced, covered in part by a right-side-out epidermis and in part by an inside-out gut epithelium. (From Holtfreter, 1939.)

a mesenchyme-filled cyst was produced, covered in part by a right-side-out epidermis and in part by an inside-out gut epithelium. These two end configurations are topologically related. In both cases, the mesoderm is apposed to the basal surfaces of both ecto- and endodermal epithelia, and each of the latter has maintained a free apical surface. The endoderm has done this by producing the usual lumen in figure 9.1 but by exposing the normally adlumenal surface to the immersion medium in figure 9.3. Both configurations are outcomes to be expected if (1) all three tissues' cells resist intermixing, (2) neither epithelium's apical surface can engage in cell adhesion, and (3) both epithelia's basal surfaces strongly adhere to the mesodermal cells.

In this way, these two end results taken together suggest that cell adhesion plays a directing role in the tissue assembly process. Holtfreter recognized these facts and devoted much attention to the powerful morphogenetic effects exerted by what he called the "surface coat," which rendered the free surfaces of amphibian eggs—and of epithelia in general—nonadhesive (Holtfreter, 1943). It is worthy of note that the configuration depicted in figure 9.3 is also produced in developing, intact embryos by conditions that prevent normal involution, in which case it is referred to as "exogastrulation."

A dramatic demonstration of the morphogenetic importance of nonadhesive cell surfaces was provided by another experiment (Holtfreter, 1944b, p. 201): "If we remove the ectoderm [bearing the surface coat] from the invaginated layer of mesoderm ... the mesoderm contracts into smaller patches which sink into the underlying endoderm. Any part of the mesoderm, however, that is covered by epidermis will not invaginate." In fact, in the absence of the surface coat, the three germ layers turn themselves inside out. As Kelland and

I discovered in 1967, the uncoated endoderm envelops the uncoated mesoderm, which in turn envelops the uncoated ectoderm, making the uncoated ectoderm the innermost rather than the outermost of these three tissues (see also Phillips and Davis, 1978).

What would happen if tissues that never normally encountered each other in the course of development were placed in mutual contact? Would they even "recognize" each other? In fact, tissues placed in such "abnormal" combinations consistently adhere to each other and rearrange to give rise to a specific, if unprecedented, structure. Holtfreter reported that "a fragment of medullary plate readily invaginates into the interior of a morula. In these instances, the graft slipped in between the large cells of the animal pole region to become permanently lodged within the underlying endoderm" (Townes and Holtfreter, 1955, p. 71). Other "abnormal" cell combinations also give rise to specific structures when paired in sorting-out experiments. Commonly, the reaggregated cell mixture forms one or more spheroidal aggregates; within each aggregate, one tissue partner forms an internal "medulla" surrounded by a "cortex" formed by the other. Chick embryonic limb precartilage, for example, has been reconstructed surrounded by cells as "foreign" as those of the mesonephros (Trinkaus and Groves, 1955), liver (Moscona, 1957; Steinberg, 1963a), pigmented epithelium of the eye (Steinberg, 1962c), and heart ventricle (Steinberg, 1963a).

Cell Behavior as a Guide to Morphogenetic Mechanism

In the course of evolution, genes must have been selected whose actions assure that our limbs, for example, have load-bearing skeletal structures on the inside, a protective epidermis on the outside, and muscle in between, rather than some other, less functional arrangement. Thus, when the intermixed cells of a dissociated and reaggregated limb bud sort themselves out to give the precise arrangement of a limb, they have been genetically programmed to do so. Yet, even though there can be no adaptive significance to any structure that might be generated by a combination of tissues that have never previously encountered each other in the entire course of evolution, all such tissue combinations are nevertheless also "genetically programmed" to assemble a specific structure that has no counterpart in Nature and has never before been assembled by these cells' ancestors. The specification of such a structure cannot be rationalized as the genetic fixation of a successful experiment of Nature. Rather, it must reflect the expression of cell properties acquired for employment in a different context. But what properties are these?

Inspired by the work of Townes and Holtfreter, we set out to identify these properties by the strategy of considering the then-existing hypotheses concerning the causes of cell sorting within heterogeneous cell aggregates and devising situations in which the different hypothetical mechanisms would produce different results. Initially there were only two candidate hypotheses. As already noted, Holtfreter had favored directed migration of individual cells inward or outward within aggregates, guided by radial concentration gradients of hypothetical chemotropic agents. He had proposed that differences in intercellular adhesiveness came into play only after cells had completed their inward or outward migrations.

I suggested an alternative possibility: that intercellular adhesive differentials might cause these cell regroupings directly, in the absence of chemotaxis, the progressive association of more cohesive cells squeezing less cohesive cells to the periphery (Steinberg, 1958, 74–75). Cell dilution experiments offered a means of testing these two hypotheses. I reasoned that

> if directed migration in concentration gradients occurs, then even a few cells of the internally segregating type, incorporated into an aggregate composed chiefly of cells of the externally segregating type, will migrate into the central region of the aggregate. If, however, differences in the mutual adhesiveness of the cells are directly responsible for the sorting out and selective localization, no such translocation of the few "internally segregating" cells will occur, since they will be too sparsely distributed to encounter one another and exert group action to exclude the other cells. (Steinberg, 1962a, p. 1578)

The results of such an experiment were decisive: chick embryonic cells of a type that segregated internally to those of a different type when they represented 23 percent of the total cell number failed to move toward an aggregate's center when they represented only 1 percent of the total. They did, however, avoid remaining in the aggregate's surface, moving from surface to subsurface positions (Steinberg, 1962a). Thus their behavior was not to seek the aggregate's center but rather both to avoid its surface and to exchange heterotypic for homotypic adhesions when presented with that opportunity. When internalizing cells were abundant, the latter behavior led to their "initial clustering ... in innumerable foci throughout the interior of each aggregate," these clusters continuing "to encounter and fuse with one another, progressively building up one or more coherent, internal masses of ... tissue, the number of which reflects [their] proportion in the population" (Steinberg, 1962b, p. 762). This progression is depicted on the right in figure 9.4 and contrasted with the progression, depicted on the left, to be expected if the internalizing cells were following a radial concentration gradient of a chemokine.

Although directed migration following radial chemokine concentration gradients was now excluded as an explanation of cell sorting in these heterogeneous cell aggregates, a third explanation had in the meantime been offered by Adam Curtis, who suggested that cells of different types undergo certain surface changes at different times after their dissociation. When cells of different types were coaggregated, these changes would be such that cells experiencing them would be immobilized by contact either with the aggregate's surface or with other cells already so immobilized. Cells of the type that first experienced this change would be trapped initially at an aggregate's surface and then in sequential layers beneath it, leaving those of types that later experienced the change to be immobilized

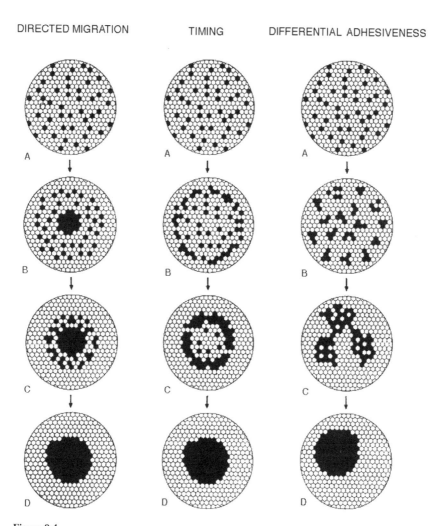

Figure 9.4
Time course of the sorting out (A–D) of two cell types within a mixed aggregate, as it would appear if brought about through three different mechanisms. Centripetal migration in a diffusion-generated radial concentration gradient ("directed migration") would produce the pathway shown on the left. Differences in the intensities of cell-cell adhesions ("differential adhesion") would produce the pathway shown at the right. Timed changes in cell surfaces following dissociation in the manner proposed by Curtis (1960), leading to a "herding" of one class of cells in from the periphery, would produce the pathway shown in the middle. (After Steinberg, 1964.)

in successive layers, the last cells to change occupying the aggregate's center (Curtis, 1960, 1961, 1962). This hypothesis, which would produce the progression of events depicted in the center column of figure 9.4, was also eliminated by establishing the progression shown to its right. Of the three mechanisms proposed to explain cell sorting, only differential adhesiveness correctly predicted the observed behavior of the segregating cells.

In 1962, the sorting out of embryonic cells from a mixture was commonly interpreted as indicating a preference of cells for association with "self," that is, with other cells of like kind (Moscona, 1962). In the course of an experiment designed as a direct test of Curtis's "timing hypothesis," we stumbled upon an observation that transformed our perspective. The object of this experiment was to mix cells from two tissues, one of which had been dissociated earlier than the other. If inside versus outside stratification really reflected the timing of a process initiated by cell dissociation, giving the later-recovering cells a sufficient "head start" should cause them to segregate externally instead of internally after coaggregation. As it happened, the flasks allowed to "recover" for the longer time periods called for by our protocol already contained sizable reaggregates before the time came to "mix" their contents. Rather than depart from the protocol, we combined these reaggregates, placing them in the same shaker flask, although it seemed pointless to do so. The next morning, many of these aggregates had adhered to one another, and it was clear that in all cases in which two fused aggregates were of different kinds, one was enveloping the other. Moreover, the enveloping tissue was always of the same kind, regardless of the timing of dissociation of either it or its partner; and this was also the same tissue that segregated externally in mixed aggregates of freshly dissociated cells of these two tissues. A follow-up experiment showed that cell dissociation could be dispensed with altogether; undissociated bits of the same tissues displayed the same mutual envelopment behavior (Steinberg, 1962c).

In the course of this envelopment of one tissue fragment by another, the two tissues were "choosing" to *increase* their mutual contacts, not to decrease them. The common feature of the phenomena of tissue spreading and cell sorting was therefore not a selection for self-association at the level of cell pairs but a drive on the part of the cell population as a whole to approach a particular configuration. The cells that engaged in these behaviors were clearly both mutually adhesive and motile. Not only was the "timing hypothesis" eliminated by these findings, but it now became possible to account for all of the observed reorganizational behavior on the basis of a single, plausible premise: that the cell rearrangements in all of these circumstances were guided by the exchange of weaker for stronger adhesions, which would drive the system toward that configuration in which total cell-cell adhesive bonding energy was maximized. We came to call this principle the "differential adhesion hypothesis" (DAH; Steinberg, 1970).

Adhesion-Specified Most Stable Structures Given a combination of two cell types which might, in principle, cohere (to the same cell type; "homotypic") and adhere (to the other cell type; "heterotypic") with any set of relative intensities whatsoever, what would be the correspondence between particular sets of relative adhesive intensities and the configurations specified as "most stable"? In precomputer 1961, we approached this question by modeling the behavior of such a system in two dimensions, representing the two cell types as equal numbers of black (A) and white (B) squares on a graph paper grid representing a cell aggregate. Various sets of numerical values were assigned to the black-black, white-white, and black-white cell-cell interfaces. Ten different cell distributions were modeled, including random mixing, one or more square black (or white) islands in a white (or black) sea and configurations in which black (or white) regions partially enveloped or lay side by side with their opposites. For each of the ten configurations, the number of black-black (AA), white-white (BB) and black-white (AB) interfaces was counted. These numbers were multiplied by the numerical values ("adhesive strengths") assigned to each category of interface, and the sum of the products, representing the total adhesiveness at all cell-cell boundaries within the aggregate, was calculated. The higher this number, the greater would be the "total adhesiveness" or "adhesive stability" of the cell aggregate.

Each set of relative adhesive "strengths" examined generated a different "most stable" configuration. For heterotypic adhesions exceeding the average strength of the two kinds of homotypic adhesions, stability was greatest when the cells were intermixed and least when the more cohesive A cells totally enveloped B cells. Decreasing the strength of the heterotypic adhesions to a value that was still intermediate between the two kinds of homotypic adhesions but below the average of their strengths radically changed matters: stability was least when the two types of cells were intermixed, greater when they were segregated, and greatest when a B cell mass totally enveloped a mass of A cells. When heterotypic adhesions were the weakest of the three kinds, a still different configuration was specified: stability was now greatest when A cells occupied one side of the square and B cells occupied the other side.

These results demonstrated graphically that differences in the strengths of adhesion between motile cells of different types were sufficient to specify a particular most stable anatomical configuration that would be approached from any initial cell distribution that provided the necessary cell-cell contacts. Cell sorting, tissue spreading, the cell translocations themselves, and the self-ordering of segregating tissues into strata with a specific arrangement were all natural consequences of the existence of particular sets of adhesive differentials among motile cells.

Our graphical models were presented at the December 1967 meeting of the American Association for the Advancement of Science, in a symposium to honor Johannes Holtfreter as he approached retirement. Proceedings of the symposium were not published, and I never submitted our simulations for publication because, unfortunately, it never occurred to me, as an experimentalist, that such things were actually publishable. On the other hand, much more sophisticated computer simulations of cell sorting and tissue spreading have since been published by others, with conclusions that both agree with ours and extend them (e.g., Goel and Rogers, 1978; Rogers and Goel, 1978; Glazier and Graner, 1993; Glazier et al., 1995; Mombach et al., 1995; Palsson, 2001).

A Syndrome of Liquid Tissue Behaviors The described behaviors of self-organizing embryonic cell populations, although regarded as manifestations of the remarkable self-assembly capacities of living systems, are in fact not unique to them. The same behaviors are displayed by ordinary liquids. Thus, in heterogeneous mixtures of embryonic cells and in dispersions of immiscible liquid droplets alike, cells or droplets

1. sort out (demix) to approach a specific configuration;
2. sort out by a coalescent pathway (smaller "islands" coalesce to form larger ones);
3. spread over the surface of those of another type;
4. approach the same specific configuration by spreading as by sorting out;
5. round up (irregularly shaped masses assume a spherical shape);
6. tend to envelop or be enveloped by those of another cohesive type in a transitive series of envelopment tendencies.

The sixth behavior served as the final behavioral test of the differential adhesion hypothesis (DAH). When two immiscible liquid droplets are apposed, it is always the less cohesive one (of lower surface tension) that tends to envelop its partner. This implies that if droplets representing a large set of mutually immiscible liquids were combined in all possible pairs, their mutual envelopment tendencies would form a transitive series, the least cohesive droplet enveloping—and the most cohesive droplet being enveloped by—all of the others. Such a series of mutually immiscible phases, though not readily available among ordinary liquids, is readily found among embryonic tissues, almost all of which have proven to be mutually immiscible. When six different chick embryonic tissues were combined in all fifteen possible pairings, their mutual envelopment tendencies established a single, hierarchical ranking (Steinberg, 1963b, 1970). This established a sixth behavior common to embryonic cell populations and immiscible liquids. (The syndrome of liquid tissue behaviors is illustrated in figure 9.5).

Analytical and Synthetic Confirmations of the Differential Adhesion Hypothesis

Direct Physical Measurements: Tissue Surface Tensions Direct Mutual Tissue Spreading

Although, in the tests noted above, the actual behavior of confronted cell populations conformed in detail to the behavior predicted by the DAH, those tests were correlative in

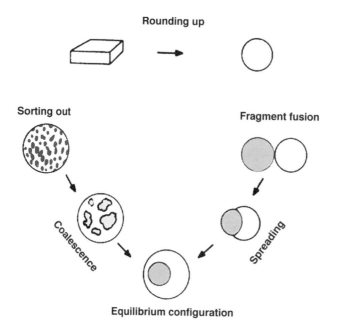

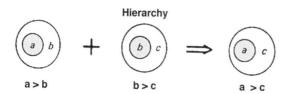

Figure 9.5
Syndrome of behaviors displayed both by many embryonic cell populations and by immiscible liquid pairs. (*Top*) A mass of arbitrary shape rounds up to form a sphere, minimizing its surface area. (*Middle*) Intermixed phases sort out by a process of coalescence, forming a continuous externalizing phase that envelops, to greater or lesser degree, a discontinuous internalizing phase. When placed in mutual contact as separate masses, the same two phases spread, one over the other, to approach the same (equilibrium) configuration approached by sorting out. (*Bottom*) In a set of mutually immiscible phases, the tendencies of one phase to spread over another are transitive: if b tends to spread over a and c tends to spread over b, then c will tend to spread over a. (After Phillips, 1969.)

nature. A direct proof of the hypothesis requires that embryonic tissues be actually shown to possess liquidlike surface tensions whose measured values consistently predict their mutual spreading tendencies: a tissue that envelops another tissue must have the lower surface tension of the two. Moreover, these tissue surface tensions should arise from the intensities of adhesion between the tissues' component cells.

The late Michael Abercrombie pointed me to the relevant physical literature on surface and interfacial tensions and the derived "works of adhesion" within and between liquid phases which correspond with one's subjective sense of "adhesive intensity." How to make the proper measurements of "intercellular adhesiveness" eluded me and all others attempting such measurements until a biophysics graduate student with a background in physics, Herbert Phillips, pointed out that the relevant thermodynamics required this to be done through measurements of the shapes of liquidlike cell aggregates at shape equilibrium under a measured distorting force. This was the crucial insight, and the effort to make such measurements became the subject of Phillips's doctoral dissertation.

In our first efforts to make such measurements, Phillips and I devised an incubator centrifuge to apply a sustained deforming force to living embryonic cell aggregates of chick embryonic limb bud mesoderm, heart ventricle, and liver (sessile droplet method). We established that these three tissues possess liquidlike surface tensions whose relative values decline in the sequence cited, in agreement with their mutual spreading tendencies (Phillips, 1969; Phillips and Steinberg, 1969). However, the development of a density gradient in the serum-containing culture medium during prolonged centrifugation prevented us from obtaining reliable numerical values of these tissue surface tensions. To avoid this complication, Phillips and his student, Grayson Davis, subsequently applied sustained deforming forces to spheroidal cell aggregates, not by centrifugation, but by compressing them between parallel plates to which they did not adhere (see below), calculating the tissue surface tensions from the Laplace equation.

We subsequently devised a parallel plate tissue surface tensiometer that continuously records both the force applied to a living cell aggregate and the aggregate's profile shape, allowing the approach to shape equilibrium to be constantly monitored in real time. Using this device, we initially reported numerical surface tension values for chick embryonic heart ventricle and liver (Foty et al., 1994) that confirmed the sequence established earlier (Phillips and Steinberg, 1969). We then extended these measurements to a total of five chick embryonic tissues. The resulting surface tension hierarchy (table 9.1) corresponds precisely with the hierarchy in these tissues' mutual envelopment preferences (Foty et al., 1996). The probability that this correspondence is fortuitous being 0.0083, these measurements demonstrated that the preferential envelopment behavior of embryonic tissues does indeed result from the tissues' relative surface tensions.

Table 9.1
Segregation hierarchy of chick embryonic tissue

	Tissue	Surface tension (dyne/cm)
Internally segregating	Limb bud mesenchyme	20.1
↕	Pigmented epithelium	12.6
	Heart ventricle	8.5
	Liver	4.6
Externally segregating	Neural retina	1.6

Note: Each tissue shown segregates internally to all others of lower surface tension and externally to all others of higher surface tension.

Tissue Surface Tensions Arise from Intercellular Adhesiveness

Surface and interfacial tensions in ordinary liquids being global reflections of the intensities of cohesion and adhesion between their component molecules, we have assumed by analogy that tissue surface tensions are global reflections of the intensities of cohesion and adhesion between their component cells. The ability to genetically engineer originally non-cohesive cells to express particular cadherins in regulated amounts has now opened this assumption to experimental examination. We have measured the number of cadherin molecules expressed per cell in clones of L cells transfected and selected to express a particular cadherin in different amounts. We have then measured the surface tensions of aggregates of cells from each clone. If tissue surface tension is a pure reflection of intercellular cohesive intensity, then the surface tensions of these aggregates should be a linear expression of the numbers of cadherins per cell, extrapolating to a surface tension near zero at zero cadherin expression. The experimental demonstration of precisely this relationship (Foty and Steinberg, 1998, and in preparation) shows unequivocally that tissue surface tensions are indeed pure and quantitative reflections of intercellular cohesive intensities.

Engineered Differences in Cadherin Expression Level Produce Tissue Segregation

The above studies have established that embryonic tissues can have certain of the physical attributes of liquids; and that, like ordinary liquids, they can flow, one over the surface of another, and rearrange in response to surface tensions arising from the energies of adhesion between their subunits—in this case, their constituent cells. However, cells are collections of molecules, and a great body of information has now been amassed concerning the molecules that mediate cell adhesions. Some of these molecules are proteins protruding from cell surfaces; others are secreted by cells to form a part of the extracellular scaffoldings to which they bind. Considering the complexity and cellular heterogeneity of the structures embryos assemble, it is to be expected that the developing animal's many kinds of cells must utilize many kinds of adhesion molecules to participate in specifying the diverse structures that comprise the animal. And so they do. Moreover, it has been held to be common sense that the ability of cells to distinguish "self" from "nonself" and to segregate from one another on that basis itself requires the action of cell-type-specific, molecular recognition markers. Nevertheless, to examine this putative requirement, one may ask whether a simple difference in the *quantity* of identical "homophilic" (self-associating) adhesion molecules expressed at their respective cell surfaces would be sufficient to render two cell populations "immiscible."

In 1963, I approached this question mathematically by calculating what the equilibrium configuration of such a pair of cell populations should be. To do this, I considered the theoretical behavior of two model cell populations in which (1) the adhesive sites are randomly distributed on the cell surfaces, (2) the areal frequency of bonds formed between these sites is proportional to the probability of their apposition when the cells bearing them are apposed, and (3) the work of adhesion between two cells is proportional to the areal frequency of bonds formed between them. The configuration of minimal free energy for the combined model cell populations turned out to be a sphere within a sphere, in which the cells with the smaller number of binding sites are segregated from and totally surround those with the greater number (Steinberg, 1963a, 1964). Thus it was shown in theory that, for two cell populations to be immiscible, they do not have to use different molecular adhesion systems, although of course they can. "Specificity" in this case arises not from chemistry but from the simple mathematics of collision probabilities.

In 1963, no cell adhesion molecules were yet known. Since then, many different cell adhesion molecules have been identified and classified into families and means have been developed, as we have noted, to genetically engineer cells to express particular adhesion proteins and even to control this expression quantitatively. Thus whether two cell populations would be rendered immiscible simply by expressing the same homophilic adhesion molecule in different amounts, and what the equilibrium configuration of such a system would be, advanced from the realm of theory to that of experiment. Friedlander and colleagues (1989) transfected weakly cohesive mouse sarcoma 180 cells with a complementary DNA (cDNA) coding for N-cadherin, a calcium-dependent cell adhesion molecule first found in neural cells. Mixing two clones expressing high versus low levels of this cell adhesion molecule, they reported a degree of segregation of the two cell populations.

However, their studies did not address the specification of higher-order structure. Does one of the two transfected cell populations segregate internally or externally to the other? Would either of the two cell populations tend to envelop the other after mutual confrontation? Would the two transfected cell populations tend to arrange themselves in a specific anatomical configuration? To answer these questions, we combined two populations of

L cells transfected with P-cadherin cDNA and expressing this homophilic adhesion molecule in substantially differing amounts. When the two cell populations were intermixed, they segregated to approach a sphere-within-a-sphere configuration, the cell population expressing more P-cadherin forming islands that fused to become an internal "medulla." When the two cell populations were first prepared as separate aggregates that were subsequently allowed to fuse, the cell population expressing more P-cadherin was enveloped by its partner, which formed an external "cortex" (Steinberg and Takeichi, 1994).

Thus it was shown empirically that mere quantitative differences in the expression of a single, homophilic adhesion molecule could at the same time render two cell populations immiscible—"self-preferring"—and encode the self-assembly of these two cell populations into an organlike structure, layered in a specific, reproducible sequence. Not only the adhesive affinities and the association constants of cell adhesion molecules but also their mere abundance contribute to a thermodynamic address code specifying both the morphogenetic behavior of cells and the specific anatomical structures into which they will tend spontaneously to assemble.

Differential Adhesion as a Determinant of Normal Embryonic Morphogenesis

Stratification of the Amphibian Primary Germ Layers and Neural Primordium

The first numerical values of tissue surface tensions were determined (Davis, 1984) for subsurface samples of amphibian embryonic ectoderm, mesoderm, and endoderm (whose self-assembly is represented in figure 9.1). As we noted earlier, in the absence of the surface coat, the three germ layers turn themselves inside out, subsurface endoderm taking up the external and subsurface ectoderm the internal position. Although the number of cases was small, the measured surface tension values of these subsurface amphibian germ layers fell in the sequence required to explain their stratification (Davis, 1984; Phillips and Davis, 1978). The results of this study were extended in 1997. The self-assembly of ectoderm, mesoderm, and endoderm into the trilaminar gastrula was found to conform with expectations based on these tissues' surface tensions, taking into account both adhesive and nonadhesive cell surface domains; moreover, the internalization of the neural plate at the neurula stage was found to follow a sharp increase in its surface tension, again in conformity with thermodynamic expectations that a tissue of higher surface tension will tend to become enveloped by a tissue of lower surface tension (Davis, Phillips, and Steinberg, 1997).

Posterior Positioning of the Drosophila Oocyte within the Follicle

In *Drosophila* oogenesis, a germ-line cell divides four times to produce a cyst of sixteen germ cells, one of which will subsequently become the oocyte while the other 15 become trophic nurse cells. The oocyte moves posteriorly within the cyst, coming to occupy the most posterior position—a position that in turn determines the oocyte's own anterior-posterior polarity. At the same time, the cyst becomes enveloped by follicle cells. Recently, two groups have independently demonstrated that the oocyte's movements are brought about by quantitative differences in the expression of *Drosophila* E-cadherin (Gonzalez-Reyes and Saint Johnston, 1998; Godt and Tepass, 1998; see also Peifer, 1998). E-cadherin, the principal *Drosophila* epithelial cadherin, is expressed on all germ-line cells, but the oocyte expresses more E-cadherin than the nurse cells. All follicle cells also express E-cadherin, but more cadherin is expressed by the posterior and anterior follicle cells than by the lateral ones. In either germ-line or follicle cell clones lacking E-cadherin, the oocyte is incorrectly localized. In chimeric follicles where some but not other follicle cells express E-cadherin, the oocyte attaches itself to cadherin-expressing follicle cells, regardless of their location. Oocytes lacking E-cadherin expression show no such attachment preference. In the absence of E-cadherin-expressing posterior follicle cells, the oocyte preferentially attaches itself to the anterior follicle cells, which also have a higher cadherin-expression level than the lateral follicle cells.

The data show that the oocyte, expressing more E-cadherin than its fifteen sister cells, moves and attaches itself to follicle cells expressing higher amounts of E-cadherin. Thus quantitative differences in expression level of a single cell-cell adhesion molecule, earlier shown to be sufficient to cause tissue segregation and to specify a unique morphology (Steinberg and Takeichi, 1994), have now been shown to specify the positioning of the *Drosophila* oocyte that in turn specifies the future embryo's anteroposterior axis.

The Sources of Specificity in Tissue Assembly

From Gene to Behavior

It is widely presumed that the root cause of every developmental process is genetic. A set of genes, operating in a complex regulatory environment that includes specific transcription factors and signal transduction pathways, gives rise to corresponding varieties of messenger RNA (mRNA), which in turn give rise to proteins, which in turn become localized, interact, and specify the developmental process in question. As these genes, their descendent proteins, and the relevant regulatory pathways are identified, one hears the confident expression that "we are finally getting down to the molecular level." Although this is of course true, it does not always tell us much about the mechanisms involved, particularly those of morphogenesis, which involves the rearrangement of cells in space. Because the movements of objects are brought about by forces, morphogenesis inherently requires a *physical* analysis.

In the case of localization of the *Drosophila* oocyte within the follicle, the molecular analysis ends with E-cadherin on the surfaces of both the oocyte and the follicle cells it seeks out. But E-cadherin is not restricted to these surfaces; it is normally present on every germ cell, on every follicle cell, and on a great many other cells in the organism-to-be. Morphogenetic understanding, in this case, comes with the realization that what determines oocyte placement are the presence of E-cadherin on germ cell surfaces, its presence and graded expression on follicle cell surfaces, and a thermodynamic principle: that the intercellular adhesive intensities of cells in a mobile population tend to be maximized. It is this principle, operating in the context of the geometry of the E-cadherin gradient on the follicle cells, that guides germ cell movement. As for the importance of E-cadherin itself, presumably any other self-recognizing adhesion molecule could have served in its place.

Cadherin Specificity and Cell-Cell Adhesive Recognition

Following Holtfreter's seminal discoveries, repeated efforts were made to explain "cell-cell recognition" in terms of an underlying and corresponding molecular recognition. It was first maintained that cells release tissue-specific materials, "factors" or "ligands" whose interactions are responsible for the "tissue-specific cell adhesion" assumed to underlie cell sorting (Moscona, 1960, 1962, 1963, 1968; Moscona and Hausman, 1977). Indeed, cell adhesion was held to be tissue specific even though sorting-out cells of different types, like demixing liquid phases, were consistently found to adhere to each other at points of mutual contact from the moment of their initial coaggregation, throughout the sorting-out process, and at the tissue or phase boundaries when segregation is complete. It is both remarkable and instructive to note the wide and enduring acceptance this misconceived hypothesis received. However, with the discovery of genuine cell-cell adhesion molecules, especially the cadherins, and the finding that most are shared by many tissues (reviewed in Takeichi, 1988), putative tissue-specific adhesion factors silently fell from grace.

In newly differentiating tissue, initiation of a morphogenetic movement, such as the formation of the mesoderm, the lens vesicle, the neural tube, or the neural crest (Takeichi, 1988), was found to be associated with cessation of expression of a particular cadherin or by a switch in cadherin expression from one subclass to another. When cadherin function was blocked by antibodies (Matsunaga, Hatta, and Takeichi, 1988; Bronner-Fraser, Wolf, and Murray, 1992) and when cadherins were ectopically expressed (Detrick, Dickey, and Kintner, 1990; Fujimori, Miyatani, and Takeichi, 1990), morphological defects resulted.

These findings demonstrated that normal tissue segregation is associated with proper cadherin expression and function. E-cadherin's vital role in the creation of an epithelium was demonstrated using a mouse E-cadherin null mutant; homozygous mutant embryos were unable to form a trophectoderm epithelium and were nonviable at a very early stage of development (Larue et al., 1994). To explain why cadherin type switching causes tissue segregation, it was proposed that cadherin interaction is "homophilic" not only in the sense that cadherins on apposed cells bind to each other but in the additional and more restrictive sense that this binding is cadherin subtype specific as well (reviewed in Takeichi, 1990). Cadherin type switching would then cause cessation of adhesion between tissues newly expressing different cadherins, effecting their segregation.

This hypothesis was tested by examining the ability of initially noncohesive L-cells, transfected to express E- versus P-cadherin, to coaggregate in gyrated suspension cultures. The E- and P-cadherin-expressing cell lines were reported to aggregate largely independently, with little cross-adhesion, at least initially (Nose, Nagafuchi, and Takeichi, 1988; Miyatani et al., 1989). These reports led to the conclusion that heterophilic interaction between E- and P-cadherin is absent or weak, even though fusion between aggregates of E- and P-cadherin-expressing cells was actually observed when the cells were cocultured for a longer time period (Nose, Nagafuchi, and Takeichi, 1988). As other cadherins were discovered, their interactions were examined and characterized using similar assays. Consequently, the majority of the cadherins have been classified as being homophilic adhesion molecules in the more restricted, cadherin-type-specific sense. It has been widely accepted that, with a few exceptions, cadherin-mediated cell-cell adhesion is restricted largely to interactions between identical cadherin molecules on apposed cells (e.g., Shapiro and Colman, 1998).

Cadherin binding specificity has almost universally been assessed through the use of short term assays of initial cell-cell binding events in stirred suspensions (Nose, Nagafuchi, and Takeichi, 1988, Murphy-Erdosh et al., 1995), first used to demonstrate "tissue selectivity of adhesion" by Roth and Weston (1967; Roth, 1968). In such assays, colliding cells in a sheared cell suspension have only a brief instant in which to adhere. This kind of assay presents at least two serious problems. First, it is now known that physiological, cadherin-mediated cell adhesions are not produced instantly upon cell-cell contact but require an assembly process involving the clustering of cadherins, their connection to catenins, and de novo actin polymerization, during which initially weak adhesions are greatly strengthened (Adams, Nelson, and Smith, 1996; Angres, Barth, and Nelson, 1996; Brieher, Yap, and Gumbiner, 1996; Hinck et al., 1994; Lotz et al., 1989; Takeda et al., 1995; Yap et al., 1997; Yap, Brieher, and Gumbiner, 1997; Yap, Niessen, and Gumbiner, 1998).

Because this process has been shown to take about an hour following initial cell-cell contact (Adams et al., 1998), it cannot be reflected in an assay of the adhesive success of cell-cell collisions lasting at most only a few seconds. Moreover, cells colliding in a stirred suspension must overcome shear forces in order to come into adhesion. In such a circumstance, cells must establish, in the brief instant of their collision, sufficient adhesion energy to resist reseparation by the shear forces that, in the next instant, will act to pull them apart.

In effect, shear forces applied to mutually adhesive cells in suspension impose on the aggregation process an "activation energy" whose magnitude is a direct function of the shear rate—greater shear rates increase the ratio of elastic (rebounding) to inelastic (adhering) cell-cell collisions.

A consequence of using the aggregation of sheared cell suspensions to assess "intercellular adhesive specificity" is that quantitative differences in the *rates of initiation* of adhesions between mutually adhesive cells of various kinds can be magnified into seemingly qualitative differences. Our recent results have fully justified our long-standing reservations (Steinberg, 1970, 427–428; 1975, 440–441). Using adhesion assays conducted with little or no shear, we have found that "heterocadherin" adhesions are in some cases somewhat slower to form. If these more slowly forming adhesions are not prevented from forming by the application of high shear forces, however, L cells transfected to express any of the classical cadherins tested in fact adhere strongly to those expressing any of the other cadherin subtypes, even classical cadherins belonging to different homology classes (type I versus type II; Duguay, Foty, and Steinberg, submitted). Interestingly, type I and type II cadherins mediate strong cross-adhesions even though the latter do not possess the histidine-alanine-valine (HAV) motif regarded as type I cadherins' "adhesive recognition domain" (Blaschuk et al., 1990). Thus adhesions between cells expressing different classical cadherins have in the past been artificially made to appear cadherin type selective (Nose, Nagafuchi, and Takeichi, 1988; Murphy-Erdosh et al., 1995) by the high shear forces used in the assays.

Qualitative and Quantitative Determinants of Tissue Segregation

If adhesive interactions between classical cadherins are in general not homophilic in the more restrictive sense, what accounts for the tissue segregation that commonly accompanies a change in cadherin subtypes? In previous studies demonstrating segregation between cells expressing different cadherins, expression levels were generally not carefully measured. Because even a moderate difference in the expression level of a given cadherin is sufficient to cause two cell lines to sort out (Steinberg and Takeichi, 1994; Duguay, Foty, and Steinberg, submitted), might it be that cell sorting seen in the past between cells expressing different cadherins actually resulted from differences in the *amounts* of cadherin expressed by those cells, rather than from weak or absent recognition between the different cadherins?

To examine this possibility, we produced L cell populations expressing E-cadherin (LE cells) at levels greater than, equal to, or lower than the level of expression of P-cadherin by the L cells (LP cells) with which they were mixed. All mixtures of LE and LP cells coaggregated. When cadherin expression of LE cells and LP cells differed significantly, whichever cell population displayed the greater amount of cadherin segregated internally to the other, which enveloped it completely. However, when the LE and LP cells expressed their different cadherins at a similar level, they failed to sort out (Duguay, Foty, and Steinberg, submitted). The failure of these two cell lines to segregate when the expression levels of their different cadherins were approximately equalized implies that the "heterocadherin" adhesion between E- and P-cadherin is not only stable but is just as strong as the two "homophilic" adhesions. It should be noted that this is the same pair of cadherins that were first reported as cross-adhering weakly if at all when tested in a sheared suspension (Nose, Nagafuchi, and Takeichi, 1988).

Our results suggest that quantitative changes in cadherin expression or activity levels accompanying changes in the type of cadherin expressed may be a major cause of tissue segregation not only in these in vitro experiments with genetically engineered cells but during embryonic morphogenesis as well. This possibility could be put to an experimental test within the embryo by increasing or decreasing the *amount* of the *normal* cadherin newly expressed by tissues initiating a morphogenetic movement (e.g., N-cadherin in the mouse primitive streak or neural plate).

Levels of Causality in Morphogenesis

The examples given above illustrate the interplay of genetic, molecular, and physical causality in bringing about the self-organization of cell populations. In each of these cases, a heterogeneous population of motile cells reorganizes itself, adopting a new arrangement. If the existence of the component cells and their motility are taken as givens (figure 9.6), then the molecular cause of the rearrangements in each case is traced to the activities of members of the cadherin family of cell-cell adhesion molecules. Both the interactions of these cadherins with each other and with cytoplasmic proteins and their abundance on the cell surfaces contribute to the binding energies of the cells to the other cells that they encounter.

The latter factor—abundance—appears to play a much more important role here than has been generally realized. Although molecular recognition specificity promotes cadherin-cadherin interactions, it does not determine cell sorting, tissue segregation, or multicellular organization. Rather, intercellular and cell-substratum binding energies constitute a set of adhesive differentials that constantly and physically impel the cells to shift positions at every opportunity to increase their overall binding intensities. Through repetition of this process, the cell population approaches a configuration—a dynamically specified anatomical structure—in which its interfacial free energy represents at least a local minimum (Foty et al., 1996). No matter what other properties the cells may possess or how these properties may change due to cell interactions, if the cells remain mobile, their cohesive and adhesive interactions, both with each other and with extracellular materials, will constantly impel

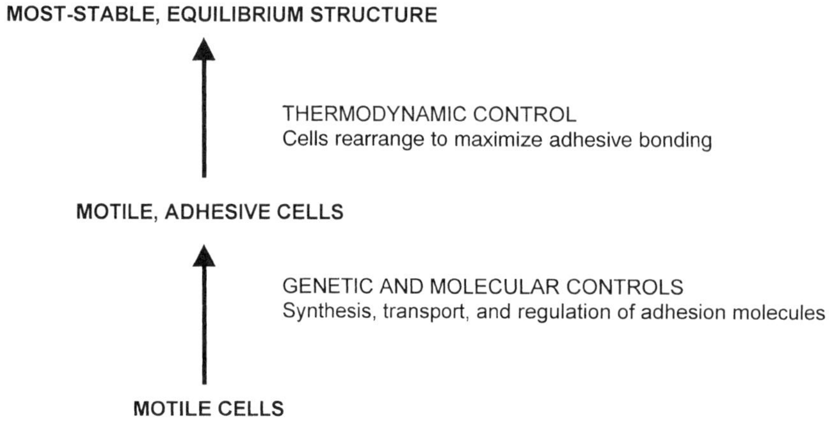

MOST-STABLE, EQUILIBRIUM STRUCTURE
↑
THERMODYNAMIC CONTROL
Cells rearrange to maximize adhesive bonding

MOTILE, ADHESIVE CELLS
↑
GENETIC AND MOLECULAR CONTROLS
Synthesis, transport, and regulation of adhesion molecules

MOTILE CELLS

Figure 9.6
Levels of causality in the specification of tissue organization by differential cell adhesion.

them toward this thermodynamically favored structure. Thus evolution has harnessed genetic information and the molecular machinery it encodes to the principles of physics to generate self-organizing, anatomical structures.

Acknowledgments

My recent research has been supported by research grants HD 30345 from the National Institute of Child Health and Human Development, and GM52009 from the National Institute of General Medical Sciences, National Institutes of Health.

References

Adams CA, Nelson WJ, Smith SJ (1996) Quantitative, single-cell analysis of cadherin-catenin-actin reorganization during development of cell-cell adhesion. J Cell Biol 135: 1899–1911.

Adams CA, Chen Y-T, Smith SJ, Nelson WJ (1998) Mechanisms of epithelial cell-cell adhesion and cell compaction revealed by high resolution tracking of E-cadherin/GFP. J Cell Biol 142: 1105–1119.

Angres BA, Barth A, Nelson WJ (1996) Mechanism for transition from initial to stable cell-cell adhesion: kinetic analysis of E-cadherin-mediated adhesion using a quantitative adhesion assay. J Cell Biol 134: 549–558.

Blaschuk O, Sullivan R, David S, Pouliot Y (1990) Identification of a cadherin cell adhesion recognition sequence. Dev Biol 139: 227–229.

Brieher WM, Yap AS, Gumbiner BM (1996) Lateral dimerization is required for the homophilic binding activity of C-cadherin. J Cell Biol 135: 487–496.

Bronner-Fraser M, Wolf J, Murray B (1992) Effects of antibodies against N-cadherin and NCAM on the cranial neural crest and neural tube. Dev Biol 153: 291–301.

Conklin EG (1932) The embryology of *Amphioxus*. J Morphol 54: 69–151.

Curtis ASG (1960) Cell contacts: some physical considerations. Am Nat 94: 37–56.

Curtis ASG (1961) Timing mechanisms in the specific adhesion of cells. Exp Cell Res, Suppl 8: 107–122.

Curtis ASG (1962) Cell contact and adhesion. Biol Rev Camb Philos Soc 37: 82–129.

Davis GS, Phillips HM, Steinberg MS (1997) Germ-layer surface tensions and "tissue affinities" in *Rana pipiens* gastrulae: quantitative measurements. Dev Biol 192: 630–644.

Davis GS (1984) Migration-directing liquid properties of embryonic amphibian tissues. Am Zool 24: 649–655.

Dean B (1896) The early development of *Amia*. QJ Micr Sci, ns, 38: 413–444.

deSelys-Longchamps M (1910) Gastrulation et formation des feuillets chez *Petromyzon planeri*. Arch Biol, Paris 25: 1–75.

Detrick R, Dickey D, Kintner C (1990) The effects of N-cadherin misexpression on morphogenesis in *Xenopus* embryos. Neuron 4: 493–506.

Duguay DR, Foty RA, Steinberg MS (submitted) Cadherin-mediated cell adhesion and tissue segregation: Qualitative and quantitative determinants. Submitted to Dev Biol.

Foty RA, Forgacs G, Pfleger CM, Steinberg MS (1994) Liquid properties of embryonic tissues: Measurement of interfacial tensions. Physiol Rev Lett 72: 2298–2301.

Foty RA, Pfleger CM, Forgacs G, Steinberg MS (1996) Surface tensions of embryonic tissues predict their mutual envelopment behavior. Development 122: 1611–1620.

Foty RA, Steinberg MS (1998) Measuring intercellular cohesivity: Surface tensions of transfected L-cell aggregates are directly proportional to cadherin expression. Mol Biol Cell 9: 460a.

Friedlander DR, Mege RM, Cunningham BA, Edelman GM (1989) Cell sorting-out is modulated by both the specificity and amount of different cell adhesion molecules (CAMs) expressed on cell surfaces. Proc Natl Acad Sci USA 86: 7043–7047.

Fujimori T, Miyatani S, Takeichi M (1990) Ectopic expression of N-cadherin perturbs histogenesis in *Xenopus* embryos. Development 110: 97–104.

Glazier JA and Graner F (1993) Simulation of the differential adhesion-driven rearrangement of biological cells. Physiol Rev E 47: 2128–2154.

Glazier JA, Raphael RC, Graner F, Sawada Y (1995) The energetics of cell sorting in three dimensions. In: Interplay of Genetic and Physical Processes in the Development of Biological Form (Beysens D, Forgacs F and Gaill F, eds), 62–73. Singapore: World Scientific.

Godt D, Tepass U (1998) *Drosophila* oocyte localization is mediated by differential cadherin-based adhesion. Nature 395: 387–391.

Goel NS, Rogers G (1978) Computer simulation of engulfment and other movements of embryonic tissues. J Theoret Biol 71: 103–140.

Gonzalez-Reyes A, Saint Johnston D (1998) The *Drosophila* AP axis is polarized by the cadherin-mediated positioning of the oocyte. Development 125: 3635–3644.

Hatschek B (1881) Studien über Entwicklung des *Amphioxus*. Arb Zool Inst Wien 4: 1–88.

Hinck L, Nathke IS, Papkoff J, Nelson, WJ (1994) Beta-catenin: A common target for the regulation of cell adhesion by Wnt-1 and src signaling pathways. Trends Biochem Sci 19: 538–543.

Holtfreter J (1939) Gewebeaffinität: Ein Mittel der embryonalen Formbildung. Arch Exp Zellforsch Besonders Gewebezücht 23: 169–209. Revised and reprinted in English, 1964, in: Foundations of Experimental Embryology (Willier BH, Oppenheimer JM, eds), 186–225. Englewood Cliffs, N.J.: Prentice-Hall.

Holtfreter J (1943) Properties and functions of the surface coat in amphibian embryos. J Exp Zool 93: 251–323.

Holtfreter J (1944a) Experimental studies on the development of the pronephros. Rev Canad Biol 3: 220–250.

Holtfreter J (1944b) A study of the mechanics of gastrulation. Part 2. J Exp Zool 95: 171–212.

Kanki JP, Ho R (1996) The development of the posterior body in zebrafish. Development 124: 881–893.

Keller RE (1986) The cellular basis for amphibian gastrulation. In: Developmental Biology: A Comprehensive Synthesis. Vol. 2. The Cellular Basis of Morphogenesis. (Browder LW, ed.) 241–327. New York: Plenum Press.

Larue L, Ohsugi M, Hirchenhain J, Kemler R (1994) E-cadherin null mutant embryos fail to form a trophectoderm epithelium. Proc Natl Acad Sci USA 99: 8263–8267.

Lenhoff SG, Lenhoff HM (1986) Hydra and the Birth of Experimental Biology, 1744: Abraham Trembley's Memoirs Concerning the Polyps with Arms Shaped like Horns. Pacific Grove, Calif.: Boxwood Press.

Lotz MM, Burdsal CA, Erickson HP, McClay DR (1989) Cell adhesion to fibronectin and tenascin: Quantitative measurements of initial binding and subsequent strengthening response. J Cell Biol 109: 1795–1805.

Matsunaga M, Hatta K, Takeichi M (1988) Role of N-cadherin cell-adhesion molecules in the histogenesis of neural retina. Neuron 1: 289–295.

Miyatani S, Shimamura K, Hatta M, Nagafuchi A, Nose A, Matsunaga M, Hatta K, Takeichi M (1989). Neural cadherin: Role in selective cell-cell adhesion. Science 245: 631–635.

Mombach JCM, Glazier JA, Raphael RC, Zajac M (1995) Quantitative comparison between differential adhesion models and cell sorting in the presence and absence of fluctuations. Physiol Rev Lett 75: 2244–2247.

Moscona A (1957) The development in vitro of chimeric aggregates of dissociated embryonic chick and mouse cells. Proc Natl Acad Sci USA 43: 184–194.

Moscona A (1960) Patterns and mechanisms of tissue reconstruction from dissociated cells. In: Developing Cell Systems and Their Control (Rudnick D, ed), 45–70. New York: Ronald Press.

Moscona A (1962) Analysis of cell recombinations in experimental synthesis of tissues in vitro. In: Symposium on Specificity of Cell Differentiation and Interaction. J Cell Compar Physiol 60 (Suppl. 1): 65–80.

Moscona A (1963) Studies on cell aggregation: Demonstration of materials with selective cell-binding activity. Proc. Natl Acad Sci USA 49: 742–747.

Moscona A (1968) Cell aggregation: Properties of specific cell ligands and their role in the formation of multicellular systems. Dev Biol 18: 250–277.

Moscona AA, Hausman RE (1977) Biological and biochemical studies on embryonic cell-cell recognition. In: Cell and Tissue Interactions (Lash JW, Burger MM, eds), 173–185. New York: Raven Press.

Murphy-Erdosh C, Yoshida CK, Paradies N, Reichardt LF (1995) The cadherin binding specificities of B-cadherin and LCAM. J Cell Biol 129: 1379–1390.

Nose A, Nagafuchi A, Takeichi, M (1988) Expressed recombinant cadherins mediate cell sorting in model systems. Cell 54: 993–1001.

Palsson E (2001) A three-dimensional model of cell movement in multicellular systems. Future Generat Comp Syst 17: 835–852.

Peifer M (1998) Birds of a feather flock together. Nature 395: 324–325.

Phillips HM (1969) Equilibrium measurements of embryonic cell adhesiveness: Physical formulation and testing of the differential adhesion hypothesis. Ph.D. diss., Johns Hopkins University, Baltimore.

Phillips HM, Davis GS (1978) Liquid-tissue mechanics in amphibian gastrulation: Germ-layer assembly in *Rana pipiens*. Am Zool 18: 81–93.

Phillips HM, Steinberg MS (1969) Equilibrium measurements of embryonic chick cell adhesiveness: 1. Shape equilibrium in centrifugal fields. Proc Natl Acad Sci USA 64: 121–127.

Rogers G, Goel NS (1978) Computer simulation of cellular movements: Cell sorting, cellular migration through a mass of cells and contact inhibition. J Theoret Biol 71: 141–166.

Roth S (1968) Studies on intercellular adhesive selectivity. Devel Biol 18: 602–631.

Roth S, Weston, JA (1967) The measurement of intercellular adhesion. Proc Natl Acad Sci USA 58: 974–980.

Schoenwolf GC (1991) Cell movements in the epiblast during gastrulation and neurulation in avian embryos. In: Gastrulation: Movements, Patterns and Molecules (Keller R, Clark Jr. WH, Griffin F, eds), 1–28. New York: Plenum Press.

Shapiro L, Colman D (1998) Structural biology of cadherins in the nervous system. Curr Opin Neurobiol 8: 593–599.

Steinberg MS (1958) On the chemical bonds between animal cells: A mechanism for type-specific association. Am Nat 92: 65–82.

Steinberg MS (1962a) On the mechanism of tissue reconstruction by dissociated cells: 1. Population kinetics, differential adhesiveness, and the absence of directed migration. Proc Natl Acad Sci USA 48: 1577–1582.

Steinberg MS (1962b) Mechanism of tissue reconstruction by dissociated cells: 2. Time course of events. Science 137: 762–763.

Steinberg MS (1962c) On the mechanism of tissue reconstruction by dissociated cells: 3. Free energy relations and the reorganization of fused, heteronomic tissue fragments. Proc Natl Acad Sci USA 48: 1769–1776.

Steinberg MS (1963a) Reconstruction of tissues by dissociated cells. Science 141: 401–408.

Steinberg MS (1963b) Hierarchical order in the anatomical patterns established by heteronomic combinations of chick embryonic cells and tissues. Am Zool 3: 512.

Steinberg MS (1964) The problem of adhesive selectivity in cellular interactions. In: Cellular Membranes in Development (Locke M, ed), 321–366. New York: Academic Press.

Steinberg MS (1970) Does differential adhesion govern self-assembly processes in histogenesis? Equilibrium configurations and the emergence of a hierarchy among populations of embryonic cells. J Exp Zool 173: 395–434.

Steinberg MS (1975) Adhesion-guided multicellular assembly: A commentary upon the postulates, real and imagined, of the differential adhesion hypothesis, with special attention to computer simulations of cell sorting. J Theor Biol 55: 431–443.

Steinberg MS (1996) Adhesion in development: An historical overview. Dev Biol 180: 377–388.

Steinberg MS, Takeichi M (1994) Experimental specification of cell sorting, tissue spreading and specific spatial patterning by quantitative differences in cadherin expression. Proc Natl Acad Sci USA 91: 206–209.

Takeda H, Nagafuchi A, Yonemura S, Tsukita S, Behrens J, Birchmeier W, Tsukita S (1995) V-src kinase shifts the cadherin-based cell adhesion from the strong to the weak state and beta-catenin is not required for the shift. J Cell Biol 131: 1839–1847.

Takeichi M (1988) The cadherins: cell-cell adhesion molecules controlling animal morphogenesis. Development 102: 639–655.

Takeichi M (1990) Cadherins: A molecular family important in selective cell-cell adhesion. Annu Rev Biochem 59: 237–252.

Townes PL, Holtfreter J (1955) Directed movements and selective adhesion of embryonic amphibian cells. J Exp Zool 128: 53–120.

Trinkaus JP, Groves PW (1955) Differentiation in culture of mixed aggregates of dissociated tissue cells. Proc Natl Acad Sci USA 41: 787–795.

Wilson HV (1899) The embryology of the sea bass (*Serranus atrarius*). Bull US Fish Comm 9: 209–277.

Wilson HV (1907) On some phenomena of coalescence and regeneration in sponges. J Exp Zool 5: 245–258.

Yap AS, Brieher WM, Pruschy M, Gumbiner BM (1997) Lateral clustering of the adhesive ectodomain: A fundamental determinant of cadherin function. Curr Biol 7: 308–315.

Yap AS, Brieher WM, Gumbiner BM (1997) Molecular and functional analysis of cadherin-based adherens junctions. Annu Rev Cell Dev Biol 13: 119–146.

Yap AS, Niessen CM, Gumbiner BM (1998) The juxtamembrane region of the cadherin cytoplasmic tail supports lateral clustering, adhesive strengthening, and interaction with p120. J Cell Biol 141: 779–789.

10 Gradients, Diffusion, and Genes in Pattern Formation

H. Frederik Nijhout

Pattern formation, the process by which an orderly structure arises in a previously featureless field, represents a fundamental problem in developmental biology that has attracted the attention of theoretical and experimental biologists alike. Both groups have made unique contributions to our understanding of self-organizing processes in development. Theoreticians have been largely concerned with the principles of how structure emerges within a homogeneous system. Experimentalists, by contrast, have focused largely on the mechanisms that control gene expression and on how such expression patterns vary in time and space. Although the potential for mutual illumination between the two research traditions is great, they appear to exist largely in parallel, with surprisingly little collaboration between them. In pointing out some of the sources of misunderstanding and miscommunication between the two groups, this chapter suggests areas where fruitful collaboration might take place in the future.

Diffusion and Lateral Inhibition in Pattern Formation

Interest in reaction-diffusion models to understand and explain how orderly pattern emerges during development stems from the pioneering work of Turing (1952). Theorists of developmental biology were intrigued by Turing's finding that organized and stable patterns could emerge spontaneously in an initially homogeneous and featureless field. Spontaneous self-organization requiring the interaction of no more than two freely diffusing substances seemed like a most parsimonious system worthy of further investigation, especially so at a time when the search for the hypothetical inducers of development was producing few if any results. Of the many mechanisms for pattern formation, reaction-diffusion has attracted the most attention from theoreticians: as the most parsimonious of mechanisms, it lends itself best to the elucidation of the principles of pattern formation.

The general applicability of reaction-diffusion mechanisms to a wide range of problems in organismal development has been explored most thoroughly by Gierer and Meinhardt (1972) and Gierer (1981). Meinhardt 1982 and 1998 are excellent surveys of the general principles of pattern formation by reaction-diffusion, illustrating the great diversity of biologically realistic patterns that can be produced by a very small set of interactions among two or three substances. More general mathematical treatments of reaction-diffusion theory are given by Segel (1984), Edelstein-Keshet (1988) and Murray (1989). These works have firmly established the necessary and sufficient conditions for diffusion-driven pattern formation. What is needed is an *activator,* a substance that somehow catalyzes its own synthesis (a condition known as "positive feedback" or "autocatalysis"), and an *inhibitor* of this synthesis, a substance that keeps the concentration of the activator from growing without bounds. In addition, the inhibitor must have a greater spatial range than the activator (a condition known as "lateral inhibition") to keep activator production from spreading. The basic components of a reaction-diffusion system are illustrated in figure 10.1. Given a certain array of parameter values (e.g., the relative diffusivities of the activator and inhibitor, the rates of synthesis and breakdown of the activator and the inhibitor, the reaction constants of their interactions), such a system can produce a highly stable spatial pattern of local activator synthesis. Moreover, a small change in one or more parameter values can alter this pattern to a different stable configuration.

All models of reaction-diffusion driven pattern formation require some kind of initial spatial inhomogeneity in the concentration of activator or inhibitor, say, a small random variation or a simple gradient in the concentration of one or the other substance. The

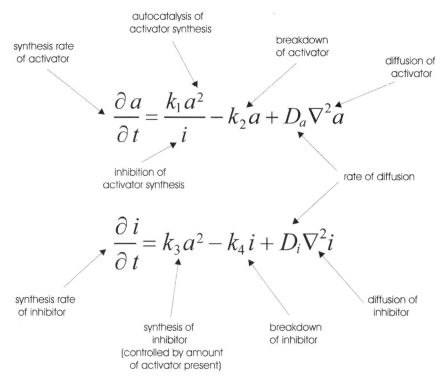

Figure 10.1
General equations of reaction-diffusion for pattern formation. Local activation (upper equation) and long-range (lateral) inhibition (lower equation) are accomplished by a small diffusion coefficient (D_a) the activator, and a much larger coefficient (D_i) for the inhibitor.

Pattern Formation

kinetics of reaction and diffusion favor and amplify certain wavelengths of this inhomogeneity and represses others. In time, a fairly homogeneous pattern of peaks of local activator production emerges, a pattern that can remain dynamically stable, or that can oscillate or move about in an orderly way across the field. By adjusting the reaction kinetics, the shape of the field, and the boundary conditions, it is possible to produce patterns that resemble the initiation sites of tentacle formation in *Hydra,* the phyllotactic patterns on plants, the patterns of imaginal disk initiation in *Drosophila,* compartment formation in *Drosophila* imaginal disks, the entire diversity of color patterns on seashells and mammalian coats, the branching patterns of veins and tracheae, and the chemotaxis of individual cells (Meinhardt, 1982, 1998; Murray, 1981a,b; 1989).

Perhaps the most important insight to emerge from the theoretical analysis of pattern formation in development is that reaction-diffusion is a special case of a virtually ubiquitous mechanism called "local activation and lateral inhibition," according to which a self-enhancing event or process, once initiated, tends to grow (mostly simply, through a positive feedback mechanism), directly or indirectly inhibiting the same process in its immediate surroundings (figure 10.2). Although reaction-diffusion happens to be a particularly simple and direct way of implementing such a mechanism, as we will see below, biological systems have many other ways of accomplishing the same end.

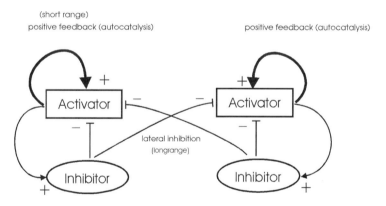

Figure 10.2
Local activation and lateral inhibition provide a general mechanism for self-organizing pattern formation. The actual mechanisms through which activation and inhibition are exercised can be extremely diverse and include molecular diffusion, action potentials, and mechanical traction. Local self-activation overwhelms inhibitory reactions, but because inhibition spreads farther and more rapidly than the activation it inhibits self-activation in adjoining areas. At a greater distance, where the inhibitory signal has decayed, activation can once more take place. Such a process can produce regular spatial patterns of activation. Whether such patterns are pointlike, linelike, or traveling waves depends on the details of the dynamics of activation and inhibition.

Notwithstanding the richness and power of pattern formation theory, many experimentalists have criticized it as largely unhelpful to their research and have therefore largely ignored it. This chapter highlights several specific critiques of pattern formation theory that illustrate the absence of reciprocal illumination between experimentalists and theorists; it discusses how pattern formation theory can and must be expanded to take account of recent findings in developmental biology and genetic regulation, and what kind of new data will need to be collected to fully integrate experimental and theoretical work in pattern formation.

Experiment versus Theory in Pattern Formation

Reaction-diffusion mechanisms can produce a wealth of realistic biological patterns, yet, despite the well-developed insights into the fundamental principles of such patterns that pattern formation theory has provided (Murray, 1989), it has had almost no influence on experimental work in the field of pattern formation. Experimentalists have found models based on the theory to be deficient in several respects. First, many models are not sufficiently robust to adequately represent development: they produce the desired results only under narrowly specified parameter values (Murray, 1982). Developmental systems, by contrast, can tolerate a great deal of environmental and genetic variation without significant changes in the final pattern. Second, because early experiments in the genetics of pattern formation indicated that communication was by nearest neighbor interaction rather than by long-range diffusion (Martinez Arias, 1989), experimentalists have questioned the relevance of diffusion-based pattern formation models to any but a few exceptional cases such as the syncytium of the early *Drosophila* embryo. Finally, the models are compatible with too many mechanisms, and, regardless of the specifics of a particular mechanism, produce essentially the same outcome (Oster, 1988). As such, they cannot be used to guide an experimental approach to the identification of specific genes.

These criticisms are accurate as far as they go, but they reveal a fundamental misunderstanding of the purposes of theoretical models in pattern formation: they are intended to examine the *principles,* not the genetic mechanisms, of pattern formation. The models show, for instance, that positive feedback is essential, but they do not specify which gene should experience positive feedback. Interestingly, the necessity of many genetic regulatory interactions that are today being "discovered" were actually predicted by earlier theoretical work (Meinhardt, 1983, 1994). Almost all theoretical work on pattern formation has been done for the purpose of uncovering general principles, and explicating how these might apply to real situations.

Experimentalists as a rule have little use for principle-oriented theory. In this regard, developmental genetics is quite different from evolutionary biology, where theory guides almost every aspect of experimental research. In evolutionary biology, research programs are designed to prove or disprove the theoretical models and principles that form the foundation of the field. Progress in the molecular genetics of development, by contrast, depends in large measure on the development of new technology. Instead of a unifying theory, there is an abundance of conceptual models to describe, for instance, how genes are regulated and how a gene activates or inhibits a specific process. A fairly general qualitative agreement of experimental results with the prediction of a conceptual model is usually accepted as satisfactory support for a hypothesis about mechanism. Conceptual models are not intended to be a rigorous statement about how a system operates, and are seldom phrased in unambiguous quantitative terms that can be rigorously tested.

A few investigators have developed mathematical models for specific processes such a the control of *eve* stripe formation in *Drosophila* (Reinitz and Sharp, 1995), Notch-Delta signaling in pattern formation (Collier et al., 1996), and the control of the cell cycle in yeast and *Xenopus* (Novak and Tyson, 1995; Tyson et al., 1995; Goldbeter, 1993). By examining the quantitative consequences of various assumptions about the physical processes by which gene products (enzymes, transcription factors) give rise to the observed feature, these models aim to discover whether the known facts adequately account for the process of interest. Such modeling can usually highlight misconceptions and point to interactions that may have been overlooked. But they do not generally get at the principle behind the process and therefore cannot say whether the process of interest is a special case of something more general, or whether it defines a truly new mechanism. Indeed, few if any researchers are investigating the principles that underlie the new molecular genetics of development.

Communication in Pattern Formation

All systems for chemical pattern formation require a mechanism that produces a graded or discontinuous distribution of an activator or inhibitor in space. From a modeler's perspective, the simplest and most parsimonious way of doing this is by means of diffusion, a random process that minimizes the potential energy of a system and therefore occurs spontaneously, unless physical barriers prevent passage of the diffusing substance. The modeler's assumption that molecules move by diffusion is merely an application of Occam's razor: diffusion is the only mechanism for moving chemicals that requires no additional factors or special assumptions. Thus, unless other mechanisms are already known to operate, investigations of chemical signal propagation should start with the assumption that diffusion is the transport mechanism. The dynamics of diffusion are well known; it is therefore possible to test empirically and quantitatively whether the dynamics of signal propagation in an experimental system support diffusion as the mechanism.

It is essential, however, to recognize that "diffusion" is not intended to be taken literally as the only, or even the primary, mechanism of transport in pattern-forming systems. In principle, diffusion means nothing more than *communication* (Harrison, 1993). Thus patterns that are produced by reaction-diffusion can also be produced by mechanisms that use other means for getting a signal from one place to another (though with different dynamics). This is well illustrated by the work of Young (1984), who simply assumed that activator and inhibitor had different ranges of constant signal strength, with sharp cutoffs at the end of that range. Without using diffusion at all, Young's mechanisms produced patterns identical to those generated by classical reaction-diffusion mechanisms. Swindale (1980) likewise proposed a model for ocular dominance stripes that relied on a static pattern of local activation and long-distance inhibition, this one mediated through neuronal signals. The critical components of pattern-forming systems, local activation and long-range inhibition, can be implemented in many ways: reaction-diffusion and Swindale's and Young's discrete mechanisms (e.g., figure 10.3) are just three of them. Mechanical force transmission and neural action potential provide yet additional means of signal propagation in pattern formation (Oster, Murray, and Maini, 1985; Ermentrout, Campbell, and Oster, 1986; Newman and Comper, 1990). When vastly different signal transmission mechanisms such as action potentials and mechanical stress are modeled, they can produce patterns that are indistinguishable from those produced by reaction-diffusion models (Oster, 1988).

That different mechanisms can produce identical patterns implies that the pattern itself cannot be used to deduce anything about the actual mechanism that gave rise to it. Pattern formation theory can be used to deduce the general properties of the mechanism, for instance, that the lateral inhibition must be weak (or strong) relative to the local activation, or that the rate of production of the inhibitor must be large (or small) relative to its rate of decay. But it cannot specify the means by which activation and inhibition are implemented. This "many-to-one" problem of process and pattern has discouraged many experimental biologists from taking models of pattern formation seriously.

This problem applies only to the *final pattern,* that is, the final product of the process that is being modeled, however. Different models make different predictions about the *dynamics* of pattern formation and about the *intermediate stages* by which a system arrives at the final pattern. There is an unfortunate paucity of experimental data on the dynamics of pattern formation, which precludes testing alternative models rigorously. Most studies on patterned gene expression, for instance, present only a single snapshot of an interesting but arbitrary point in time. What is really needed are finely timed series, showing how the patterns of all interacting species change in space and time. An example that approximates this ideal is the work of Carroll and colleagues (1994) on the dynamics of expression of *distalless* in the wing imaginal disk of the butterfly *Pecis coenia.* This pattern goes through a rather unusual series of intermediate stages before becoming restricted to a small group of cells halfway between two wing veins. These cells will, at a later stage in development, become the inducers for the formation of an eyespot pattern (Nijhout, 1994). This dynamical pattern of *distal-less* expression had been predicted many years earlier based on a rather

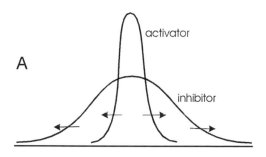

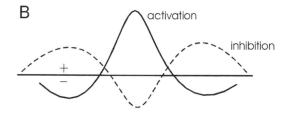

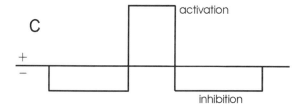

Figure 10.3
Three mechanisms for local activation and lateral inhibition. (*A*) Reaction-diffusion, in which activation and inhibition propagate from a common center but with different rates and different ranges. (*B*) Swindale (1980) model, in which static but graded complementary patterns of activation and inhibition are centered on a locus. (*C*) Young (1984) model, in which a discrete local activation is replaced at a fixed distance with a discrete inhibition. All three models can produce equivalent final patterns, but with different intermediate dynamics. The reaction-diffusion mechanism is the more versatile because different reaction schemes can cause qualitatively different patterns to develop (Nijhout, 1997).

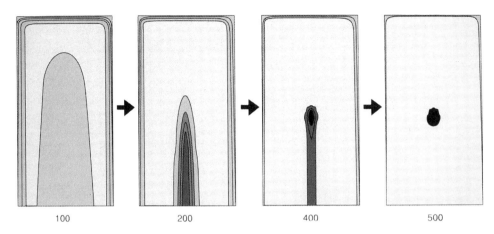

Figure 10.4
Simulation of the dynamical pattern of *distal-less* expression in a butterfly wing based on a local activation and lateral inhibition mechanism (after Carroll et al., 1994; Nijhout, 1994). In this case, the final pattern of *distal-less* expression marks the location of cells that will emit a signal that induces an eyespot of the color pattern on the adult wing. *Distal-less* expression in butterfly wings is not controlled by the same regulatory cascade that controls *distal-less* expression in *Drosophila* disks (see figure 10.6). Numbers represent arbitrary time units.

simple reaction-diffusion model (figure 10.4), indeed, the only model mechanism currently known to produce the specific sequence of patterns that precedes the final expression of *distal-less* in a small compact group of cells (Nijhout, 1990, 1994).

Oster (1988) has pointed out that different model mechanisms, however equivalent they may be in the final patterns they generate, are not equivalent in the experimental approach they suggest. For example, a pattern formed by a neural mechanism and one formed by a diffusion-based mechanism would respond differently to the introduction of a neurotropic agent. And a pattern formed by either of these mechanisms and one formed by a mechanical mechanism would respond differently to a surgical disruption. Experimentally obtained information is thus critical to uncovering the mechanism. Once the mechanism is known and some of its components have been elucidated (as in figure 10.4), theory can point the way to what additional interaction will make the system perform in the observed manner.

Cellular Mechanisms of Chemical Communication in Development

Leaving aside neural and mechanical mechanisms of communication, let us briefly review the diversity of mechanisms by which a chemical signal can be transmitted form cell to cell in development (figure 10.5A). The simplest mechanism, by which ionic coupling and dye coupling occur between cells in insect epidermis (Caveney, 1974; Safranyos and

Pattern Formation

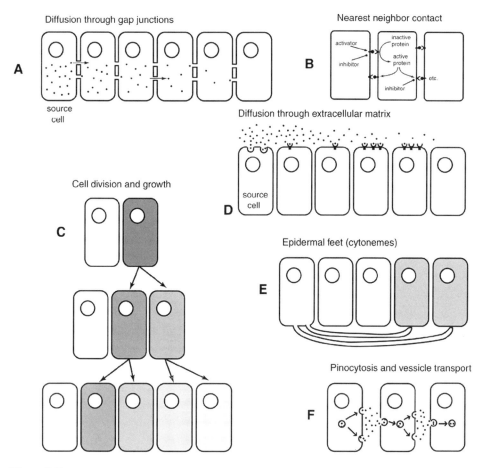

Figure 10.5
Five mechanisms (A, B, D, E, F) of cellular communication in biological pattern formation.

Caveney, 1985), is passive diffusion via gap junctions. Such communication is limited to molecules small enough to pass through the gap junctions, which restricts this mode of communication to molecules having a molecular weight of less than about one to a few thousand. Larger molecules, including most transcription factors, cannot pass through gap junctions and must be secreted into the extracellular matrix (figure 10.5D). Passive diffusion (or advection, if there is a directional flow within the extracellular matrix) can carry these molecules to the surrounding cells. Only cells that have receptors for the molecule will respond to it. It is possible that a patchy or graded distribution of receptors also plays an important role in pattern formation in such a system. Thus two gradients, one of a moving ligand and one of a static receptor, interact to form a new pattern. In order to understand the properties of such a system, we must discover by what dynamics the ligand moves, and what process sets up the prepattern of receptors.

Communication that is based on the stimulation of novel activities in a cell through the exchange of insoluble signals between immediate neighbors is another mechanism that can propagate signals over distances (figure 10.5B). Notch-Delta signaling appears to be such a mechanism (Collier et al., 1996). Here ligands that are bound to the cell surface interact with receptors bound to the surface of a neighboring cell. When a ligand is activated by an intracellular process, it transmits this information to a neighbor, which can repeat this process and transmit the same signal to cells further down the line. If the signals also include inhibition, (as they appear to in Notch-Delta signaling), then the fundamental conditions for pattern formation exist (Collier et al., 1996). Local interactions among only a few neighboring cells can set up small patterns, which can subsequently enlarge by cell division (Martinez-Arias, 1989). If during cell division some quantitative property of the cells becomes diluted, then new macroscopic gradients can be set up for further patterning (figure 10.5C). In epidermal sheets, cells can communicate over long distances by physically reaching out and touching distant cells by means of filopodia (figure 10.5E), processes discovered in the epidermal cells of several species of insects by Locke and Huie (1981a,b), who named them "epidermal feet." These processes can extend to cells as many as five to ten cell diameters removed and can thus be used for medium distance communication. Their role in morphogenesis in the developing wing was studied by Nardi and Magee-Adams (1986). Similar processes, called "cytonemes," have recently been reported in *Drosophila* imaginal disks (Ramirez-Weber and Kornberg, 1999). The filopodia in *Drosophila* have been shown to change over time and appear to reach out from the recipient cells to the signaling cells, rather than vice versa. Although obviously not a simple communication mechanism insofar as they involve rather complex and nonrandom behavior of cell membranes and cytoskeletal elements, such processes have the potential for carrying both stimulatory and inhibitory signals and thus readily fulfill the general needs for pattern formation. Finally, Moline, Southern, and Bejsovec (1999) have recently reported that the *wingless* gene product in *Drosophila* appears to be transported via pinocytotic vesicles. Presumably, cells can emit the *wingless* signal, which is then taken up by adjoining cells and transported within cytoplasmic vesicles to the other side of these cells, where it is released again (figure 10.5F). With such a mechanism, cells are in control of the directionality of transport. Moreover, cells could act to block the progress of a signal (by failing to transmit it) and could fine-tune the local timing and rate of transport of signals.

Preexisting Patterns Are Ubiquitous, but Are They Universal?

To discover the principles of pattern formation and to establish the necessary and sufficient conditions that can account for a particular phenomenon, pattern formation theories make a number of simplifying assumptions about the structure and behavior of a system, taking care, however, to make the fewest possible ad hoc assumptions about the system's structure. We have already discussed one of these simplifying assumptions, namely, that communication is by diffusion. Another simplifying assumption is that the pattern originates in a field that is *initially unpatterned*. Most models assume that the system is initially in some dynamic steady state and that pattern formation begins in response either to random fluctuations in this steady state or to simple changes in the boundary conditions.

Unfortunately, because evolution of complex systems occurs by elaboration and modification of previously existing systems, biological systems do not, as a rule, operate in the most parsimonious way possible: they are not designed from the beginning in the most optimal fashion. Development likewise proceeds by the gradual modification of preexisting structures. Pattern formation in development seldom occurs in a completely unpatterned homogeneous field. Even the egg has gradients of maternal gene products that define its polarity and that control the early events of embryonic pattern formation. The ability to visualize these previously hidden patterns is perhaps one of the greatest contributions of modern molecular biology to our understanding of development.

The best-understood pattern formation systems from a genetic viewpoint are those that control embryonic axiation, segmentation, and imaginal disk patterning in *Drosophila* (e.g., figure 10.6). The mechanisms appear to be quite simple. In all these systems a concentration gradient in a transcription factor or signaling substance induces the expression of a new gene wherever the concentration is above (or below) some threshold value. The product of the activated gene, in turn, diffuses and becomes distributed as a gradient that activates another gene in a threshold-dependent manner. Pattern specification in development thus appears to be due to a cascade of gradient threshold events of gene activation in which one gene sets up the conditions for the expression of the next one, and so on. All of which raises the question of how such a system gets started in the first place. In the case of the *Drosophila* egg, the initial conditions of embryonic pattern formation are set up in the ovary of the mother: the first gradient is due, among others, to the deposition of maternal messenger RNA (mRNA) for *bicoid* at one pole of the egg. In species of insects that do not use *bicoid*, other maternal gradients may play a role in the control of early embryonic patterning. In most cases of patterned gene expression, it is not known how the first-detectable gradient is set up. It is an open question whether there is always an infinite regression of prior gradients and patterns, or whether there exist cases in which a gradient is set up from initially homogeneous conditions. Only empirical data will answer this question.

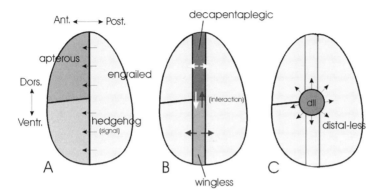

Figure 10.6
The localized expression of *distal-less* in a *Drosophila* imaginal disk is based on a series of coupled diffusion-threshold events. In each phase, relatively short-range signals at the boundaries between regions with different gene expression patterns induce the expression of new genes. Three compartments of the disk are identified by the expression of *apterous* and *engrailed* (A). A *hedgehog* signal from the posterior compartment induces the expression of *decapentaplegic* and *wingless* in the dorsal and ventral portions, respectively, of the anterior compartment (A and B). Signals of *decapentaplegic* and *wingless* then interact to induce the expression of *distal-less* at their joint boundary (C). *Distal-less* marks the location where the disk will grow outward to form a tubular limb.

Reaction-Diffusion and Genetic Circuits

The interaction of activator and inhibitor in a local activation–long-range inhibition system can be easily visualized in terms of a gene regulatory network. The genes involved in pattern formation in *Drosophila* produce transcription regulators and ligands that diffuse (or are transported; figure 10.5) to distant cells, where they affect the expression of other genes. Some of these regulators stimulate transcription, whereas others inhibit it. Many work in a concentration-dependent manner. The *bicoid* protein (a transcription regulator), for instance, inhibits transcription of the *Krüppel* gene (which codes for another transcription factor) at high concentration, but activates *Krüppel* transcription at low concentrations. Many genes are controlled through the interaction of several positive and negative transcription regulators; many others have, in addition, multiple regulatory sites through which they are regulated in different tissues under different chemical contexts. Although the dynamics of these interactions are poorly understood, their overall interplay is very similar to the interaction of activators and inhibitors in reaction-diffusion systems. The expression of *engrailed*, for instance, is regulated by positive feedback from its own product (Heemskerk et al., 1991), and *engrailed, wingless,* and *hedgehog* are all involved in a mutually stimulatory pathway. New cases of positive feedback in genetic regulation are reported almost monthly. Thus local self-enhancement (positive feedback), a critical feature of self-organizing pattern formation, appears to be a very common aspect of gene regulation. Genes are also inhibited by transcription factors produced by other genes. If these inhibitors have a longer range than the activators, then two of the fundamental conditions for pattern formation by reaction-diffusion are met. Even the diffusion part can be taken quite literally: diffusion does indeed appear to be the main mechanism for transport of many of these transcription regulators.

What is lacking in our current knowledge of the gene regulatory circuits involved in pattern formation is an explicit dynamic model of such a system that takes both the reaction kinetics and the diffusion (or whatever mode of communication is used) into account. A simple static picture of development and pattern formation as a genetic regulatory circuit, with activators and inhibitors pointing at each other, is inadequate: unless the values of activation and inhibition are specified, it is impossible to know whether such a system will actually behave in the anticipated manner. If, however, the equations by which activators and inhibitors bind to their sites and the dynamics by which they compete or interfere with each other can be specified, then it is possible to design a mathematical model that can rigorously test whether present knowledge adequately describes the system. Such a model should also be able to predict how the system would perform under novel genetic and environmental conditions.

The data necessary to develop such models are within easy reach of today's molecular biological techniques, although their collection will require a shift in research emphasis from the identification of the component of a system to the analysis of the dynamics of their interactions. From a theoretical viewpoint, the identities of the players are less important than the mechanism by which they interact, particularly in evolved and diversified biological systems, where the players may change, while the general principles by which these systems operate stay the same.

Although most of the pattern-forming systems analyzed experimentally consist of cascades of diffusion gradient threshold systems, we have at present no theoretical framework to deal with such systems. Because simple diffusion threshold systems have some surprising genetic properties (Nijhout and Paulsen, 1997), coupled systems may also have rather surprising and nonintuitive behaviors. The dynamics of much simpler coupled systems, such as biochemical reaction pathways (in which the product of one reaction is the substrate for the next one), have a host of interesting emergent properties such as the dominance relationships among alleles and the epistatic relationships among genes. (Kacser and Burns, 1981; Keightley, 1989; Keightley and Kacser, 1987). These findings have led to a well-developed body of theory and methodology called "metabolic control theory." Attempts to build up a similar body of work for static genetic circuit analysis are in their early stages (McAdams and Arkin, 1998). Little has yet been done on investigating the formal properties of genetic circuits whose kinetics vary spatially, as would be the case in systems of pattern formation.

Some Questions about Development That Arise from Models of Pattern Formation

The general principles of pattern formation are reasonably well understood. Further progress in the theoretical understanding of biological patterns will very likely depend on two new approaches, both stimulated by the need to incorporate the new findings of developmental genetics into the body of theory on pattern formation: the development of a theory of coupled dynamical systems and the acquisition of actual parameter values for the interactions, so that the models can be made to fit real-life cases. What follows is a brief list of the questions that need to be answered, and of the kinds of experimental data that need to be collected to this end:

1. What is the rate of gene product synthesis, and the rate of its decay? This is important because the gene product, the protein, is the active principle in genetic developmental regulation. The timing and rate of transcription, translation and breakdown need to be specified in quantitative terms.

2. Do synthesis and decay vary with time? The models produce different results if they assume proteins are synthesized as a single pulse, which then decays, than if they assume that protein concentration is actively maintained at a constant concentration.

3. How much protein is present? It seems that many regulatory systems require rather small numbers of proteins. A simple calculation suggests that if a large protein is present at a concentration of 10^{-6} M in a cell with a volume of 1 μm^3, only about 600 molecules are present. If numbers are any smaller, then stochastic processes are likely to be important determinants of activity, and a Monte Carlo approach rather than one based on differential equations may be more appropriate (e.g., McAdams and Arkin, 1997).

4. What controls the activity of a protein? Some proteins may always be active, but many require cofactors for activation or inhibition of whatever effect they have on a system. The kinetics of this interaction and the parameter values of this interaction (binding constants, equilibrium constants, cooperativity), need to be established. Even if only a few aspects of the reaction kinetics were known, this would place severe constraints on the types of models that could be fit to the data. The ability of Sharp and Reinitz (1998) to predict the expression patterns of *bicoid* and *hunchback* in certain mutants illustrates the power of this approach.

5. What is the time course of evolution of a given pattern? In a dynamical system, it is critical to have accurate information on the rates of processes.

6. What is a threshold? There are various mechanisms that can translate a continuous stimulus gradient into a more or less discontinuous response. Some of these are based on cooperativity or some variant on it (e.g., Lewis, Slack, and Wolpert, 1977; Reinitz, Mjolsness,

and Sharp, 1995). Here the Hill equation provides a way to make a transition that can be arbitrarily sharp, depending on the number of cooperating species. Other models are based on mutual inhibition among genes, and Meinhardt (1978, 1982) has shown how such an interaction among genes can give sharp discontinuities of gene expression along a gentle gradient of an inducing chemical. Most gradient-mediated gene activation mechanism in *Drosophila* embryos and imaginal disks are believed to be responsive to threshold concentrations of either a ligand or a transcription factor, but the mechanism and the kinetics of these thresholds have not been quantified.

7. Is the spatial pattern of gene expression at steady state or is it in flux? And a related question: Is a gradient at steady state when it is read? If not, then what determines the timing at which a gradient is read? This is important because if the system is not at steady state, small differences in the timing of events will have significant effects on the outcome. Whether a system reaches steady state will have important consequences for the evolution of robustness in development and the role of heterochrony in evolution.

Acknowledgment

Work in my laboratory is supported in part by grants from the National Science Foundation.

References

Carroll SB, Gates J, Keys DN, Paddock SW, Panganiban GEF, Selegue JE, Williams JA (1994) Pattern formation and eyespot determination in butterfly wings. Science 265: 109–114.

Caveney S (1974) Intercellular communication in a positional field: Movement of small ions between insect epidermal cells. Dev Biol 40: 311–322.

Collier JR, Monk NAM, Maini PK, Lewis JH (1996) Pattern formation by lateral inhibition with feedback: A mathematical model of Delta-Notch intercellular signalling. J Theor Biol 183: 429–446.

Edelstein-Keshet L (1988) Mathematical Models in Biology. New York: Random House.

Ermentrout B, Campbell J, Oster GF (1986) A model for shell patterns based on neural activity. Veliger 28: 369–388.

Gierer A (1981) Generation of biological pattern and form: Some physical, mathematical, and logical aspects. Prog Biophys Mol Biol 37: 1–47.

Gierer A, Meinhardt H (1972) A theory of biological pattern formation. Kybernetik 12: 30–39.

Goldbeter A (1993) Modeling the mitotic oscillator driving the cell division. Commun Theor Biol 3: 75–107.

Harrison LG (1993) Kinetic Theory of Living Pattern. Cambridge: Cambridge University Press.

Heemskerk J, DiNardo S, Kostriken R, O'Farrell PH (1991) Multiple modes of *engrailed* regulation in the progression towards cell fate determination. Nature 352: 404–410.

Kacser H, Burns JA (1981) The molecular basis of dominance. Genetics 97: 639–666.

Keightley PD (1989) Models of quantitative variation of flux in metabolic pathways. Genetics 121: 869–876.

Keightley PD, Kacser H (1987) Dominance, pleiotropy and metabolic structure. Genetics 117: 319–329.

Lewis J, Slack JMW, Wolpert L (1977) Thresholds in development. J Theor Biol 65: 579–590.

Locke M, Huie P (1981a) Epidermal feet in insect morphogenesis. Nature 293: 733–785.

Locke M, Huie P (1981b) Epidermal feet in pupal segment morphogenesis. Tissue Cell 13: 787–803.

Martinez-Arias A (1989) A cellular basis for pattern formation in the insect epidermis. Trends Genet 5: 262–267.

McAdams HH, Arkin A (1997) Stochastic mechanisms in gene expression. Proc Natl Acad Sci USA 94: 814–819.

McAdams HH, Arkin A (1998) Simulation of prokaryotic genetic circuits. Annu Rev Biophys Biomol Struct 27: 199–224.

Meinhardt H (1978) Space-dependent cell determination under the control of a morphogen gradient. J Theor Biol 74: 307–321.

Meinhardt H (1982) Models of Biological Pattern Formation. London: Academic Press.

Meinhardt H (1983) Cell determination boundaries as organizing regions for secondary embryonic fields. Dev Biol 96: 375–385.

Meinhardt H (1994) Biological pattern formation: New observations provide support for theoretical predictions. BioEssays 16: 627–632.

Meinhardt H (1998) The Algorithmic Beauty of Sea Shells (2d ed). Berlin: Springer.

Meinhardt H, Gierer A (1974) Application of a theory of biological pattern formation based on lateral inhibition. J Cell Sci 15: 321–346.

Moline MM, Southern C, Bejsovec A (1999) Directionality of *wingless* protein transport influences epidermal patterning in the *Drosophila* embryo. Development 126: 4375–4384.

Murray JD (1981a) On pattern formation mechanisms for lepidopteran wing patterns and mammalian coat markings. Philos Trans R Soc Lond B Biol Sci 295: 473–496.

Murray JD (1981b) A pre-pattern formation mechanism for animal coat markings. J Theor Biol 88: 161–199.

Murray JD (1982) Parameter space for Turing instabilities in reaction-diffusion mechanisms: A comparison of models. J Theor Biol 98: 143–163.

Murray JD (1989) Mathematical Biology. New York: Springer.

Nardi JB, Magee-Adams SM (1986) Formation of scale spacing patterns in a moth wing: 1. Epithelial feet may mediate cell rearrangement. Dev Biol 116: 278–290.

Newman SA, Comper WD (1990) "Generic" physical mechanisms of morphogenesis and pattern formation. Development 110: 1–18.

Nijhout HF (1990) A comprehensive model for colour pattern formation in butterflies. Proc R Soc Lond B Biol Sci 239: 81–113.

Nijhout HF (1994) Genes on the wing. Science 265: 44–45.

Nijhout HF (1997) Pattern formation in biological systems. In: Pattern Formation in the Physical and Biological Sciences (Nijhout HF, Nadel L, Stein D, eds), 269–297. Reading, Mass.: Addison-Wesley.

Nijhout HF, Paulsen SM (1997) Developmental models and polygenic characters. Am Nat 149: 394–405.

Novak B, Tyson JJ (1995) Quantitative analysis of a molecular model of mitotic control in fission yeast. J Theor Biol 173: 283–305.

Oster GF (1988) Lateral inhibition models of developmental processes. Math Biosci 90: 265–286.

Oster GF, Murray JD, Maini PK (1985) A model for chondrogenic condensation in the developing limb: The role of extracellular matrix and traction. J Embryol Exp Morphol 89: 93–112.

Ramirez-Weber F-A, Kornberg TB (1999) Cytonemes: Cellular processes that project to the principal signaling center in *Drosophila* imaginal disks. Cell 97: 599–607.

Reinitz J, Sharp DH (1995) Mechanisms of *eve* stripe formation. Mech Dev 49: 133–158.

Reinitz J, Mjolsness E, Sharp DH (1995) Model for cooperative control of positional information in *Drosophila* by *bicoid* and maternal *hunchback*. J Exp Zool 271: 47–56.

Safranyos RGA, Caveney S (1985) Rates of diffusion of fluorescent molecules via cell-to-cell membrane channel in a developing tissue. J Cell Biol 100: 736–747.

Segel LA (1984) Modeling Dynamic Phenomena in Molecular and Cellular Biology. Cambridge: Cambridge University Press.

Sharp DH, Reinitz J (1998) Prediction of mutant expression patterns using gene circuits. BioSystems 47: 79–90.

Swindale NV (1980) A model for the formation of ocular dominance stripes. Proc R Soc Lond B Biol Sci 208: 243–264.

Turing A (1952) The chemical basis of morphogenesis. Philos Trans R Soc Lond B Biol Sci 237: 37–72.

Tyson JJ, Novak B, Chen K, Val J (1995) Checkpoints in the cell cycle from a modeler's perspective. In: Progress in Cell Research, vol 1 (Meijer L, Guidet S, Tung HYL, eds), 1–18. New York: Plenum Press.

Young DA (1984) A local activator-inhibitor model of vertebrate skin patterns. Math Biosci 72: 51–58.

11 A Biochemical Oscillator Linked to Vertebrate Segmentation

Olivier Pourquié

Embryogenesis follows a precise and specific schedule of events shared by all individuals in any given animal species. Although this precise developmental sequence might result from the natural kinetics of the cascades of biochemical reactions required for constructing the organism, an accumulation of evidence indicates that time-measuring devices must exist to ensure that developmental processes in the embryo operate at the right time (Cooke and Smith, 1990; French-Constant, 1994; Snow and Tam, 1980).

One type of mechanism for measuring time is thought to rely on oscillatory devices (as does a clock). Well-described examples include oscillations in calcium concentration (Jones, 1998; McDougall and Sardet, 1995; Swanson, Arkin, and Ross, 1997), the cell division cycle (King et al., 1996; Murray and Kirschner, 1989), and other rhythmic processes such as the one recently identified in somitogenesis (Palmeirim et al., 1997). How these oscillations are converted into temporal information, however, remains poorly understood.

A second type of mechanism employs rate-limiting processes to measure time by the accumulation or loss of a product (as does an hourglass). At a defined time point, a threshold of the product is reached, which acts as a switch to trigger a particular developmental process. Such a mechanism has been implicated in, for example, controlling the onset of midblastula transition or the temporal control of oligodendrocyte progenitor differentiation in the rat optic nerve (Durand et al., 1998; Newport and Kirschner, 1982). These biological time-measuring mechanisms have been called either "developmental clocks" or "timers," depending on whether they relied on oscillations or on rate-limiting processes (Cooke and Smith, 1990). This review discusses several such time-measuring mechanisms, placing particular emphasis on oscillatory systems.

Ionic Oscillators in Development

The role of calcium as a second messenger within developing and differentiating cells is well established. In many signaling processes, an increase of calcium concentration within the cell occurs either as a single event or as a periodic phenomenon (Jones, 1998; Swanson, Arkin, and Ross, 1997). Calcium oscillators have been described in early ascidian and mammalian embryos, where calcium oscillations occur in response to sperm penetration. In ascidians, these oscillations correspond to rapid calcium transients that travel along the egg surface from the sperm entry point (McDougall and Sardet, 1995) and are correlated with the first cell surface contraction wave, which leads to the relocalization of the superficial cortex of the egg. They have a much shorter periodicity than the cell division cycle: a series of 12 to 25 calcium oscillations are observed over the period of 25 minutes between fertilization and expulsion of the first polar body.

In all species examined thus far, a calcium increase (whether oscillating or not) at the time of fertilization is thought to be important in promoting exit from meiosis, which allows the first cell cycle to proceed. In mammals, the protein oscillin has been implicated in triggering the periodic release of calcium from internal stores (Parrington, Lai, and Swann, 1998).

Because some species show only an increase in calcium concentration following fertilization, the advantage of an oscillating release of calcium remains unclear, although Dolmetsch, Xu, and Lewis (1998) have suggested it may be important in reducing the calcium threshold to that required for activation of transcription. Moreover, the frequency of calcium oscillations has been shown to regulate transcriptional specificity, with a particular frequency activating a given set of transcription factors. Thus the frequency of calcium oscillations might also direct cells toward specific developmental pathways. On the other hand, the number of oscillations does not appear to be tightly controlled, and it is debatable whether these calcium oscillators should be considered as clocks.

Oscillations of calcium, chloride, and potassium concentrations whose frequency correlated with that of the cell cycle have been reported during the first cell divisions following fertilization in sea urchins, fish, frog, and mouse eggs (Block and Moody, 1990; Day, Johnson, and Cook, 1998; Johnson and Day, 2000; Swanson, Arkin, and Ross, 1997). The relationship between these oscillations and the triggering of mitosis remains controversial, however: they proceed even if early cell divisions are prevented. Similar oscillations linked to cell division have been observed in cultured cells (Kao et al., 1990). The driving force of these oscillators is unknown and it remains to be established whether they are a cause or a consequence of the cell cycle oscillator.

Researchers have also found evidence of calcium oscillations in later embryos. In the zebrafish *Danio rerio,* the embryo exhibits a periodic series of intercellular, long-range calcium waves with a 5- to 10-minute frequency during gastrulation and somitogenesis (Gilland et al., 1999). Moreover, these pulses emanate from distinct loci at different developmental stages, notably from the node and tail bud regions during somitogenesis (Gilland et al., 1999). The role of these periodic waves remains to be established. Because somite formation time in the zebrafish does not correspond to the frequency time of the calcium waves described, a direct link with the somitogenesis is unlikely. Nevertheless, the cells remain phase coordinated over a long time span which, in the absence of cell communication, appears to demand an improbably precise control mechanism within each cell. The high frequency of calcium waves described in the zebrafish suggests a means of intercellular communication, possibly via gap junctions, which could provide a mechanism of coordinating the spatial and temporal regulation of highly localized processes across large cellular domains (Bagnall, Sanders, and Berdan, 1992).

Counting Early Cell Cycles?

One of the best-studied biochemical oscillators operating during development is the cell cycle. Historically, the major component of the cytoplasmic oscillator driving the cell cycle was identified by studying the first frog cell cycles, which are composed of rapidly alternating S and M phases (Hara, Tydeman, and Kirschner, 1980; Murray, Solomon, and Kirschner, 1989). This isolated component was able to induce premature mitosis when injected in an interphase cell and was therefore called "maturation-promoting factor" (MPF). Subsequently, MPF was shown to correspond to the cyclin B/cdc2 complex. In contrast to circadian oscillators, which rely on transcriptional mechanisms, the MPF oscillations are largely driven by posttranslational modifications such as phosphorylation and protein degradation (King et al., 1996). Thus entry into mitosis is promoted by the phosphorylation of the cyclin B/cdc2 complex. This complex activates the anaphase-promoting complex (APC), which is responsible for cyclin destruction by the ubiquitination pathway. This consequently allows exit from mitosis and commencement of the next cycle. These biochemical oscillations of MPF activity are coupled to and may drive the periodic cell movements called "surface contraction waves," which in turn are thought to be important for cytokinesis (Perez-Mongiovi, Chang, and Houliston, 1998; Rankin and Kirschner, 1997).

An association between early cell cycles and time measurement is important during early embryogenesis for controlling the timing of a phenomenon called "midblastula transition" (MBT). For many animal species in which the egg contains a large amount of maternal material, the first cell divisions occur in the absence of RNA synthesis. After a defined number of cell cycles (twelve in the frog and ten in the fly), MBT occurs: transcription of the zygotic genome becomes abruptly activated. It is now well established for insects and vertebrate species, that the activation of zygotic transcription starts when a threshold in the nucleocytoplasmic ratio is reached (Newport and Kirschner, 1982; Pritchard and Schubiger, 1996; Sibon, Stevenson, and Theurkauf, 1997). Because no growth occurs at these early stages, zygotic transcription is thought to be triggered when specific sets of maternal factors are titrated out as a result of cell divisions (Newport and Kirschner, 1982; Prioleau et al., 1994; Pritchard and Schubiger, 1996; Sibon, Stevenson, and Theurkauf, 1997). Thus, even though the timing of this mechanism does not involve direct counting of the cell divisions, it is nevertheless a consequence of this process.

The Hox Clock or the Einbahnstrasse

Cell cycle oscillations have been hypothesized to function as a potential time measurement device when the Hox gene cluster opens during organogenesis (Duboule, 1994). Transcription factors that are expressed in almost all embryonic tissues, including the nervous system, Hox genes are thought to specify positional information in the cells, along the anteroposterior axis in the embryo. Genes in the Hox complex are linearly arranged in four homologous clusters in amniotes and expressed colinearly such that a gene located more 3′ in the complex is activated in more anterior embryonic structures than its 5′ neighbor. In vertebrates, these genes are also expressed according to a temporal colinearity resulting in an earlier activation during development of the genes located more 3′ than those located 5′.

Studies on Hox gene activation in the developing limb bud have led to the proposal that there may be a link between cell proliferation and the progressive opening of the Hox gene complex (Duboule, 1994). During development, the limb bud extends proximodistally, as tissue is progressively laid down from a highly proliferative zone (the progress zone) located at the distal tip of the bud. Cells of the limb bud obey spatial and temporal colinearity rules, in that the first cells produced by the progress zone give rise to the most proximal part of the limb (humerus) and activate more 3′ genes of the cluster, whereas cells exiting the zone later in development contribute to more distal structures such as the hand and activate the more 5′ genes of the complex. Because the distal cells have undergone a higher number of cell divisions than those more proximal (Wolpert, Lewis, and Summerbell, 1975), it was proposed that cell division is linked to the progressive opening of the Hox genes cluster.

This model is referred to as "Einbahnstrasse" German for "one-way street" because the progressive opening of the Hox cluster, like time, is unidirectional (figure 11.1). A

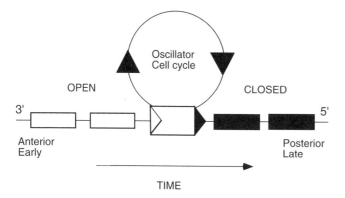

Figure 11.1
Schematic representation of the potential link between the cell cycle and the opening of the Hox gene cluster during development. Cells with a high proliferative rate progressively open the Hox cluster (progressively activate more and more 5′ Hox genes) as they divide. Cells exiting the high proliferation zones stop this process and somehow freeze their configuration of activated Hox genes. Cells remaining the longest in the highly proliferating zone will be positioned most distally in the limb bud and will display a fully opened Hox cluster with the most 5′ genes activated. (After Duboule, 1994.)

molecular explanation was proposed to account for this phenomenon. In this hypothesis, proteins aggregate at a point of high affinity in the 5′ part of the cluster, preventing access to the transcription machinery. Because of cell divisions, these proteins (which are rate-limiting) become progressively titrated out resulting in a progressive depletion from the Hox complex in a 3′ to 5′ fashion (Kondo, Zákány, and Duboule, 1998). This example constitutes a true example of a clock in which the oscillator (the cell cycle) is linked to a time-measuring device (unidirectional progression along the cluster).

Hourglass Types of Timers

In many instances in development, cells divide a defined number of times before they differentiate and eventually exit the cell cycle. It has therefore been proposed that one means by which cells know when they should undergo differentiation may be as a result of simply counting the number of cell divisions (Quinn, Holtzer, and Nameroff, 1985; Temple and Raff, 1986). No experimental demonstration of such a counting device, however, has been reported; to the contrary, many experiments suggest that the timing of several developmental process is controlled independently of cell cycle progression. For instance, blocking cell divisions during early cleavage in ascidians or in the nematode *Caenorhabditis elegans* does not prevent the expression of differentiation markers according to their appropriate schedule (Laufer, Bazzicalupo, and Wood, 1980; Satoh, 1979). Similarly, temporal control of competence windows whereby cells respond differently to inducers such as mesoderm inducers in early frog development was shown to be independent of the cell cycle (Cooke and Smith, 1990; Grainger and Gurdon, 1989). Degradation of the cyclins E and A prior to gastrulation also appears to be controlled by an intrinsic timer independent of cell cycle progression and of protein synthesis (Howe and Newport, 1996; Stack and Newport, 1997).

Another developmental system in which time measurement was shown to be independent of the cell cycle is the maturation of oligodendrocyte progenitors in the rat optic nerve (Gao, Durand, and Raff, 1997). Clonal analyses of oligodendrocyte precursor differentiation suggest that these cells are intrinsically programmed with respect to the time at which they arrest proliferation and start differentiating (Barres and Raff, 1994; Temple and Raff, 1986). The existence of a time measurement device composed of a timer, which measures the time elapsed from the moment of commitment, and of an effector, which ensures that cells adopt the correct fate at the right moment, was proposed to account for this behavior. Time measurement was shown to be independent of the cell cycle by placing cultures at 33°C and showing that, even when the cell cycle was slowed down, the time at which progenitors differentiated remained unaffected (Gao, Durand, and Raff, 1997).

One mechanism proposed for the timing of this process is the progressive accumulation of products that result in blocking the cell cycle, once a particular threshold is reached

(Durand, Gao, and Raff, 1997). Thus the CDK inhibitor p27kip1, which blocks progression from G1 to S phase, when it accumulates (even at 33°C) in the proliferating precursors, was proposed to be part of the timing mechanism (Durand et al., 1998; Durand, Gao, and Raff, 1997). Mice mutant for the p27kip1 gene are one-third larger than wild types due to increased cell proliferation in all tissues (Kiyokawa and Koff, 1998). This phenotype is in agreement with p27 playing a role in limiting cell division in vivo. When oligodendrocyte precursors from the optic nerve of p27kip1-/- mutant mice are cultured, they tend to differentiate later than wild-type cells and to complete one more cell cycle before becoming postmitotic (Durand et al., 1998). These observations indicate that levels of p27kip1 may be part of both the timer and of the effector mechanisms in this system, although the differentiation delay is rather limited, and other partners may well be involved in this process. Similar mechanisms controlling the exit from the cell cycle are likely to be used by many cell types.

A Biochemical Oscillator in Vertebrate Segmentation

Another type of oscillator involved in development has been identified during the process of vertebrate somitogenesis (Dale and Pourquié, 2000; Palmeirim et al., 1997). In vertebrate embryos, the most obvious segments are the somites, repeated epithelial structures of mesodermal origin formed in an anteroposterior sequence, in a bilateral fashion, and at regular time intervals. Somites give rise to the vertebral column and to the striated muscles of the body and provide the frame for the segmentation of the peripheral nervous system (Hirsinger et al., 2000). The study of a vertebrate homologue of the fly pair-rule gene *hairy*, *c-hairy1*, revealed a very unusual expression of its messenger RNA (mRNA) during somitogenesis in the chick embryo. The profile of *c-hairy1* exhibits a dynamic anteroposterior expression sequence which appears as a wave front sweeping along the avian presomitic mesoderm (PSM), once during the formation of each somite.

The dynamic wave front of expression is independent of cell movements and does not result from the propagation of a signal in the plane of the PSM but rather corresponds to an intrinsic property of this tissue. Between the moment a cell becomes specified to a paraxial mesoderm fate and enters the PSM and the moment this cell actually becomes incorporated into a somite, twelve new somites will be formed in the chick embryo (figure 11.2). Therefore PSM cells will undergo twelve cycles of *c-hairy1* expression before becoming incorporated into a somite. These oscillations of the *c-hairy1* mRNA in the presomitic mesoderm have been proposed to define a molecular clock linked to somitogenesis (Palmeirim et al., 1997). Strikingly, the existence of such oscillations had been predicted in several theoretical models for somitogenesis. These include Cooke's "clock and

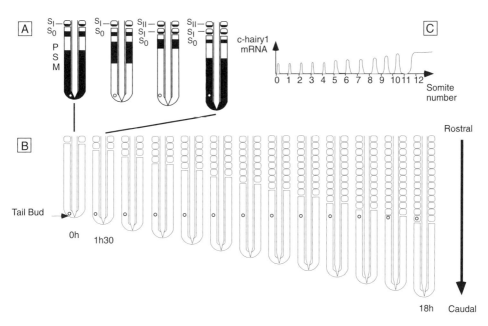

Figure 11.2
Expression of *c-hairy1* in the presomitic mesoderm defines a clock linked to vertebrate segmentation. (*A*) Expression of *c-hairy1* (black zones) appears as a wave arising from the posterior region of the embryo and progresses in a rostral direction. S_0, forming somite; S_I, newly formed somite; S_{II}, next oldest somite. (*B*) Between the time that a presomitic mesoderm (PSM) cell (open circle) becomes first specified after gastrulation (0h) and the moment it becomes incorporated into a somite (18h), twelve somites will form. Thus any cell in the PSM will experience twelve *c-hairy1* waves. (*C*) Oscillations of *c-hairy1* mRNA expression in the cell illustrated in *B* during the period it resides in the PSM. At the twelfth cycle, *c-hairy1* expression either remains on if the cell is in the caudal somitic compartment or turns off if it is in the rostral compartment. These oscillations of the *c-hairy1* mRNA occur in every cell of the PSM and define a clock linked to somite segmentation. (After Palmeirim et al., 1997.)

wavefront" model (1998), Meinhardt's model (1986), Stern's "cell cycle" model (Stern et al., 1988), and a cell-cycle biochemical oscillation model (Newman, 1993). The purpose of such a clock would be to generate a temporal periodicity, which could then be translated into the basic metameric pattern of the somites.

The *c-hairy1* homologue in the fly is a primary pair-rule gene that encodes a basic helix-loop-helix transcriptional repressor (Barolo and Levine, 1997). Similarly, the *HES1* gene in the mouse, a gene whose structure is highly related to that of the *c-hairy1* gene and one that was demonstrated to oscillate in mouse presomitic mesoderm, encodes a transcriptional repressor, shown to bind to its own promoter (Jouve et al., 2000; Takebayashi et al., 1994). Thus *c-hairy1* might be thought to regulate its own transcription, and regulation of the somitogenesis clock to resemble that of other well-studied biological clock systems, implying that *c-hairy1* might itself be acting as a crucial clock component. Because inhibiting protein synthesis does not arrest cyclic *c-hairy1* expression, however, such a transcriptionally based mechanism seems unlikely (Palmeirim et al., 1997).

These findings are also supported by genetic evidence in the mouse, where no effect on the segmentation clock is observed in the *HES1-/-* mutant. All of which suggests the existence of a mechanism acting at the posttranscriptional level, by means as yet unknown, to regulate the transcription of *hairy-like* genes.

Meinhardt's model incorporated the idea that the oscillatory mechanism used to generate the somites could serve to bestow regional identity to the somites. The number of oscillations undergone by each cell would determine the segment-specific fate of its derivatives, whether cervical, thoracic, lumbar, or other. He proposed an analogy with a grandfather clock, in which oscillations of the pendulum correspond to the oscillations in the presomitic mesoderm cells (Meinhardt, 1986). These would drive the rhythmic movements of the clock hands, which can be related to formation of somite units and the regional domains described above.

For such a mechanism to be implicated in somite determination along the anteroposterior (AP) axis, the oscillations would have to start very early in the development of somitic cells. Regional determination of the paraxial mesoderm cells along the AP axis is believed to occur immediately after they leave the primitive streak (or tail bud) and enter the presomitic mesoderm (Hogan, Thaler, and Eichele, 1992; Kieny, Mauger, and Sengel, 1972). The PSM cells in which *c-hairy1* oscillate are thus likely already determined with respect to their future AP identity and location. Consequently, for a time-measuring mechanism such as those proposed by Duboule and Meinhardt to be operational, the segmentation clock has to be functional prior to cells entering the PSM, that is, in the somitic stem cells of the primitive streak.

Studies of *c-hairy1* expression have to date not addressed the status of the clock in the presomitic territory of the streak and the tail bud, that is, before these cells enter into the presomitic mesoderm. We have observed that, in the chick, the first appearance of cycling RNA expression in the prospective paraxial mesoderm correlates with its ingression from the epiblast into the primitive streak (Jouve, Iimura, and Pourquié, 2002). Oscillations of the cycling genes are then detected in the rostral primitive streak and the forming PSM. Therefore, somite precursors in the streak undergo oscillatory expression of these genes before their release into the PSM. This suggests that the segmentation clock is already active in the somitic stem cells, which are located in the rostral streak. Consequently, somitic cells deriving from these stem cells will not only have experienced twelve oscillations, as reported in the initial study of *c-hairy1* expression, but the number of oscillations that directly corresponds to their position along in the AP axis. Thus cells that leave the stem cell zone early, and will thus be located anteriorly, will experience fewer cycles of gene expression prior to their oscillation cycles in the PSM than cells that populate more posterior somites and that continue to cycle in the streak before entering the PSM.

These data suggest that the number of oscillations experienced by presomitic mesoderm cells characterizes their specific anteroposterior position and therefore may be directly linked with the regionalization of their somitic derivatives. One possible means of linking these events is that the segmentation clock could control the cell cycle in these cells, thus directly or indirectly controlling the activation of Hox gene expression.

Resetting and Pausing the Clock

The developmental clock is set at fertilization and starts to tick in every cell. Once cells become determined, the time-measuring process becomes irreversible and cells follow their own internal developmental program (Kato and Gurdon, 1993). On the other hand, depending on the developmental mode of the embryos, the clock can be reset, made to pause, or restarted at particular developmental stages in many species. For example, although the clock is stopped in mouse embryonic stem cells (which can be cultured indefinitely), it can be restarted, because these cells retain the capacity to generate a complete embryo when injected into a blastocyst.

The same embryonic structures form at different developmental times in different animal species, a phenomenon called "heterochrony." This differential timing is thought to account for the generation of much of the diversity of body shapes found in the animal kingdom. To achieve such heterochronies, the developmental clock must therefore be independently controlled in different structures. It also means that it can be made to pause at different developmental time points, which vary between different animal species.

One way to produce these heterochronies might be to play on the timing of activation of the Hox genes. Slight variations in the timing of expression of these genes can result in the generation of different shapes from the same basic embryonic material (Duboule, 1994). In *Caenorhabditis elegans*, the heterochronic genes have been shown to play an important role in controlling when cells respond to inducing signals as well as cell cycle kinetics in the development of specific organs (Euling and Ambros, 1996; Moss, Lee, and Ambros, 1997). Mutation in these genes can either advance or delay activation of the developmental program in a specific organ while the other organs are able to maintain their appropriate pace. A high level of coordination between these different developmental programs is thus required to generate a normal organism; such coordination is achieved through a tight control of the different developmental clocks.

Conclusion

Although different examples of developmental clocks are now well described, there is no evidence to strictly associate an oscillating mechanism with the developmental measurement of elapsed time. A role in time measurement has been more clearly indicated for hourglass types of developmental timers. The measurement of time during embryonic development is also related to that of the whole lifespan. This has been demonstrated in *Caenorhabditis elegans,* where in most cases, the gene mutations that lengthen the animal's lifespan also lengthen its embryonic development (Hekimi et al., 1998).

An important difference between circadian and developmental clocks is the absence of temperature compensation. Modifying the ambient temperature of a developing embryo usually changes the speed of its development and thus the period of most its developmental clocks, whereas temperature has little influence on the period of the circadian clocks. Thus ambient temperature might play a role in the epigenetic control of morphogenesis by acting on the developmental clocks as it does, for instance, in the control of sex determination in reptiles (Ferguson and Joanen, 1982). Finally, it must be recognized that whereas studies in organisms as distantly related as algae and humans, have produced a reasonably unified picture of circadian clocks and their regulation, no such picture has yet emerged from the study of developmental clocks.

References

Bagnall KM, Sanders EJ, Berdan RC (1992) Communication compartments in the axial mesoderm of the chick embryo. Anat Embryol 186: 195–204.

Barolo S, Levine M (1997) *Hairy* mediates dominant repression in the *Drosophila* embryo. EMBO J 16: 2883–2891.

Barres BA, Raff MC (1994) Control of oligodendrocyte number in the developing rat optic nerve. Neuron 12: 935–942.

Block ML, Moody WJ (1990) A voltage-dependent chloride current linked to the cell cycle in ascidian embryos. Science 247: 1090–1092.

Cooke J (1998) A gene that resuscitates a theory: Somitogenesis and a molecular oscillator. Trends Genet 14: 85–88.

Cooke J, Smith JC (1990) Measurement of developmental time by cells of early embryos. Cell 60: 891–894.

Dale KJ, Pourquié O (2000) A clock-work somite. BioEssays 22: 72–83.

Day ML, Johnson MH, Cook DI (1998) A cytoplasmic cell cycle controls the activity of a K^+ channel in pre-implantation mouse embryos. EMBO J 17: 1952–1960.

Dolmetsch RE, Xu K, Lewis RS (1998) Calcium oscillations increase the efficiency and specificity of gene expression. Nature 392: 933–936.

Duboule D (1994) Temporal colinearity and the phylotypic progression: A basis for the stability of a vertebrate *Bauplan* and the evolution of morphologies through heterochrony. Development Suppl 135–142.

Durand B, Fero ML, Roberts JM, Raff MC (1998) p27Kip1 alters the response of cells to mitogen and is part of a cell-intrinsic timer that arrests the cell cycle and initiates differentiation. Curr Biol 8: 431–440.

Durand B, Gao FB, Raff M (1997) Accumulation of the cyclin-dependent kinase inhibitor p27/Kip1 and the timing of oligodendrocyte differentiation. EMBO J 16: 306–317.

Euling S, Ambros V (1996) Heterochronic genes control cell cycle progress and developmental competence of *C. elegans* vulva precursor cells. Cell 84: 667–676.

Ferguson MW, Joanen T (1982) Temperature of egg incubation determines sex in *Alligator mississippiensis*. Nature 296: 850–853.

French-Constant C (1994) Developmental timers: How do embryonic cells measure time? Curr Biol 4: 415–419.

Gao FB, Durand B, Raff M (1997) Oligodendrocyte precursor cells count time but not cell divisions before differentiation. Curr Biol 7: 152–155.

Gilland E, Miller AL, Karplus E, Baker R, Webb SE (1999) Imaging of multicellular large-scale rhythmic calcium waves during zebrafish gastrulation. Proc Natl Acad Sci USA 96: 157–161.

Grainger RM, Gurdon JB (1989) Loss of competence in amphibian induction can take place in single nondividing cells. Proc Natl Acad Sci USA 86: 1900–1904.

Hara K, Tydeman P, Kirschner M (1980) A cytoplasmic clock with the same period as the division cycle in *Xenopus* eggs. Proc Natl Acad Sci USA 77: 462–466.

Hekimi S, Lakowski B, Barnes TM, Ewbank JJ (1998) Molecular genetics of life span in *C. elegans:* How much does it teach us? Trends Genet 14: 14–20.

Hirsinger E, Jouve C, Dubrulle J, Pourquié O (2000) Somite formation and patterning. Int Rev Cytol 198: 1–65.

Hogan BL, Thaller C, Eichele G (1992) Evidence that Hensen's node is a site of retinoic acid synthesis. Nature 359: 237–241.

Howe JA, Newport JW (1996) A developmental timer regulates degradation of cyclin E1 at the midblastula transition during *Xenopus* embryogenesis. Proc Natl Acad Sci USA 93: 2060–2064.

Johnson MH, Day ML (2000) Egg timers: How is developmental time measured in the early vertebrate embryo? BioEssays 22: 57–63.

Jones KT (1998) Ca^{2+} oscillations in the activation of the egg and development of the embryo in mammals. Int J Dev Biol 42: 1–10.

Jouve C, Iimura T, Pourquié O (2002) Onset of the segmentation clock in the chick embryo: Evidence for oscillations in the somite precursors in the primitive streak. Development 129: 1107–1117.

Jouve C, Palmeirim I, Henrique D, Beckers J, Gossler A, Ish-Horowicz D, Pourquié O (2000) Notch signalling is required for cyclic expression of the *hairy*-like gene *HES1* in the presomitic mesoderm. Development 127: 1421–1429.

Kao JP, Alderton JM, Tsien RY, Steinhardt RA (1990) Active involvement of Ca^{2+} in mitotic progression of Swiss 3T3 fibroblasts. J Cell Biol 111: 183–196.

Kato K, Gurdon JB (1993) Single-cell transplantation determines the time when *Xenopus* muscle precursor cells acquire a capacity for autonomous differentiation. Proc Natl Acad Sci USA 90: 1310–1314.

Kieny M, Mauger A, Sengel P (1972) Early regionalization of somitic mesoderm as studied by the development of axial skeleton of the chick embryo. Dev Biol 28: 142–161.

King RW, Deshaies RJ, Peters JM, Kirschner MW (1996) How proteolysis drives the cell cycle. Science 274: 1652–1659.

Kiyokawa H, Koff A (1998) Roles of cyclin-dependent kinase inhibitors: Lessons from knockout mice. Curr Top Microbiol Immunol 227: 105–120.

Kondo T, Zákány J, Duboule D (1998) Control of colinearity in AbdB genes of the mouse HoxD complex. Mol Cell 1: 289–300.

Laufer JS, Bazzicalupo P, Wood WB (1980) Segregation of developmental potential in early embryos of *Caenorhabditis elegans*. Cell 19: 569–577.

McDougall A, Sardet C (1995) Function and characteristics of repetitive calcium waves associated with meiosis. Curr Biol 5: 318–328.

Meinhardt H (1986) Models of segmentation. In: Somites in Developing Embryos (Bellairs R, Ede DA, Lash JW, eds), 179–191. New York: Plenum Press.

Moss EG, Lee RC, Ambros V (1997) The cold shock domain protein LIN-28 controls developmental timing in *C. elegans* and is regulated by the lin-4 RNA. Cell 88: 637–646.

Murray AW, Kirschner MW (1989) Dominoes and clocks: The union of two views of the cell cycle. Science 246: 614–621.

Murray AW, Solomon MJ, Kirschner MW (1989) The role of cyclin synthesis and degradation in the control of maturation promoting factor activity. Nature 339: 280–286.

Newman SA (1993) Is segmentation generic? BioEssays 15: 277–283.

Newport JW, Kirschner MW (1982) A major developmental transition in early *Xenopus* embryos: 1. Characterization and timing of cellular changes at the midblastula stage. Cell 30: 675–686.

Palmeirim I, Henrique D, Ish-Horowicz D, Pourquié O (1997) Avian *hairy* gene expression identifies a molecular clock linked to vertebrate segmentation and somitogenesiss. Cell 91: 639–648.

Parrington J, Lai FA, Swann K (1998) A novel protein for Ca^{2+} signaling at fertilization. Curr Top Dev Biol 39: 215–243.

Perez-Mongiovi D, Chang P, Houliston E (1998) A propagated wave of MPF activation accompanies surface contraction waves at first mitosis in *Xenopus*. J Cell Sci 111: 385–393.

Prioleau MN, Huet J, Sentenac A, Mechali M (1994) Competition between chromatin and transcription complex assembly regulates gene expression during early development. Cell 77: 439–449.

Pritchard DK, Schubiger G (1996) Activation of transcription in Drosophila embryos is a gradual process mediated by the nucleocytoplasmic ratio. Genes Dev 10: 1131–1142.

Quinn LS, Holtzer H, Nameroff M (1985) Generation of chick skeletal muscle cells in groups of 16 from stem cells. Nature 313: 692 694.

Rankin S, Kirschner MW (1997) The surface contraction waves of *Xenopus* eggs reflect the metachronous cell-cycle state of the cytoplasm. Curr Biol 7: 451–454.

Satoh N (1979) On the "clock" mechanism determining the time of tissue-specific enzyme development during ascidian embryogenesis: 1. Acetylcholinesterase development in cleavage-arrested embryos. J Embryol Exp Morphol 54: 131–139.

Sibon OC, Stevenson VA, Theurkauf WE (1997) DNA-replication checkpoint control at the *Drosophila* midblastula transition. Nature 388: 93–97.

Snow MHL, Tam PPL (1980) Timing in embryological development. Nature 286: 107.

Stack JH, Newport JW (1997) Developmentally regulated activation of apoptosis early in *Xenopus* gastrulation results in cyclin A degradation during interphase of the cell cycle. Development 124: 3185–3195.

Stern CD, Fraser SE, Keynes RJ, Primmett DR (1988) A cell lineage analysis of segmentation in the chick embryo. Development 104: 231–244.

Swanson CA, Arkin AP, Ross J (1997) An endogenous calcium oscillator may control early embryonic division. Proc Natl Acad Sci USA 94: 1194–1199.

Takebayashi K, Sasai Y, Sakai Y, Watanabe T, Nakanishi S, Kageyama R (1994) Structure, chromosomal locus, and promoter analysis of the gene encoding the mouse helix-loop-helix factor HES-1. Negative autoregulation through the multiple N box elements. J Biol Chem 269: 5150–5156.

Temple S, Raff MC (1986) Clonal analysis of oligodendrocyte development in culture: Evidence for a developmental clock that counts cell divisions. Cell 44: 773–779.

Wolpert L, Lewis J, Summerbell D (1975) Morphogenesis of the vertebrate limb. Ciba Found Symp 29: 95–130.

12 Organization through Intra-Inter Dynamics

Kunihiko Kaneko

This chapter provides a new look at biological organization, based on studies by my colleagues and me of systems consisting of the intra-inter dynamics and reproduction of units. Our scenario for cell differentiation and development serves to explain (1) the robustness of developmental processes; (2) the mechanism of the stochastic differentiation of stem cells and the hierarchical organization in the stem cell system; (3) the relevance of chemicals with low concentrations to differentiation; (4) the differentiation of the equivalence group and the community effect; (5) separation of germ cell line; and (6) the origin of multicellular organisms and their life histories. Applying our results to evolution, we suggest that sympatric speciation can occur without assuming premating isolation in advance, and that it necessarily occurs when the interaction among units becomes strong. Our results also help explain why the rate of diversification of species seems to be different for each ecosystem and era, as well as the relevance of low penetrance to evolution.

Organization in Dynamic Interaction

In considering biological organization, one often tries to isolate each part and its function and to understand the whole organization by combining the processes of all the parts. One extracts the function of each enzyme in the metabolic process, the role of each cell in the tissue, and so forth. After selecting out each "part," one looks for a rule connecting the parts, such as "If this process is on, then that process starts. . . ." One focuses on the nature of an interaction-independent part first, and then studies the effect of the interaction, to describe the whole biological system as the "programmed" combination of all parts. One studies the interaction dependence only as the modification of the character of individual units (Alberts et al., 1994).

Of course, the function of each part is not rigidly predetermined: it depends on the whole process through interaction. For example, the function of G-proteins is determined by other kinases. The character of a cell is determined only after its interaction with the other cells is prescribed. The fate of a cell through development generally depends on the other cells, studied in terms of the equivalence group (Greenwald and Rubin, 1992) or community effect (Gurdon, Lemaire, and Kato, 1993). Indeed, the function of a cell in isolation can be essentially different from its function in the organism.

Nevertheless, one might think that by combining several predetermined functions, the whole organism could be understood. The process of successively turning on and off combinations of genes is believed to produce a body plan. Assuming that the embryo is a machine like a parallel processor, then all cell differentiations occurring in the development have to follow a strictly organized course.

Owing to the uncertainty of each cell differentiation, however, such preprogrammed rules cannot always work "correctly" (Kaneko and Yomo, 1999). For example, turning on and off genes is not necessarily a logical process that always works correctly like a computer. Although the expression of genes is a biochemical process often believed to be determined by the threshold of the signaling molecule concentrations, because the number of such signaling molecules is not large, this process is subject to inevitable molecular fluctuations. Typically, for 1,000 molecules, the probability of error should be $1/\sqrt{1000} \approx 0.03$. As the number of genes increases, then, it is almost impossible to have a "correct" expression pattern.

Still, one might expect that errors in the developmental process would be overcome by the interplay among the genes. For example, in Kauffman's Boolean networks (1969), some gene expression patterns are selected from several initial conditions. Each attracted pattern is assumed to give rise to a particular cell type. Even though errors occur and change the gene expression for each cell type, the expression pattern of genes can recover the original one through the interplay among genes. Thus stability might be expected. If, however, an error occurs during the differentiation of a cell to another cell type, the gene expression patterns are affected, and a "wrong" cell type is produced. Despite these molecular fluctuations, the developmental process is robust; thus the threshold mechanism alone cannot account for the whole of it.

In earlier publications, Yomo, Furusawa, and I (Kaneko and Yomo, 1994, 1997, 1999; Kaneko, 1997a; Furusawa and Kaneko, 1998a,b) have put forward a different viewpoint, where developmental rules are not given in advance, but are formed from the interplay of reproducing units with internal chemical dynamics. After outlining our approach, this chapter sets forth the cell differentiation scenario we extracted from several simulations, revealing its underlying logic in later sections. It begins, not by combining symbolic rules of the if-then type for cell differentiation, as is often the case in discussions of body plans from a genetic network perspective, but by exploring how such rules appear from the complex pattern dynamics at a lower (chemical) level. It then describes how our cell differentiation scenario accounts for the formation of the multicellular organism, giving rise to a higher-order recursive production. Finally, it discusses the evolutionary implications of our scenario, where genetic fixation takes over the initial interaction-induced phenotypic differentiation.

Our Approach

Intra-Inter Dynamics

The internal chemical dynamics of each cell are not necessarily stable; indeed, because the cell must be able to switch from one type to another in differentiation, they need to be somewhat unstable. With orbital instability, small differences between two cells can be amplified. However paradoxical it may seem, stability is realized by taking advantage of such "unstable" dynamics. Through development, the instability is weakened as some distribution of differentiated cell types is established. Rules for differentiation or reproduction of units are generated that allow for the robustness of the developmental process. The generated rule is interaction dependent: information about the whole organism (i.e., distribution of cell types) influences the rule. The robustness of the cell society is a consequence of such interaction-dependent rules.

In trying to understand how structures and rules are generated from dynamics, we adopted a dynamical systems approach in the broadest sense. All biological units have an internal structure that can change in time. Such a unit is represented by a dynamical system, which consists of time, a set of states, an evolution rule, an initial condition of the states, and a boundary condition. The state is represented by a set of k variables, the degrees of freedom. Thus the state at an instant is represented by a point in the k-dimensional space, the phase space.

These units, however, are not completely separated from the outside world. For example, isolation by a biomembrane is flexible and incomplete, allowing the units, represented by dynamical systems, to interact with each other and with the external environment. Hence, we need to model the interplay between interunit and intraunit dynamics (Kaneko, 1990, 1993, 1998). For example, the complex chemical reaction dynamics in each unit (cell) are affected by the interaction with other cells, which provides an interesting example of "intra-inter dynamics." In intra-inter dynamics, elements having internal dynamics interact with each other. This type of intra-inter dynamics is neither only the perturbation of the internal dynamics by the interaction with other units, nor is it merely a perturbation of the interaction by adding some internal dynamics.

If N elements with k degrees of freedom exist, the total dynamics are represented by an Nk-dimensional dynamical system (in addition to the degrees of freedom of the environment). Furthermore, the number of elements is not fixed in time, although they are born and die in time, which is to be expected when one considers the development of a cell society. As a result, the number of degrees of freedom, Nk, changes in time.

After the division of a cell, if two cells remained identical, another set of variables would not be necessary. If the orbits have orbital instability (such as chaos), however, the orbits of the daughters will diverge. Thus the increase in the number of variables is tightly connected with the internal dynamics. It should also be noted that in the developmental process, in general, the initial condition of the cell states is chosen so that their reproduction continues. Thus a suitable initial condition for the internal degrees of freedom is selected through interaction.

As a specific example of the scheme of intra-inter dynamics, I will focus on the developmental process of a cell society accompanied by cell differentiation. Here, the intra-inter dynamics consist of several biochemical reaction processes. The cells interact through the diffusion of chemicals or their active signal transmission. The change of dynamics is brought about by cell division and death, which also depend on the cellular state.

Our Model

My colleagues and I (Kaneko and Yomo, 1994, 1997; Kaneko, 1994, 1997a; Furusawa and Kaneko, 1998a,b) have tested several models by choosing (1) the internal variables and their dynamics; (2) the interaction type, and (3) the rule to change the degrees of freedom (e.g., cell division). For the internal dynamics, we chose autocatalytic reaction among chemicals. Such autocatalytic reactions are necessary to produce the chemicals in a cell needed for reproduction (Eigen and Schuster, 1979). Autocatalytic reactions often lead to nonlinear oscillation in chemicals; we assumed the possibility of such oscillation in the intracellular dynamics (Goodwin, 1963; Hess and Boiteux, 1971). The relevance of the oscillation of Ca^{2+} concentrations to cell differentiation has been established experimentally (Dolmetsch, Xu, and Lewis, 1998).

For the interaction mechanism, we chose the diffusion of chemicals between a cell and its surroundings. To avoid the complication of spatial pattern formation, we assumed, for the most part, that the surrounding medium was spatially homogeneous, in other words, that each cell was coupled to all the other cells. In testing some models, we assumed that a nutrient was actively transported into the cells, with the rate of active transport depending on the concentration of other chemicals in the cell. In testing others, we assumed only diffusion.

The cell divides according to its internal state. In a number of the models we studied, we assumed that some products were accumulated through biochemical reactions, and that the cell divided into two when the accumulated products rose above some threshold. In other models, we assumed that the cell volume depended on the amount of biochemicals in the cell, with a cell dividing into two when the volume became twice the original volume.

Of course, despite the variety of choices for a biochemical reaction network, the observed results do not depend on the details of a particular choice, as long as the network allows for the cell division and for the growth of the number of cells. Note that our network is not constructed to resemble an existing biochemical network. Rather, we tried to demonstrate that important features in a biological system were the natural consequences of a system with internal dynamics, interaction, and reproduction. From our studies, we extracted a universal logic underlying this class of models. I will survey this logic with regard to cell biology in a later section; we believe our scenario holds generally for all biological systems.

Remarks In his pioneering study, Turing (1952; see also Newman and Comper, 1990) proposed a pattern formation mechanism through spatial symmetry breaking induced by dynamic instability. Although this work is the classic masterpiece in dynamical systems biology and pattern formation, Turing's theory is not sufficient to explain how a differentiation rule can be generated from pattern dynamics. In our approach, we studied the generation of differentiation rules by introducing rich internal dynamics (such as chaos) and the reproduction of units, selecting initial conditions to form discrete cell types and rules for differentiation.

On the other hand, the aim of Kauffman's Boolean network (1969) is the formation of different cell types from symbolic rules, where multiple attractors coexist as different patterns of gene expressions, each of which corresponds to a differentiated cell type (for a realistic genetic network model with cell-to-cell interaction, see Mjoliness, Sharp, and Reinetz, 1991). Still, Kauffman neither clarified the rules of differentiation nor resolved the problem of the robustness of the cell society. We decided therefore to study the generation of rules from patterns where the observed cell type is not an attractor but an attracting state stabilized through cell-to-cell interaction.

Constructive Biology

Note that our model does not have any direct one-to-one correspondence in a specific cell system in nature. Rather, we tried to capture the general consequences of a system with internal dynamics, interaction, and reproduction of units. In this sense, our approach was constructive (Kaneko and Tsuda, 1994, 1996; Kaneko, 1998). We first constructed a simple "world" by combining fundamental procedures and by clarifying a universal class of phenomena. Then we tried to reveal the underlying universal logic that the life process should obey, to provide a new look at present-day organisms.

My colleagues and I believe that this "constructive" approach is essential to understand the logic of life. As long as we study only the present organisms in nature, which come to us as frozen accidents through evolution, it will be hard to distinguish the logic that organisms necessarily should obey. The logic and universality of life can be clarified only by constructing some world. Actual organisms are best understood as representatives of a universal class, to which the life-as-it-could-be also belongs.

Scenario for Differentiation to Types

From several simulations of the models starting from a single cell initial condition, Tetsuya Yomo and I (Kaneko and Yomo, 1997, 1999) proposed the following scenario ("isologous diversification"), as a general mechanism of spontaneous differentiation of replicating biological units (see figures 12.1a–12.1d for schematic representation; our results are supported by numerical simulations of several models; see Kaneko and Yomo, 1997; Furusawa and Kaneko, 1998a, for direct numerical results).

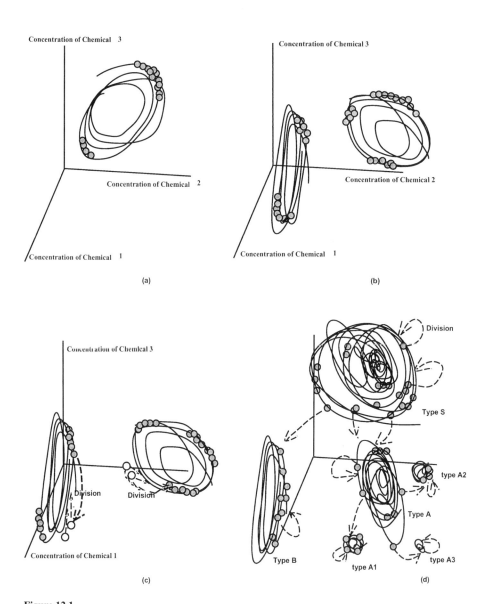

Figure 12.1
(*a*) Stage I of differentiation scenario: differentiation in the phases of oscillation. (*b*) Stage II of differentiation scenario: differentiation in chemical composition. (*c*) Stage III of differentiation scenario: determination of differentiated cells. (*d*) Stage IV of differentiation scenario: differentiation of cell groups into subgroups. Formation of stem cells ($S \to A, B; A \to A1, A2, A3$) is represented here.

Synchronous Divisions with Synchronous Oscillations of the Chemicals

Up to a certain number of cells (depending on the model parameters), the dividing cells from a single cell have the same characteristics. Although, because of the accompanying fluctuation in the biochemical composition, each cell division is not exactly symmetrical, the phase of oscillation in the concentrations and thus also the timing of cell division remain synchronous for all cells. Synchronous cell division is also observed up to the first eight cells in the embryogenesis of mammals.

Clustering in the Phases of Oscillations

When the number of cells rises above a certain threshold value, the state of identical cells is no longer stable. Small differences introduced by the fluctuation start to be amplified, until the synchrony of the oscillations is broken. Then the cells split into a few clusters, each having a different oscillation phase, identical for all cells belonging to that cluster. Because the time average of the biochemical concentrations reveals the cells to be almost identical, this diversification of phases cannot be called "cell differentiation," however. At this stage, the cells differ only in their phases of oscillation (figure 12.1a).

Which cells fall into the same cluster is highly sensitive to the small fluctuations brought about by cell division, during which small differences are amplified, and the orbit of each cell's intracellular chemical dynamics is unstable. Once these small differences are amplified, the newly divided cells produce different phases of oscillations, and the orbital instability of the intracellular dynamics is diminished. Even if further fluctuation is added, all the cells of a cluster remain in the same cluster. When the ratio of the number of cells in the various clusters falls within some range, the intracellular biochemical dynamics are mutually stabilized by cell-to-cell interactions. For example, the dynamics are stable if the numerical ratio of two type of cells is between, say, 38 : 26 and 40 : 24. In other words, the distribution is stable in the face of external perturbations. But then if all or many cells of one cluster are removed externally, the cell society is destabilized, with some cells switching to a different cluster to recover the original distribution.

Differentiation in Chemical Composition

As the number of cells increases, the average concentrations of the biochemicals over the cell cycle, the composition of biochemicals, and the rates of catalytic reactions and transport of the biochemicals become different for each group. Although the orbits of chemical dynamics plotted in the phase space of biochemical concentrations lie in a distinct region within the phase space, the phases of oscillations remain different for each group (see figure 12.1b).

Distinct groups of cells are thus formed with different chemical characters. Each group is regarded as a different cell type, and the process by which such types are formed is called "differentiation." In biological terms, this third stage is none other than the division of labor of several biochemical reactions in the cell, as can be seen in the use of biochemical resources, which is different for each group.

With the nonlinear nature of the reaction network, the difference in chemical composition between the cell groups is amplified. By the formation of groups of different biochemical composition, the intracellular biochemical dynamics of each cell are again stabilized (see figure 12.1b).

Determination of the Differentiated Cells

After the formation of cell types, the biochemical composition of each group is inherited by its daughter cells. In other words, the biochemical composition of cells is recursive over divisions. To see this recursivity, we plotted the "return map," that is, the relation between the biochemical averages of mother and daughter cells. In the return map, the recursivity is represented by points lying around the diagonal ($y = x$) line. Although, initially, biochemical composition changes with each cell division, later it settles down to almost a fixed value. The biochemical properties of a cell are inherited by its progeny; in other words, the properties of the differentiated cells are stable, fixed, or determined over the generations (see figure 12.1c).

It is important to note that the biochemical characters are "inherited" solely through the initial conditions of biochemical concentrations after cell division. We have not explicitly implemented any external mechanisms to get such cellular memory. The determination of cells occurs at this stage, with daughters of one cell type preserving their type. After several cell divisions, initial conditions of units are chosen to make the next generation of cells the same type as their mother cell. Thus a kind of memory is formed. This memory lies not only in the internal states but also in the interactions among the units, as will be shown later.

Generation of Rules for Hierarchical Organization

As the number of cells further increases, differentiation proceeds. Each group of cells differentiates further into two or more subgroups. Thus the total system consists of units of diverse behaviors, forming a heterogeneous society.

The most interesting example here is the formation of stem cells (see figure 12.1d; Furusawa and Kaneko, 1998a). This cell type, denoted as S here, either reproduces the same type or forms different cell types, denoted, for example, as types A and B. Then, after cell division, event $S \to A$ occurs. Depending on the adopted biochemical networks, the types A and B replicate, or switch to different types. For example, $A \to A1, A2, A3$ is observed in some networks. This hierarchical organization is often observed when the internal dynamics have some complexity, such as chaos.

As long as we distinguish only the cell type, the choice for a stem cell either to replicate or to differentiate appears to be stochastic (i.e., probabilistic). Of course, because our model itself is deterministic, if one completely determined the biochemical composition and the division-induced asymmetry, then the fate of the cell would also be determined. Here, however, because such deterministic dynamics lead to stochastic behavior if one cannot determine the internal cellular state in every detail (as the phenomenon of chaos demonstrates), such a possibility is practically irrelevant.

As we shall see, this stochastic differentiation is accompanied by a regulative mechanism. When some cells are removed externally during the developmental process, the rate of differentiation changes so that the final cell distribution is recovered.

Logic of Differentiation

Diversification

The change in phases of oscillation at the second stage is due to dynamic clustering, studied in coupled nonlinear oscillators (Kaneko, 1990, 1994). In a coupled dynamical system, clustering generally appears if there is both an orbital instability to amplify small differences and interaction among units to maintain synchronization. Clusters are formed through the balance between the two tendencies, with the numerical ratio of cells in the various clusters satisfying some condition. This balance results in stability in the face of perturbations.

The differentiation at the third stage is expected if the instability is related also to the amplitude of oscillations. Some internal degrees of freedom are necessary to support the difference in the phase space position. Taking advantage of internal degrees of freedom, each intracellular orbit is differentiated in the phase space.

Formation of Discrete Types That Are Recursive

Although the diversification trend can generally lead to cells whose (biochemical) characters continuously change, discrete types are formed through the (re)production process. Formation of discrete types is due to selection of initial conditions for the cell states (after cell division) that provide stability through later stages of the developmental process.

As for cell types, one might think that this selection is nothing more than a choice among basins of attraction for a multiple attractor system. If the interaction were neglected, a different type of dynamics would be interpreted as a different attractor. In our case, this is not true. In our model, cell-to-cell interactions are necessary to stabilize cell types. Hence, each intracellular state is an attracting state stabilized by means of the cell-to-cell interactions. Given cell-to-cell interactions, the cellular state is stable with respect to perturbations on the level of each cell's intracellular dynamics. It is also stable with respect to small changes of the interaction term. Even if the cell type number distribution is changed within some range, each type of intracellular dynamics remains the same. Hence discrete, stable types are formed through the interplay between intracellular dynamics and interaction. The recursive production is attained through the selection of initial conditions of the intracellular dynamics of each cell, so that it is rather robust to the change of interaction terms as well.

In real organisms, cells tend to form several types, whose number is much smaller than the number of units. As the number of cells increases, newborn cells fall into a few discrete types. Although cells of the same type are not completely identical, the differences between them are much smaller than those between cells of different types. In the course of development, cells are initially undifferentiated, and change their states with time. Later, in a process called "determination," fixed cell types appear (Alberts et al., 1993). These findings agree with the scenario my colleagues and I obtained from model simulations.

Rule Generation

Rules for transition between cell types—differentiation—formed after several generations are classified as follows (see figure 12.2):

- "Undifferentiated cell" class (U): Cells change their biochemical composition with each generation (figure 12.2a).
- "Stem cell" class (S): Cells either reproduce the same type or switch to different types with some rate (figure 12.2b).
- "Determined cell" class (D): Biochemical composition is preserved in each cell division (figure 12.2b).
- "Tumor-type" cells: Appearing in a few examples with a large diffusion coupling, these cells differentiate in an extraordinary way, destroy the cooperativity attained in the cell society, and selfishly multiply faster. Their biochemical composition loses diversity, and their ongoing chemical pathways are simpler than in other cell types (Kaneko and Yomo, 1997).

Differentiation rules are observed to be at a higher level than the rules for chemical reactions. A differentiation rule is not necessarily determined by cell type alone; it also depends on the intracellular state and the interaction.

Stem Cell

In our model, probabilistic differentiation from a stem cell originates in chaotic intracellular dynamics. Because such stochasticity is not due to external fluctuation but is a result of the internal state, the probability of differentiation can be regulated by the intracellular state. This is the origin of regulation mentioned in the previous section.

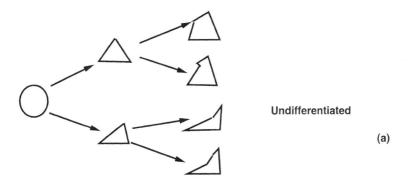

Undifferentiated

(a)

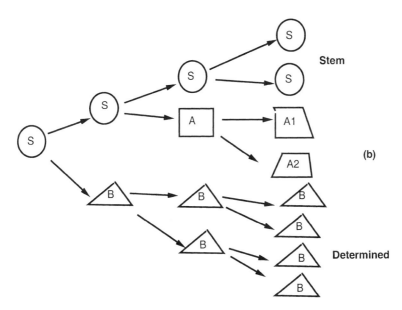

Stem

Determined

(b)

Figure 12.2
Schematic representation of the differentiation rules observed in several models.

The state of a stem cell has to be stable for reproduction, and unstable for differentiation. Indeed, the cell's stability has to change depending on its interaction with other cells. In this sense, the intracellular state is weak with respect to external perturbation. Such weak attracting states have recently been found in a class of dynamical systems of coupled units (Kaneko, 1997b), where orbits starting from a large number of initial conditions are attracted to such states, although any small perturbation to the states can kick the orbit away from them. In this sense, they are a candidate for a system having both the stability and instability of state, that is, the propensity to switch to a different state. Furthermore, the transition from one such weak attracting state to another caused by any small perturbation is not completely random, but has some constraints. The switching dynamics among the states, known as "chaotic itinerancy," emerge out of high-dimensional chaotic dynamics (Tsuda, 1992; Ikeda et al., 1989; Kaneko, 1990). Accordingly, there is a basis in dynamical systems theory for the nature of a stem cell observed in our model.

In biological systems such as the hematopoietic system stem cells either replicate or differentiate into different cell types. The differentiation rule is often hierarchical (Alberts et al., 1993; Ogawa, 1993), and differentiation is often thought to be stochastic (Till, McCulloch, and Siminovitch, 1964; Ogawa, 1993). The probability of differentiation to one of the several blood cell types is presumed to depend on the interaction; otherwise, it is hard to explain why the developmental process is robust. For example, when the number of some terminal cells decreases, some mechanism is needed to increase the rate of differentiation from the stem cell to the terminal cells. This need suggests the existence of interaction-dependent regulation of the type we demonstrated in our results.

Relativity in Determination

Note that determination also depends on interaction with other cells. One effective method to clearly demonstrate the intra-inter nature of the determination is the transplantation experiment. Transplantation experiments are carried out by choosing determined cells (obtained from the normal differentiation process) and putting them into a different set of surrounding cells to make a cell society that does not appear through the normal course of development.

When a determined cell is transplanted to another cell society, the daughters of the cell keep the same type, unless the cell type distribution of the society is strongly biased (e.g., a society consisting only of the same cell type as the transplanted cell). When a cell is transplanted into a biased society, differentiation from a "determined" cell occurs. For example, a homogeneous society only of one determined cell type is unstable, and some cells start to switch to a different type. Thus, even though the cell memory is preserved mainly in each individual cell, suitable intercellular interactions are also necessary to maintain it.

Internal Representation

How is the interaction-dependent rule mentioned earlier generated? Note that, depending on the distribution of the other cell types, the orbit of internal cellular state is deformed. Such modulation is made possible by our model's dual memory system: a state of a cell is characterized mainly by its discrete type. However, there remains analog modulation even between the same cell types, carrying global information about cell type distribution.

For the stem cell case, the rate of the differentiation or the replication (e.g., the rate of $S \to A, B$) depends on the cell type distribution. For example, when the number of type A cells is reduced, the orbit of a type S cell is shifted toward the orbits of type A, and the rate of switching to type A is enhanced. The information of the cell type distribution, represented by the internal dynamics of type S cells, is essential to the regulation of the differentiation rate (Furusawa and Kaneko, 1998a, 2001).

Robustness

Our simulation and theory account for robustness at two different levels. First the developmental process in our example is robust to molecular fluctuations. Second, it is robust to macroscopic perturbations such as somatic mutations or the removal of some cells.

In general, when units with unstable dynamics interact with each other, there is stability at the ensemble level (Kaneko, 1992; Kaneko and Ikegami, 1992). In the present case, macroscopic stability is sustained by the change in the rate of differentiation, which is achieved by the change in the internal dynamics through the interaction (as already shown). But why is regulation oriented to preserving stability instead of undermining it? In the example of figure 12.1d, the rate of differentiation from S to A is increased when some of the type A cells are removed. Assume that regulation worked in the opposite way (i.e., that the rate of $S \to A$ was *decreased* by the removal of type A cells). Then, at the initial stage when type A cells were to be produced, their number would decrease to zero. Hence the type A cell would not appear from the beginning. In other words, only the cell types that have a regulatory mechanism to stabilize their coexistence with other types can appear in our developmental process.

In our scenario, global robustness has a higher priority than the local differentiation rule. Let us consider an example from real life. Whereas, in the newt, the cells of a triploid type are three times as large as those of the wild type, the total number of its cells is only one-third the normal number and its body does not appear to be much affected by the condition. In other words, the local rule of cell division (here, the number of divisions) is modified so that the global body pattern remains undamaged (Alberts et al., 1993).

It should be stressed that our model's dynamical differentiation process is always accompanied by this kind of regulation, without any sophisticated programs implemented in advance. This autonomous robustness provides novel insights into the stability of the cell society in multicellular organisms.

Irreversibility in Development

In general, stability of a state and dynamical irreversibility are tightly interrelated. For a state to have stability, perturbation to it has to be damped, which implies the existence of irreversibility in the dynamics of the cellular state. Because, in actual organisms and in our model alike, there is a clear temporal flow (e.g., $U \to S \to D$) as development progresses, with the degree of determination normally increasing, how can one quantify such irreversibility? Thus far, no decisive answer is available, although we have found the following common tendencies in our model simulations of the process of $U \to S \to D$ in time (Kaneko and Furusawa, 2000):

1. Stability of intracellular dynamics increases;
2. Diversity of chemicals decreases;
3. Temporal variation of intracellular chemical concentrations is smaller, and intracellular dynamics become less chaotic.

The degree of tendency 1 could be determined by a minimum change in the interaction to switch a cell state, by properly extending the "attractor strength" introduced in Kaneko, 1997b. A cell in state U spontaneously changes its state without a change in the interaction term, whereas the S state can be switched by a tiny change in the interaction term. The degree of determination in the D state is roughly measured as the minimum perturbation strength required for a switch to a different state.

The diversity of chemicals tendency 2 can be measured, for example, by

$$-\sum_{j=1}^{k} p(j) \log p(j),$$

with $p(j)$ as the fraction of the concentration of chemical j. As cells are differentiated from stem cells and determined, this measure for the diversity decreases successively. Tendency 3 is also numerically confirmed by measuring an index characterizing the temporal variation of the chemical concentrations (Furusawa and Kaneko, 2001).

Origin of Recursive Units

In our studies, we took as given the existence of the cell itself, a unit separated from the outside by a membrane and dividing to produce daughters. At the next level, there is the problem of the origin of multicellular organisms, where an ensemble of cells forms a unit, which produces a similar unit, namely, the problem of recursive production at a higher level than the cell replication.

To study this problem, Chikara Furusawa and I (Furusawa and Kaneko, 1998b) extended the intra-inter dynamics to incorporate spatial structure. First, we assumed that units were located in two-dimensional space and interacted locally through the environment, which could change in space and in time. Second, we assumed that units moved in space passively and randomly and that they were subject to fluctuation. Third, we assumed that adhesion forces between the cells kept cell aggregates from disintegrating and that these adhesion forces could depend on the character of the cells (i.e., chemical composition). In terms of cellular process, we assumed only diffusion of chemicals through the cell membrane, Brownian motion, and cell adhesion force.

We carried out several numerical experiments for this class of models. Again, up to some number, cells were identical. With the division process, cells formed an aggregate. Then the cells started to differentiate according to the mechanism mentioned already, forming some spatial structure, such as a ring or a stripe pattern. For example, one type of cells was located at the middle, and two different types of determined cells lay at the opposite edge. As this stripe pattern was formed, positional information was given as gradients of some chemical concentrations (Furusawa and Kaneko, 2000, 2002). In contrast with the conventional positional information theory (Wolpert, 1969), this information was not externally given but was instead generated spontaneously through the interplay between intracellular dynamics and cell-cell interaction. Note also that the pattern here was formed only with an increase of cell numbers, in contrast with the Turing pattern.

When all the cell types adhered to each other, the cell aggregate could not split; the inner core no longer had access to resources. The growth rate of the cell cluster decreased and finally the cluster stopped growing (in most cases).

With the adhesion force dependent on cell type, the cell aggregate could arrive at a novel stage (see figure 12.3 for schematic representation). Because some types of cells did not adhere to each other, a few type S cells were released from the aggregate. Due to Brownian motion, the released cells moved away from the original aggregate. The chemical interaction between the released and the original cells became negligible. The released cells encountered a new environment with rich chemical substances, and started to divide actively. They formed an aggregate with a similar spatial pattern and with the same differentiated cell types as the original cell aggregate. Again, some type S cells were released from this "daughter aggregate." Hence a new generation of a multicellular unit was formed from the mother aggregate. At the same time, after some releases of cells, the cell division process at the original cell aggregate grew slower, until only inactive cells without division were

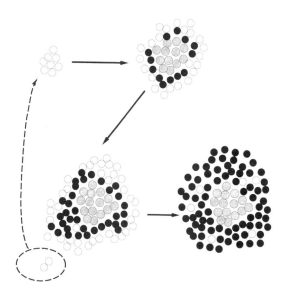

Figure 12.3
Schematic illustration of the emergence of life cycle and the multicellular organism. For simulation results, see Furusawa and Kaneko, 1998b.

left. In this sense, the aggregate arrived at a halting (or dead) state, determining the lifetime of the replicating multicellular unit. Because the new generation always started from type S cells, the other D cell types (such as types A and B) did not affect the next generation. Thus separation of germ cells and a closed life cycle emerged together.

Hence a life cycle of multicellular replicating units emerged without explicit implementation. Note that this emergence of replicating cell societies was a natural consequence of a system with internal cellular dynamics with nonlinear oscillation, cell-to-cell interaction through a medium, and adhesion dependent on cell type.

At the first stage of multicellularity in evolution as we observed it in our model simulations, two daughter cells failed to separate after division, and a cluster of identical cell types was formed. The differentiation of cells generally followed if they continued dividing. Then the multicellular cluster had to release its active cells before the cells lost the capability of cell division. Thus the simultaneous emergence of germ cells and a closed life cycle was a natural consequence of the system described here.

Of course, there can be other scenarios for the origin of multicellular organisms. We have studied other possibilities by introducing cell death and active motion, instead of passive Brownian motion. From some preliminary studies, these processes also lead to the emergence of life cycles. In evolution, there are at least three routes for multicellularity, each of which may involve a different splitting mechanism.

Evolution with Developmental Plasticity

Relevance of Phenotype Differentiation to Evolution

One might wonder what the role of genes is in our cell differentiation scenario. Because genes are chemicals, our scenario does not contradict the conventional view of differentiation by genetic switches. In fact, we have often observed that some chemicals exhibit an on-off type behavior in our model.

Still, genes are generally believed to have a rather special role among the intracellular chemicals, working as controllers over other chemicals. They change to a much smaller extent than other chemicals. In terms of dynamical systems, they act as the control parameter on the other variables. Assuming this distinction between parameters and variables, let us discuss the inevitable consequences of the interaction-induced phenotype differentiation (isologous diversification) for the problem of evolution.

According to the standard view of evolution in contemporary biology (see Futuyma, 1986; Alberts et al., 1994): (1) each organism has a genotype and a phenotype; (2) fitness for survival is given for a phenotype, and the Darwinian natural selection process acts for the survival of organisms, having a higher fitness; (3) only the genotype is transferred to the next generation (Weismann's doctrine); and (4) Finally, there is unidirectional flow of information from the genotype to phenotype (the central dogma of molecular biology), i.e., the phenotype is determined through developmental processes, given a particular genotype and environment. With reservations about points 3 and 4 in certain cases, our scenario follows this standard view.

Note, however, that point 4 does not necessarily mean that the phenotype is uniquely determined from a given genotype and environment. Although standard population genetics assumes that it is, the standard view summarized above does not. Indeed, there are three reasons to doubt the assumption of uniqueness, one theoretical and two experimental.

First, and according to our isologous diversification theory as discussed thus far, two or more groups with distinct phenotypes can appear from the same genotype. Because orbital instability in the developmental process amplifies small differences (or fluctuations) to the macroscopic level, the dynamical state of two organisms (cells) can be different, even if they have the same set of genes. The organisms are differentiated into discrete types through the interaction, where the existence of each type is necessary to eliminate the dynamic instability in the developmental process that exists when one of the types is isolated. Hence the existence of each type is required for the survival of the others, even though every individual has an identical or only slightly different genotype.

Second, it is well known experimentally that in some mutants, various phenotypes may arise from a single genotype (Holmes, 1979), a phenomenon known as "low" or "incomplete penetrance" (Opitz, 1981). One of the more prominent headaches of genetics, alleles of low penetrance exist even in *Caenorhabditis elegans*.

Third, in an experimental demonstration of interaction-induced phenotypic diversification in specific mutants of *Escherichia coli* having identical genes, Ko, Yomo, and Urabe (1994) showed (at least) two distinct coexisting types of enzyme activity in the well-stirred environment of a chemostat. The coexistence of each type was supported by each other; indeed, when one type of *E. coli* was removed externally, the remaining type differentiated to recover the coexistence of the original two types.

Modeling Phenotypic Differentiation

How does the problem of interaction-induced phenotypic differentiation from the same genotype bear on evolution? To explore the relationship between genotype and phenotype, we need to consider a developmental process that maps a genotype to a phenotype. Let us consider an abstract model consisting of several biochemical processes. Each organism possesses internal dynamic processes that transform external resources into certain products depending on the internal dynamics. Through these processes, organisms mature and eventually become ready for reproduction.

Here, the phenotype is represented by a set of state variables. For example, each individual i has several cyclic processes $j = 1, 2, \ldots, k$, whose state at time t is denoted by $X_t^j(i)$. With k such processes, the state of an individual is given by the set $(X_t^1(i), X_t^2(i), \ldots X_t^k(i))$, which defines the phenotype. This set of variables can be regarded as concentrations of chemicals, rates of metabolic processes, or some quantity corresponding to a higher function. The state is not fixed in time, but changes temporally according to a set of deterministic equations with some parameters.

Genes, which are nothing more than information encoded in DNA, could in principle be included in the set of variables. However, according to the central dogma of molecular biology (see point 4 of standard view of evolution summarized above), the gene has a special role among such variables. Genes can affect phenotypes, the set of variables, but the phenotypes cannot change the code of genes. During the life cycle, changes in genes are negligible compared with those of the phenotypic variables they control. The set corresponding to genes can be represented by parameters $\{g^1(i), g^2(i), \ldots g^m(i)\}$ that govern the dynamics of phenotypes. Because the parameters are not changed through the developmental process and control the dynamics of phenotypic variables, let us represent the genotype by a set of parameters, which changes slightly by mutation, however, when an individual organism is reproduced.

For example, let us adopt a model where each unit represents an abstract cell, whose variables change according to some dynamical equation governed by parameters. (Although our model is directed to unicellular organisms, the core idea here can be applied to multicellular organisms). Indeed, Tetsuya Yomo and I (Kaneko and Yomo, 2000, 2002) have carried out several model simulations with the following specifications:

1. *Dynamical change of states generating a phenotype.* The temporal evolution of the state variables $(X_t^1(i), X_t^2(i), \ldots X_t^k(i))$ was given by a set of deterministic equations, which were described by the state of the individual, and parameters $\{g^1(i), g^2(i), \ldots g^m(i)\}$ (genotype), and the interaction with other individuals.

2. *Interaction between the individuals.* Through the set of variables $(X_t^1(i), X_t^2(i), \ldots X_t^k(i))$, individuals interacted with all others through competition for resources, which were used by all the individuals. To show sympatric speciation clearly, we took this extreme all-to-all interaction in a well-stirred soup of resources, excluding any spatially localized interaction.

3. *Mutation.* Each individual split into two when a given condition for growth was satisfied. When the unit reproduced, the set of parameters $\{g^j(i)\}$ changed slightly by mutation, by adding a random number with a small amplitude δ, corresponding to the mutation rate.

4. *Death.* To avoid a population explosion and to ensure competition, individuals died according to a certain death rate.

Scenario for Sympatric Speciation

From several simulations, we obtained a scenario for a sympatric speciation process, as schematically shown in figure 12.4, which, in general terms, represents the results we obtained when a phenotypic variable (P) and a genotypic parameter (G) were plotted at every cell division event in several simulations (Kaneko and Yomo, 2000, 2002). The scenario is summarized as follows.

Stage I: Interaction-Induced Phenotypic Differentiation When there are many individuals interacting for finite resources, the phenotypic dynamics starts to be differentiated even though the genotypes are identical or differ only slightly. Phenotypic variables split into two or more types. This interaction-induced differentiation is an outcome of the mechanism presented in previous sections. Slight phenotypic differences between individuals are amplified by the internal dynamics, whereas through the interaction between organisms, the differences in phenotypic dynamics tend to be clustered into two or more types. Note that the differences are fixed at this stage neither at the genotypic nor phenotypic level. After reproduction, an individual's phenotype can switch to another type. With this interaction-induced differentiation, the phenotypes of individuals split into two groups, called "upper" and "lower" groups, in (figure 12.4b).

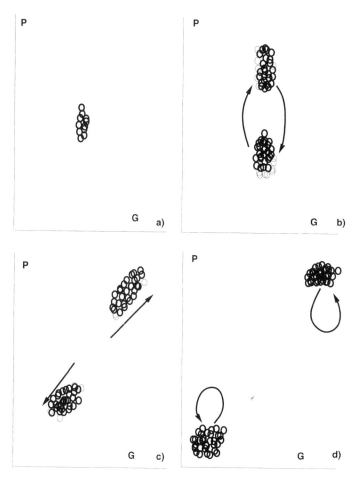

Figure 12.4
Schematic representation of genetic speciation through interaction-induced phenotypic differentiation. For simulation results, see Kaneko and Yomo, 1999.

Stage II: Amplification of the Difference through Genotype-Phenotype Relationship

At the second stage, the differences between the two groups are amplified both at the genotypic and at the phenotypic level. This is realized by a kind of positive feedback process between the change of genotypes and phenotypes.

First the genetic parameters separate as a result of the phenotypic change. This occurs if the parameter dependence of the growth rate is different for the two phenotypes: there are one or several parameters g^j, such that the growth rate increases with g^j for the upper group and decreases for the lower group (or the other way around).

Such parameter dependence is by no means exceptional. As a simple illustration, assume that the use of metabolic processes is different between the two groups. If the upper group uses one metabolic cycle more than the lower group, then the mutational change of the parameter g^l to enhance the cycle favors the upper group, and vice versa. Indeed, several numerical results support the existence of such parameters.

This dependence of growth rate on the genotypes leads to the genetic separation of the two groups, as long as there is competition for survival, to keep the population numbers limited.

Genetic separation is often accompanied by a second process, the amplification of the phenotypic difference by the genotypic difference. In the situation illustrated in figure 12.4, as parameter g^l increases, phenotype variable X (i.e., a characteristic quantity for the phenotype) tends to increase for the upper group, and to decrease or remain the same for the lower group. It should be noted that this second stage is always observed in our model simulation when phenotypic differentiation occurred at the first stage.

Stage III: Genetic Fixation After the two groups have separated, each phenotype and each genotype start to be preserved by the offspring, in contrast to the situation at the first stage. However, up to the second stage, the two groups with different phenotypes cannot exist in isolation. When isolated, offspring with the phenotype of the other group start to appear. The two groups coexist interdependently.

Only at this third stage does each group start to exist on its own. Even if one group of units is isolated, no offspring with the phenotype of the other group appear. Now the two groups exist on their own. Such a fixation of phenotypes is possible through the evolution of genotypes (parameters). In other words, the differentiation is fixed into the genes (parameters). Now each group exists as an independent "species," separated both genetically and phenotypically. The initial phenotypic change introduced by interaction is fixed in genes, and the "speciation process" is completed.

Speciation as Reproductive Isolation in the Presence of Sexual Recombination

The speciation process is defined by both genetic differentiation and by reproductive isolation (see Dobzhansky, 1937). Although simulating evolution through stages I–III led to genetically isolated reproductive units, the term *speciation* should not be used unless these units formed isolated reproductive groups under sexual recombination. Indeed, because the parental genotypes are mixed by recombination, one might wonder whether our proposed differentiation scenario works under sexual recombination.

On the other hand, because our scenario showed itself to be robust to perturbations, such as the removal of one phenotype group, it might be stable in the face of sexual recombination, which mixes the two genotypes and may bring about a hybrid between the two genotypes. To determine whether this was so, we extended the previous model to include the mixing of genotypes by sexual recombination. To be specific, when the threshold condition for reproduction of two individuals, i_1 and i_2, was satisfied, the two genotypes were mixed, producing two offspring, $j = j_1$ and j_2, with intermediate parameter values such that $g^l(j) = g^l(i_1)r_j + g^l(i_2)(1 - r_j) + \delta$, with a random number $0 < r_j < 1$ to represent mixing the parents' genotypes and a random mutation term, δ.

Even if two separated groups were formed according to our scenario, the above recombination could form "hybrid" offspring with intermediate parameter values g^l between the two groups. Despite this mixing by recombination, we found in our numerical simulations that two distinct groups were again formed, and that it became harder for an individual offspring of parents from the two distinct groups to breed true.

Of course, the mating between the two groups could produce an individual with parameters in the middle of the two groups. However, an individual with intermediate parameters between the two groups had a lower reproduction rate; indeed, before the reproduction condition was satisfied, there was a high probability that the individual would die.

As the two groups were formed with the split of the parameter values, the average number of offspring of an individual having the control parameter between those of the two groups began to decrease, and soon went to zero, implying that the hybrid between the two groups was sterile. Because, sterility or low reproduction of the hybrid appeared as a result, it is proper to call stages I–III "speciation" because they satisfy the criterion of genetic differentiation and reproductive isolation under sexual recombination.

Note that we did not assume any preference in mating choice. Rather, according to our proposed scenario, mating preference in favor of similar phenotypes "naturally" evolves because it is disadvantageous for individuals to produce a sterile hybrid. In other words, the present mechanism also provides a basis for the evolution of sexual isolation through mating preference (see also Lande, 1981; Turner and Burrows, 1995; Dieckmann and Doebeli, 1999; Kondrashov and Kondrashov, 1999). The premating isolation can evolve as a consequence of postmating isolation.

Implications for Evolution

Although our scenario is based on Darwinian selection and on the central dogma of modern biology (i.e., the parameters—the genes—can change the variables, but the variables cannot directly change the parameters), it makes clear that the gene cannot be a "cause" for the evolutionary process. Indeed in our model simulations, the speciation into two or more groups occurs *if and only if* the phenotypes are separated into two or more groups by interaction. Genes (parameters) become important later, to amplify and fix the differentiation.

Note that the interaction-induced phenotypic differentiation is deterministic in nature: the initial parameters of the model determine whether such differentiation will occur. The model's speciation process is therefore also deterministic in nature despite its stochastic mutation process. Once the condition for phenotypic differentiation is satisfied, speciation proceeds rapidly. Indeed, in the simulations we carried out, it is completed in about the first fifty generations. Our scenario may give a new and plausible explanation for the substantial variation in the rates of evolutionary change, specifically for, punctuated equilibrium (Gould and Eldredge, 1977).

In our scenario's speciation process, the potentiality for a single genotype to produce several phenotypes decreases. After the phenotypic diversification of a single genotype, each genotype again appears through mutation and assumes one of the diversified phenotypes in the population. Thus the one-to-many correspondence between the original genotype and phenotypes eventually ceases to exist. As a result, one might expect that a phenotype would be uniquely determined for a single genotype in wild types, because most present-day organisms have gone through several speciation processes. One might also expect that a mutant would tend to have a higher potentiality to produce various phenotypes from a single genotype. Hence our scenario suggests an explanation for why low or incomplete penetrance (Holmes, 1979; Opitz, 1981) is more frequently observed in mutants than in wild types.

Conclusion

Cell differentiation and the formation of cell societies are a natural consequence of reproducing, interacting units with internal dynamics, a universal phenomenon in a system having intra-inter dynamics. The multicellular organism with a life history is also a consequence of producing units with internal dynamics, which have mechanical and chemical interactions locally in space. The speciation process is also a natural consequence of a system having intra-inter dynamics and mutating parameters, where the interaction-dependent phenotypic differentiation is fixed in the genes.

To determine the validity of "necessity" claims, the study of existing organisms is not sufficient: it is hard to distinguish "chance" from "necessity." Only by constructing a biological system under artificial conditions, can we show that fine-tuning by evolution is not necessary to produce biological organization. We are hopeful that such experimental constructive biology (see Ko, Yomo, and Urabe, 1994; Xu et al., 1996; Prijambada et al., 1996) will confirm our scenario for biological organization.

It is often taken for granted that symbolic representations such as genes are fundamental, and that the combination of such representations leads to a complex organization of behaviors. This chapter has put forward another viewpoint. As the evolutionary scenario discussed above clearly demonstrates, interaction-dependent differentiation of behaviors can come first, and the Symbolic representations (types) and rules can be generated and fixed later.

Acknowledgments

This chapter is based on the studies undertaken in collaboration with Tetsuya Yomo and Chikara Furusawa. I am grateful to them for stimulating discussions. I am indebted to Roeland Merks for his critical reading of the manuscript. I thank the volume editors, Gerd Müller and Stuart Newman, for their kind invitation to contribute and their careful reading of my manuscript. Work on this chapter is supported in part by Grant-in-Aids for Scientific Research from the Ministry of Education, Science, and Culture of Japan (Komaba Complex Systems Life Science Project).

References

Alberts B, Bray D, Lewis J, Raff M, Roberts K, Watson JD (1994) The Molecular Biology of the Cell. New York: Garland.

Dieckmann U, Doebeli M (1999) On the origin of species by sympatric speciation. Nature 400: 354–356.

Dobzhansky T (1951) Genetics and the Origin of Species (2d ed). New York: Columbia University Press.

Dolmetsch R, Xu K, Lewis RS (1998) Calcium oscillations increase the efficiency and specificity of gene expression. Nature 392: 933–936.

Eigen M, Schuster P (1979) The Hypercycle. Berlin: Springer.

Furusawa C, Kaneko K (1998a) Emergence of rules in cell society: Differentiation, hierarchy, and stability. Bull Math Biol 60: 659–687.

Furusawa C, Kaneko K (1998b) Emergence of multicellular organisms: Dynamic differentiation and spatial pattern. Artif Life 4: 79–93.

Furusawa C, Kaneko K (2000) Origin of complexity in multicellular organisms. Phys Rev Lett 84: 6130–6133.

Furusawa C, Kaneko K (2001) Theory of robustness of irreversible differentiation in a stem cell system: Chaos hypothesis. J Theor Biol 209: 395–416.

Furusawa C, Kaneko K (2002) Generation of positional information and robust pattern formation. Submitted to J Theor Biol.

Futuyma DJ (1986) Evolutionary Biology. Sunderland, Mass.: Sinauer.

Goodwin B (1963) Temporal Organization in Cells. London: Academic Press.

Gould SJ, Eldredge N (1977) Punctuated equilibria: The tempo and mode of evolution reconsidered. Paleobiology 3: 115–151.

Greenwald I, Rubin GM (1992) Making a difference: The role of cell-cell interactions in establishing separate identities for equivalent cells. Cell 68: 271–281.

Gurdon JB, Lemaire P, Kato K (1993) Community effects and related phenomena in development. Cell 75: 831–834.

Hess B, Boiteux A (1971) Oscillatory phenomena in biochemistry. Annu Rev Biochem 40: 237–258.

Holmes LB (1979) Penetrance and expressivity of limb malformations. Birth Defects Orig Artic 15: 321–327.

Ikeda K, Otsuka K, Matsumoto K (1989) Maxwell-Bloch Turbulence. Prog Theor Phys 99 (Suppl): 295–324.

Kaneko K (1990) Clustering, coding, switching, hierarchical ordering, and control in networks of chaotic elements. Physica D 41: 137–172.

Kaneko K (1992) Mean field fluctuation in network of chaotic elements. Physica D 55: 368–384.

Kaneko K (1994) Relevance of clustering to biological networks. Physica D 75: 55–73.

Kaneko K (1997a) Coupled maps with growth and death: An approach to cell differentiation. Physica D 103: 505–527.

Kaneko K (1997b) Dominance of Milnor attractors and noise-induced selection in a multi-attractor system. Phys Rev Lett 78: 2736–2739.

Kaneko K (1998) Life as complex systems: Viewpoint from intra-inter dynamics. Complexity 3: 53–60.

Kaneko K (ed) (1993) Theory and Applications of Coupled Map Lattices. New York: Wiley.

Kaneko K, Furusawa C (2000) Robust and irreversible development in cell society as a general consequence of intra-inter dynamics. Physica A 280: 23–33.

Kaneko K, Ikegami T (1992) Homeochaos: Dynamic stability of a symbiotic network with population dynamics and evolving mutation rates. Physica D 56: 406–429.

Kaneko K, Tsuda I (1994) Constructive complexity and artificial reality: An introduction. Physica D 75: 1–10.

Kaneko K, Tsuda I (1996) Complex Systems: Chaos and Beyond (in Japanese). Tokyo: Asakura. (Revised English version, New York: Springer, 2000.)

Kaneko K, Yomo T (1994) Cell division, differentiation, and dynamic clustering. Physica D 75: 89–102.

Kaneko K, Yomo T (1997) Isologous diversification: A theory of cell differentiation. Bull Math Biol 59: 139–196.

Kaneko K, Yomo T (1999) Isologous diversification for robust development of cell societies. J Theor Biol 199: 243–256.

Kaneko K, Yomo T (2000) Sympatric speciation: Compliance with phenotype diversification from a single genotype. Proc Soc Lond B Biol Sci 267: 2367–2373.

Kaneko K, Yomo T (2002) Genetic diversification through interaction-driven phenotype differentiation. Evol Eco Res 4: 317–350.

Kauffman S (1969) Metabolic stability and epigenesis in randomly connected nets. J Theor Biol 22: 437–467.

Ko PE, Yomo T, Urabe, I (1994) Dynamic clustering of bacterial population. Physica D 75: 81–88.

Kondrashov AS, Kondrashov AF (1999) Interactions among quantitative traits in the course of sympatric speciation. Nature 400: 351–354.

Lande R (1981) Models of speciation by sexual selection on phylogenic traits. Proc Natl Acad Sci USA 78: 3721–3725.

Maynard-Smith J (1966) Sympatric speciation. Am Nat 100: 637–650.

Mjolsness E, Sharp DH, Reinitz J (1991) A connectionist model of development. J Theor Biol 152: 429–453.

Newman SA, Comper WD (1990) Generic physical mechanisms of morphogenesis and pattern formation. Development 110: 1–18.

Ogawa M (1993). Differentiation and proliferation of hematopoietic stem cells. Blood 81: 2844–2853.

Opitz JM (1981) Some comments on penetrance and related subjects. Am J Med Genet 8: 265–274.

Prijambada ID, Yomo T, Tanaka F, Kawama T, Yamamoto K, Hasegawa A, Shima Y, Negoro S, Urabe I (1996) Solubility of artificial proteins with random sequences. FEBS Lett 382: 21–25.

Till JE, McCulloch EA, Siminovitch L (1964) A stochastic model of stem cell proliferation, based on the growth of spleen colony-forming cells. Proc Natl Acad Sci USA 51: 29–34.

Tsuda I (1992) Dynamic link of memory: Chaotic memory map in nonequilibrium neural networks. Neural Networks 5: 313–326.

Turing AM (1952). The chemical basis of morphogenesis. Philos Trans R Soc Lond B Biol Sci 237: 37–72.

Turner GF, Burrows MT (1995) A model for sympatric speciation by sexual selection. Proc R Soc Lond B Biol Sci 260: 287–292.

Wolpert L (1969) Positional information and the spatial pattern of cellular differentiation. J Theor Biol 25: 1–47.

Xu W, Kashiwag A, Yomo T, Urabe I (1996) Fate of a mutant emerging at the initial stage of Evolution. Res Popul Ecol 38: 231–237.

13 From Physics to Development: The Evolution of Morphogenetic Mechanisms

Stuart A. Newman

Materials of the nonliving world take on forms dictated by an interplay between their inherent physical properties and external forces to which they are susceptible. Water, for example, forms waves and vortices when it is mechanically agitated, whereas clay bears the record of physical impressions long after they have been made. Living metazoans—multicellular animals—seem to obey different rules: their forms appear to be expressions of intrinsic genetic "programs." Aside from the exchange of energy and matter with the external world that organisms require to stay alive, the general plans and fine details of the forms they assume are taken to be independent of the external environment.

This chapter advances the hypothesis that the forms of the earliest multicellular organisms were not generated in such a rigid programmatic fashion, that they were more like certain materials of the nonliving world than are the forms of their modern, highly evolved counterparts, and that they were therefore almost certainly molded by their physical environments to a much greater extent than contemporary organisms. Only later, with the evolution of integrating, stabilizing, and canalizing biochemical circuitry, would generation of organismal form have become more autonomous and programmatic.

This hypothesis has important implications for morphological evolution: it affirms that whatever internal, programmed bases exist for form generation in modern organisms, the templates for these forms must have been established by conditional organism-environment interactions earlier in the history of the Metazoa. Stated simply, tissue forms emerged early and abruptly because they were physically inevitable—they were not acquired incrementally through cycles of random genetic change followed by selection. What was acquired through these classic Darwinian means were stabilizing mechanisms, which eventually moved biological morphogenesis beyond the realm of determination by generic physical processes. A "physicoevolutionary" approach (Newman, 1998a) thus provides a rational explanation for the complexity and "overdetermination" of morphogenetic processes in modern organisms. Discussions and examples of this approach can be found in Newman and Comper, 1990; Newman, 1992, 1993, 1994; Müller and Newman, 1999; and Newman and Müller, 2000.

Multicellular organisms first arose more than 700 million years ago. By approximately 540 million years ago, at the end of the "Cambrian explosion," virtually all the body plans seen in modern organisms already existed (Conway Morris, 1989, and chapter 2, this volume). In particular, metazoan bodies are characterized by axial symmetries and asymmetries, multiple tissue layers, interior cavities, segmentation, and various combinations of these properties. The organs of these organisms are organized in similar ways, on a smaller scale. Although the early world contained many unoccupied niches within which new organismal forms could flourish, this alone can neither account for the rapid profusion of body plans once multicellularity was established, nor for the particular forms these plans assumed.

Just as we recognize that nonliving materials such as liquids, clays, taut strings, and soap bubbles can take on only limited, characteristic arrays of shapes and configurations, it is reasonable to ask what the characteristic spectrum of forms would have been for ancient cell aggregates. Although such ancient aggregates would have resembled modern early-stage embryos in being composed of cells producing numerous proteins, some used intracellularly, others at the cell surface (providing the means for cell aggregation) and still others released, they would have lacked the highly integrated genetic mechanisms of pattern formation and morphogenesis by which modern organisms coordinate the generation of their forms. Rather, as we will see, many of the body and tissue forms we have come to associate with modern multicellular organisms were inherent in the physical makeup of their less rigidly programmed ancestors.

The Earliest Metazoa: Excitable Soft Matter

The hallmark of a liquid is the ability of its component parts to readily move past one another and thereby dissipate stresses, a property known as viscous flow. Elastic solids have the ability to recover their shape after deformations. Living tissues may exhibit both viscous flow and elasticity. In this sense they are similar to viscoelastic physical materials such as clay, rubber, lava, and jelly—what physicists refer to as "soft matter" (de Gennes, 1992). In many mature tissues, however, the ability to behave in a liquidlike fashion is curtailed by complexities of tissue architecture. Mature epithelial and epithelioid tissues, for example, contain cells that are firmly attached to one another by means of highly structured specializations such as desmosomes and adherens junctions, and therefore exhibit little or no fluidity; the fibrous or mineralized extracellular matrices of mature connective tissues permit little cell rearrangement.

In contrast, morphogenetically active tissues (those participating in embryonic development, regeneration, and neoplasia) often exhibit liquid behavior. The cells of morphogenetically active epithelioid tissues, for instance, are attached by relatively weak protein-protein interactions, and thus readily slip past one another. As a result, they display properties characteristic of liquids, such as fluidity, viscosity, and surface tension (Foty et al., 1996; Steinberg and Poole, 1982; see Steinberg, chapter 9, this volume). Connective tissues that are morphogenetically active also differ from their mature counterparts: their matrices are devoid of mineral, and relatively poor in structured fibrous materials. In such "mesenchymes," local rearrangements of both cells and matrix can occur, and these tissues may also exhibit liquidlike properties (Newman and Tomasek, 1996).

The single cells of the premetazoan world were of a composition, internal organization, and spatial scale for which the physics of soft matter was essentially inapplicable. But once cell aggregates appeared on the scene things changed. Lacking the complex specializations that restrict cell arrangement in mature tissues of modern organisms, these macroscopic parcels of matter would have behaved like liquids, spreading and exhibiting surface tension, diffusion gradients, and other properties of soft matter.

Physical scientists characterize materials that actively respond to their environment mechanically, chemically, or electrically (such as liquid crystal displays on electronic devices) as "excitable media" (Gerhardt, Schuster, and Tyson, 1990; Mikhailov, 1990; Starmer et al., 1993; Winfree, 1994). Because even the earliest multicellular organisms were composed of chemically active cells with the ability to respond to their microenvironments by exhibiting motile activity and expressing new arrays of gene products, metazoan progenitors were excitable media from the start. The cells of these primitive aggregates were highly evolved metabolically, with complex biochemical and genetic networks; they were open and responsive to the external environment; and they were capable of "self-organizing" dynamical activities under the appropriate circumstances.

For instance, positive and negative feedback loops of chemical reactivity, when confined to the interior of an individual cell, will often lead to temporal oscillations in one or more chemical component (Goldbeter, 1996). Oscillations of glycolytic intermediates, for example, were identified as dynamical curiosities in yeast more than forty years ago. Functional roles for these periodic activities are obscure, and may not even exist. It is clear, however, that they can readily arise as self-organizing "side effects" of the metabolic circuitry (Elowitz and Leibler, 2000), rather than as the expression of an evolved program.

Although soft matter and excitable media do not, in general, coincide, in the parcels of tissue that constituted the most primitive metazoans the defining properties of both were found together in the same material. And because excitable soft matter has "generic" (characteristic, physically inevitable) properties that can be predicted and analyzed by standard physical theories and methods, we can make a number of general statements about the kinds of forms that would have been readily assumed by this new kind of material.

The Physics of Cell Aggregates and the Origin of Body Plans

The following subsections briefly explore the consequences of some of the physical attributes of cell aggregates as excitable soft matter and suggest how the morphological characteristics of metazoan body plans may have emerged from these properties.

Diffusion and the Formation of Spatial Gradients

The advent of multicellularity opened up possibilities for the molding of biological form that were unavailable to single-celled organisms. The primary reason for this is that different sets of physical forces predominate at different spatial scales—the shapes and forms of macroscopic objects, such as multicellular aggregates, are influenced by physical determinants different from those that noticeably affect microscopic objects, such as individual cells.

On the scale of a cell aggregate, the formation of gradients of released molecules is fostered, rather than undermined, by diffusion (Crick, 1970). A group of cells in one region of an aggregate that releases a product at a higher rate than its neighbors—either by spontaneous, stochastic effects, or because it is induced by an external cue to do so—can take on a privileged, organizing role in the aggregate, particularly if an effect of the product is to inhibit surrounding cells from making the same thing. The "organizer" cells do not need to be predetermined (although in certain embryos maternal determinants will bias their location): once they are established, the diffusion gradients they set up will have global patterning consequences. Although generation of molecular gradients by diffusion-dependent processes is used widely in contemporary metazoan patterning, the basic ingredients were present even in the most primitive cell aggregates.

Differential Adhesion and Compartment Formation

Unlike modern organisms, which have numerous highly evolved regulatory mechanisms devoted to controlling the precise strength of intercellular adhesion, the earliest cell aggregates would have been novices in the regulation of cell-cell interactions. Cell stickiness, with little evolutionary history behind it (by some accounts, it may even have arisen as a result of changes in the ionic composition of seawater; Kazmierczak and Degens, 1986) is likely to have been less stringently regulated in the earliest metazoans than it is at present. But we know from experiments in which cell types with different amounts of adhesion molecules on their surfaces are mixed together that they will sort out into islands of more cohesive cells within lakes composed of less cohesive neighbors. Eventually, by random cell movement, the islands will coalesce and an interface become established, across which cells will not intermix (Steinberg, 1998; chapter 9, this volume). When two or more differentially adhesive cell populations are present within the same tissue mass (as they would have been in primitive metazoan ancestors), multilayered structures can form automatically, comprising nonmixing "compartments," distinct spatial domains within a single tissue, in which no interchange or mixing of cells occurs across the common boundary (Crick and Lawrence, 1975; Garcia-Bellido, 1975).

What is observed is similar to what happens when two immiscible liquids, such as oil and water, are poured into the same container. As long as the molecules that make up one of the liquids have a greater binding affinity for one another than they do for the molecules of the other liquid, phase separation will take place. This takes the form of an interface within the common fluid that neither type of molecule will cross. When oil and water are shaken together the molecules will move randomly by Brownian motion and eventually settle into the same phase-separated equilibrium configuration. For cells, the analogous source of random motion is undirected locomotion. When sufficient differential adhesion exists between two cell populations, not only will each type of cell keep to its own side of the interface, but, when dissociated and randomly mixed, the two populations will "sort out," much like the shaken mixture of oil and water, and for the same thermodynamic reasons (figure 13.1A).

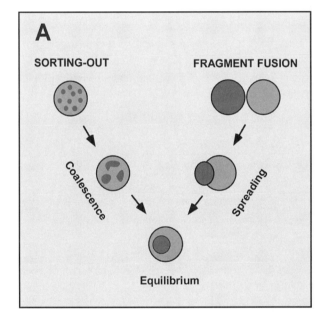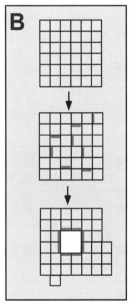

Figure 13.1
Differential adhesion can drive the sorting out of cells, as well as the formation of boundaries and lumina. (A) Schematic representation of the behavior of intermixed cells and corresponding tissue fragments in two differentially adhesive cell populations. The cell mixture will sort out as the more adhesive cells establish more stable bonds with one another than with cells of the other population. Random motion leads to the formation of cohesive islands of these cells, and these will ultimately coalesce into a separate tissue phase or compartment. The equilibrium configuration of the cell mixture is identical to that which would be formed by fusion and spreading of fragments of tissue consisting of the same differentially adhesive cell populations. (B) Schematic view of formation of a lumen or internal cavity by differential adhesion in an epithelioid tissue consisting of polarized cells. In the original state (*top*), the cells are uniformly adhesive, and make contacts around their entire peripheries. Upon expression of an antiadhesive protein (gray bars) in a polarized fashion in a random subpopulation of cells (*middle*), followed by random movement of the cells, bonds between adhesive surfaces are energetically favored over those between adhesive and nonadhesive surfaces, resulting in lumen formation (*bottom*). (*A* adapted from Steinberg, 1998; *B* adapted from Newman, 1998a.)

It is important to recognize that, whereas compartments in the developing embryos of modern organisms are typically allocated with precision by spatially distributed signals based on molecular gradients, because the sorting-out process will generally bring the cells of similar adhesive state to one side or another of a common boundary, even random assignment of cells to distinct adhesive states can result in a compartmentalized tissue. Thus compartmentalization could have arisen rather early in multicellular evolution as a result of stochastic processes within cell aggregates, and only later come under the control of biochemically sophisticated regulatory processes.

The presence of nonmixing tissue layers within a common cell aggregate would have represented a primitive form of gastrulation, achieved by physical, rather than genetically programmed means. In such early forms, the spontaneous multilayering would have provided a template upon which different cell lineages within early metazoans (originally differing only by adhesivity), can have taken independent evolutionary trajectories, leading to cell type specialization and differentiation. Such a central role of generic physical properties in the early evolution of the metazoan body plan clearly contrasts with scenarios based on adaptations resulting from competition between specialized cell populations and among the organisms that contain them (Buss, 1987).

Cell Polarity and Lumen Formation

Although some epithelioid cells, such as the blastomeres of the early mammalian embryo, have uniform adhesive properties around their entire surfaces, many epithelioid and all epithelial cells are polarized in the expression of several functions, notably adhesion (Rodriguez-Boulan and Nelson, 1989).

The targeting of adhesive or antiadhesive molecules to specific regions of the cell surface can have dramatic consequences. A tissue mass consisting of motile epithelioid cells that are nonadhesive over portions of their surfaces will readily develop cavities or lumina (figure 13.1B). As a result of random cell movement or the loss of cells next to the nonadherent surfaces of neighboring cells, such spaces will come to adjoin one another. Lumen formation could therefore have originated as a simple consequence of differential adhesion in cells that expressed adhesive properties in a polarized fashion.

The notion that lumen formation is a consequence of a delicate balance of adhesive interactions between cells, their extracellular substrata, or both is supported by experimental studies and genetic analyses of various developmental and pathological conditions. Recent studies of human autosomal dominant polycystic kidney disease (ADPKD) indicate that the formation of cysts, rather than tubules, in the kidneys of severely affected patients, involves the expression of mutated forms of polycystin-1 (Qian et al., 1996), an integral membrane glycoprotein that normally forms complexes with the cell-cell adhesive protein E-cadherin in kidney epithelial cells (Huan and van Adelsberg, 1999). Mammary epithelial cells when grown on tissue culture plastic in the absence of extracellular matrix adopt a flat "cobblestone" appearance. In the presence of laminin, however, they round up and cluster, and depending on the culture conditions, may form hollow, alveolar structures with well-defined apical and basal surfaces (Li et al., 1987). Here formation of lumina in polarized cells depends, not on a sorting process per se, but on apoptosis in the interior of the aggregate, apparently brought on by abrogated adhesion to the polarized surface layer (Lund et al., 1996; Bissell et al., chapter 7, this volume).

Although cell polarity would have had functional consequences even for free-living cells, only with multicellularity would its morphological consequences have been manifested. As a potential disrupter of the cell-cell adhesive interactions that define the multicellular state, cell polarity set the stage for lumen or cavity formation. The conjunction of cell polarity and differential adhesion can properly be identified as the basis of the critical epithelioid-epithelial transition.

In contemporary organisms, other cellular mechanisms, such as the contraction of apical actin filaments in a group of cells in a localized domain of an epithelial sheet, undoubtedly contribute to and may even initiate lumen formation. But this mechanism also requires polarized cells as well as a global pattern formation system to specify the position of the contracting domain. The latter feature most likely arose later in evolution than the emergence of the first hollow tissue formations.

Excitability and Segmentation

Because individual cells are metabolically active, thermodynamically open systems, tissues composed of them are excitable media, as noted above. The capacity to generate oscillations of gene products or metabolites, and to have such oscillations triggered by external effects, is a prime example of such excitability, which can be exhibited at the level of a single cell or of a cell aggregate.

The cell division cycle is a temporally periodic process driven by a chemical oscillation, but one that has no *morphological* consequence in the world of single cells: for free-living cells, cell division simply leads to more cells, no matter what its temporal dynamics may be. Even in a multicellular entity the division cycle typically acts only to increase the mass of the undifferentiated aggregate. But in a multicellular entity which contains an additional biochemical oscillation with a period different from the cell cycle, populations of cells will be generated periodically with distinct, recurrent, initial compositions. The morphological consequences of such metabolic excitability becomes clear when we look at what happens if the periodic expression of some molecule becomes linked to the cell's adhesive properties.

Let us consider a synchronized population of cells that divide at regular intervals, with the number of adhesive molecules on the surfaces of these cells set at the time of mitosis as a function of the cellular concentration of a regulatory molecule, R, and with each cell retaining its adhesive state throughout its lifetime. If the concentration of R were to oscillate with the same period as that of the cell cycle, each cell would be born with the same adhesive state: the tissue so generated would be adhesively uniform. If, however, the period of the R oscillation were different from that of the cell cycle, successive populations of cells would be born with different adhesive states: the phase relation between cell cycle and regulatory oscillator would ensure that adhesive states periodically recurred, and a spatial array of alternating tissue compartments or segments would form (figure 13.2A; Newman, 1993).

It is therefore of considerable interest that *c-hairy1,* an avian homologue of the *Drosophila* segmentation gene *hairy,* is expressed in the paraxial mesoderm of avian embryos in cyclic waves whose temporal periodicity, though different from the cell cycle, corresponds to the formation time of one somite (Palmeirim et al., 1997; see Pourquié, chapter 11, this volume). Because the *hairy* molecular clock is upstream of the Notch-Delta signaling pathway (McGrew et al., 1998), which in turn is implicated in the regulation of cell adhesion (Lieber et al., 1992) and compartmental boundary formation (Micchelli and Blair, 1999) in *Drosophila,* it corresponds to the factor R in the segmentation model described above. Clearly, a periodic, adhesivity-regulating factor of this type could have been involved in the origin of segmental organization during any of the the multiple times it arose in the history of the Metazoa.

In this regard, it has seemed puzzling that evolutionarily related organisms such as beetles ("short germ band" insects) and fruit flies ("long germ band" insects) have apparently different modes of segment formation. For example, in short germ band insects (Patel, Kornberg, and Goodman, 1989) and crustaceans (Itow, 1986), segmental primordia are added in sequence from a zone of cell proliferation. In contrast, in long germ band insects, such as *Drosophila,* a series of "chemical stripes," consisting of alternating evenly spaced bands of transcription factors in the syncytial embryo, ensures that when cellularization finally takes place, the cells of the blastoderm have periodically distributed identities. When these covert cell identities are later transformed into states of differential adhesivity (Irvine and Wieschaus, 1994), overt morphological segments are formed.

Treating tissues as excitable media can unify our understanding of these phenomena (Newman, 1993). The kinetic properties that give rise to a chemical oscillation (which most simply arises from what mathematicians refer to as the "Hopf instability") can also, when one or more of the components are diffusible, give rise to standing or traveling spatial periodicities of chemical concentration via the "Turing instability" (Boissonade, Dulos, and DeKepper, 1994). Such reaction-diffusion coupling represents another means by which the earliest metazoans could have acquired complex patterns (see Nijhout, chapter 10, this volume).

In the syncytial embryo of long germ band insects such as *Drosophila,* several of the pair-rule gene products that become organized into the early chemical stripes freely diffuse

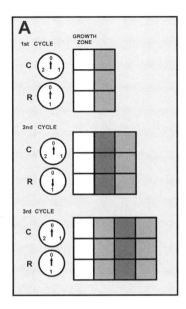
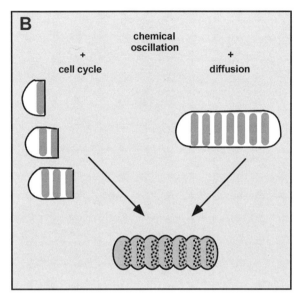

Figure 13.2
Chemical oscillation can drive sequential or simultaneous segmentation. (A) Model for the generation of segments in a zone of synchronized cell multiplication, by the temporal oscillation of the concentration of a molecule (e.g., en, ftz, hairy) that regulates expression of a cell adhesion molecule. The clock faces represent the phase of the cell cycle (C) and that of the periodically varying regulatory molecule (R). The duration of the cell cycle is assumed to be three hours, the period of the chemical oscillation two hours, with both cycles starting together. During the first cell cycle, newly formed cells have a level of cell adhesion specified by the initial value of R (light gray). During the second cell cycle, R is in midcycle, and the newly formed cells have a different level of cell adhesion (dark gray). During the third cell cycle, R is again at its initial concentration, and the new cells have the first level of cell adhesion. The alternation of adhesive states can proceed indefinitely. The assumption of cell synchrony is for simplicity; the mechanism would also give rise to segments in a zone of asynchronous cell multiplication with local cell sorting. (B) Schematic representation of two modes of tissue segmentation that can arise when the tissue's cells contain a biochemical circuit that generates a chemical oscillation or "molecular clock," and the oscillating species directly or indirectly regulates the strength or specificity of cell adhesivity. (*Left*) The periodic change in cell adhesivity occurs in a growth zone in which the cell cycle has a different period from the regulatory oscillator; as in (A). (*Right*) One or more of the biochemical species diffuse, leading to a set of standing waves of concentration of the regulatory molecule by a reaction-diffusion mechanism, and then to the simultaneous formation of bands of tissue with alternating cohesive properties. (*A* adapted from Newman, 1993; *B* adapted from Newman and Tomasek, 1996.)

among the cell nuclei that synthesize their messenger RNA (mRNA). Some of these also positively regulate their own synthesis (Harding et al., 1989; Ish-Horowicz et al., 1989), a sine qua non of both chemical oscillators and Turing pattern-forming systems.

Once a successful morphological motif had been established in a particular taxonomic group, the developmental mechanisms by which it had been ontogenetically achieved would have undergone stabilizing evolution (see below), becoming more complex on the molecular level. That striped expression of pair-rule genes in modern *Drosophila* can involve ornate systems of multiple promoter elements responsive to preexisting, nonuniformly distributed molecular cues (e.g., maternal and gap gene products; Goto, MacDonald, and Maniatis, 1989; Small et al., 1991) is thus not inconsistent with this pattern having originated as a Turing-type process (figure 13.2B). Although, absent any direct evidence about ancient mechanisms of segmental pattern formation, we can only hypothesize that such was the case, it should be noted that, with the reaction-diffusion mechanism, chemical stripe formation is achieved simply, with a minimum of molecular ingredients and physical processes. Because periodicities are a virtually inevitable consequence of such systems, reaction-diffusion processes represent a plausible basis for the origination of segmentation in a syncytial setting. Indeed, recent computer simulations of the evolution of segmentation suggest that morphogenetically prolific genetic mechanisms would likely have been replaced by less prolific but more stable ones as evolution proceeded (Salazar-Ciudad, Newman, and Solé, 2001; Salazar-Ciudad, Solé, and Newman, 2001). In related organisms that did not have syncytial embryos, as we have seen, segmentation could have arisen from the phase offset between the cell cycle and a molecular clock (Newman, 1993; Salazar-Ciudad, Solé, and Newman, 2001).

Physical Processes and the Origins of Organogenesis

Once the major body plans were established, similar formative processes were played out on a less global scale in the generation of organ primordia and the establishment their characteristic structures. Pattern-forming mechanisms (such as the sorting-out and reaction-diffusion process discussed above) were probably first deployed when cells had only limited ability to differentiate and adhesion differences were among their few distinguishing properties, and could later have affected other tissue features when repertoires of cell types were more extensive.

Epithelial Tissues

A Turing-type reaction-diffusion mechanism apparently alters local adhesive properties in chicken skin epithelium (Jiang et al., 1999), giving rise to the epithelial placodes that provide the primordia for developing feathers (Chuong and Edelman, 1985a,b; Sengel, 1976). In fish skin, a similar mechanism controls pigment production, resulting in the formation of stripes or spots of color (Kondo and Asai, 1995). Comparative studies of epithelial development in butterflies and fruit flies have shown that the same set of underlying morphogens can regulate integumentary color patterning or tissue shape depending on the species (Carroll et al., 1994).

Depending on local and polar expression of adhesive molecules, epithelial sheets can be bent, everted, invaginated, or formed into placodes, cysts, or tubules (Newman, 1998a). Moreover, the physics of fluids (confined to a plane in this case) can account for many details of epithelial sheet morphogenesis (Gierer, 1977; Mittenthal and Mazo, 1983). For epithelial tissue primordia in which the cells have relative mobility (i.e., in early embryos), the physics of soft matter must contribute to the molding of epithelial organ form (Newman, 1998a). In ancient metazoans before the evolution of complex structures mediating cell-cell attachment, physical determinants likely set the structural templates for organs, as they did for body plans in an earlier era.

Mesenchymal and Connective Tissues

In contrast to epithelioid and epithelial tissues, in which cells directly adhere to one another over a substantial portion of their surfaces, mesenchymal and other connective tissues consist of cells suspended in an extracellular matrix (ECM). They are therefore subject to additional morphogenetic mechanisms that depend on changes in the distance between cells, the effects of cells on the organization of the ECM, and the effects of the ECM on the shape and cytoskeletal organization of cells, changes that do not occur in epithelioid tissue types (Newman and Tomasek, 1996).

As in epithelioid tissues, boundaries of immiscibility can occur in connective tissues: for example, at the interface between the flank of a developing vertebrate embryo and a limb bud emerging from the flank (Heintzelman, Phillips, and Davis, 1978). Because differential adhesion per se is not relevant to cell populations in which cells do not contact one another directly, a different explanation must be sought for immiscibility of various embryonic mesenchymes. Here the in vitro phenomenon known as "matrix-driven translocation" is helpful. Detailed investigation of this morphogenetic effect in nonliving ECM-based colloids has shown that subtle differences in the organization of molecular fibers can define distinct physical phases (Newman, 1998b; Newman et al., 1985, 1997; Forgacs et al., 1989, 1994). In principle, these phases can provide the basis for compartment formation in mesenchymal tissues.

In mesenchymal condensation, often a transient effect in development, mesenchymal cells initially dispersed in a matrix move closer to one another. Condensations generally progress to other structures, such as feather germs (Chuong, 1993), cartilage or bone (Hall and Miyake, 1992, 2000), or, after conversion to epithelium, kidney tubules (Ekblom, 1992).

Various cellular mechanisms have been suggested for condensation formation, including local loss of matrix materials, centripetal, chemotactically driven movement through the matrix, cell traction, and absence of cell movement away from a center (see Newman and Tomasek, 1996 for further references). In vitro studies of limb bud precartilage mesenchyme have provided evidence that mesenchymal condensation can be accounted for by local increases in adhesion proteins and random cell movement (Frenz et al., 1989; Frenz, Jaikaria, and Newman, 1989), in line with predictions from theoretical models (Glazier and Graner, 1993; Graner, 1993). Moreover, recent experiments have provided direct evidence that the pattern of condensations in vitro arises from a reaction-diffusion process leading to a patterned distribution of adhesion proteins (Miura, Komori, and Shiota, 2000; Miura and Shiota, 2000a,b) in agreement with earlier suggestions (Newman and Frisch, 1979; Leonard et al., 1991; Newman, 1996). It is therefore plausible that this self-organizing process was at the origin of the vertebrate limb skeletal plan, whose basic features are mathematically predictable from a reaction-diffusion model (Newman and Frisch, 1979).

The formation of elaborate cell-ECM adhesive structures in developmentally mature connective tissues permits physical forces originating within cells to contribute to tissue morphogenesis (reviewed in Newman and Tomasek, 1996). In particular, intracellular forces necessary for cell shape changes and migration can be imparted to the surrounding cells and ECM, resulting in mechanical stress in the tissue as a whole (Beloussov, Dorfman, and Cherdantzev, 1975; Grinnell, 1994; Ingber et al., 1994). Such cell-generated stresses can cause extracellular fibers or cytoskeletal filaments to contract, orient, or assemble (Forgacs, 1995; Halliday and Tomasek, 1995; Harris, Stopak, and Wild, 1981; Mochitate, Pawelek, and Grinnell, 1991; Nogawa and Nakanishi, 1987; Stopak, Wessells, and Harris, 1985; Sumida, Ashcraft, and Thompson, 1989; Tomasek et al., 1992). In such cases, because the cells are not independently mobile, the tissue no longer exhibits liquid-like behavior, acting instead like an (excitable) elastic medium.

The linking of cells in an elastic medium, which converts individual cell contractility into global mechanical stress, is another example of new physical processes coming into play in a multicellular setting. And because the resulting elastic media are also biochemically excitable, the stresses and strains generated within them can also regulate the active behavior of the subunits. Cells recognize and respond to mechanical stresses by changing their shape, growth, expression of specific gene products, and cytoskeletal organization (Grinnell, 1994; Ingber et al., 1994), as well as by remodeling their extracellular matrix (Lambert et al., 1992; Unemori and Werb, 1986). This response occurs in part by the transduction of mechanical forces into chemical signals within specialized structures such as the focal adhesion complex (Kornberg et al., 1992; McNamee, Ingber, and Schwartz, 1993; Schaller and Parsons, 1994; Seko et al., 1999).

The importance of mechanical stress in morphogenesis is seen in the role of exogenous mechanical loads in determining tissue pattern. It has long been known, for instance, that the organization of the cardiovascular system is influenced by mechanical forces arising from blood pressure and flow (Russell, 1916). In tendons, the fibroblasts and the extracellular matrix they produce appear to be modulated according to their mechanical status (Vogel and Koob, 1989). For example, in regions of tendons that wrap around bone, compressive forces lead to the expression of the proteoglycans aggrecan and biglycan and the formation of fibrocartilage (Evanko and Vogel, 1993). Similarly, the ECM is deposited along lines of eventual tension and compression in bone (Koch, 1917). Interestingly, however, the ECM organization first arises during embryonic development, before substantial mechanical stresses are placed on the bones. While some stress undoubtedly is generated during embryonic growth (Herring, 1993), it is also possible that this is an example of the phenomenon alluded to earlier, in which biochemical circuitry evolved that stabilized or reinforced an outcome originally generated by external forces. Such pathways may come to be triggered earlier in the life history of the organism than the stage at which the external forces originally acted.

Because connective tissue and tendons have the capacity to react to biomechanical stimuli by forming cartilage and bone, mechanical stress can also provide the basis for morphological innovation during evolution. For example, altered stresses on embryonic connective tissue and tendon insertions arising as a consequence of changes of bone proportions can generate novel, sesamoid skeletal structures (Müller and Streicher, 1989; Streicher and Müller, 1992). Such structures represent interaction-dependent morphological templates that, with subsequent genetic evolution, can become assimilated into the developmental repertoire, which is to say, "autonomous" (Müller and Newman, 1999; Müller, chapter 4, this volume).

Interplay of "Generic" and "Programmatic" Mechanisms of Development

The foregoing discussion supports the hypothesis that metazoan organisms look the way they do in a general sense, not because of specific evolved adaptations, but because of the combined effects of the various physical properties that were generic to the earliest multicellular aggregates, presumed to be chemically excitable, viscoelastic "soft matter." These properties and the associated physical processes made a profusion of multilayered, hollow, segmented forms virtually inevitable early in the history of metazoan life. During later stages of evolution, these same and additional physical processes decreed the formation of organs and appendages consisting of lobes and lobules, tubules, sacs, nodules, and rods.

Thus, even though the somatic organization of ancient organisms resembled in many respects that of their modern counterparts, their developmental modes and mechanisms were, by this hypothesis, profoundly different. In particular, many of the earliest organisms are likely to have exhibited multiple, interconvertible forms by virtue of the conditional and interactive nature of the physical forces that molded them (Newman and Müller, 2000). Only with the subsequent evolution of genetic redundancy and biochemical integration (in the course of which forms that originated by physical processes were co-opted by "hardwired" genetic circuitry) would organisms of the more familiar modern variety have emerged: entities in which bodily form is achieved with decreased participation of external physical forces and increased dependence on routinized genetic control. Scenarios in which biological traits and properties originating in the interaction of the organism with its environment were later incorporated into its developmental repertoire through natural selection, were discussed by Baldwin (1902) and referred to by Simpson (1953) as the "Baldwin effect."

Certain theoretical considerations support the idea that morphological evolution can proceed from physically determined, and hence plastic, form generation to increased reliance on routinized expression of gene activities—what is sometimes simplistically referred to as "genetic programs." For example, dynamical systems theory, a branch of physics that describes the generic properties of complex networks of interacting components, has been used to demonstrate that cells comprising identical biochemical networks can exhibit differentiated states whose existence and stability depend on interactions among neighbors (Kaneko and Yomo, 1997; 1999; Kaneko, chapter 12, this volume). Although phenotypic plasticity and interaction-dependent stabilization could thus reasonably have provided a basis for cell type diversification among early multicellular forms, it is clear that modern organisms use a host of sophisticated chemical and structural mechanisms, some of them—"epigenetic" in the sense of Jablonka and Lamb, 1995—to stabilize the differentiated state. As opportunities for adaptation drove selection for functional integration and reproducibility, biochemical differentiation originally dictated by physical processes and interactions would over time have become transformed into developmentally reliable generation of sets of functionally coherent cell types.

In a parallel approach, my colleagues and I (Salazar-Ciudad, Newman, and Solé, 2001; Salazar-Ciudad, Solé, and Newman, 2001) have considered the pattern-generating capabilities of dynamical networks of interacting genes using computer simulation techniques. When such systems were allowed to undergo an evolutionary search for spatiotemporal patterns by mutation of parameters defining gene interactions, "emergent networks," which exhibited reciprocity of interactions and self-organizing behavior arrived at complex patterns earlier than "hierarchic networks," which were organized with unidirectional gene interaction (Salazar-Ciudad, Garcia-Fernandez, and Solé, 2000). However, complex patterns generated by hierarchically organized networks were evolutionarily more stable than the same patterns generated by self-organizing networks, indicating a tendency for genetic "rewiring" (Szathmary, 2001) to take place over the course of the evolution of development. A possible example of this is the case described above in which the striped pattern of expression of certain pair-rule genes in *Drosophila*, hypothesized to have originated as a self-organizing reaction-diffusion system in an ancestral form, evolved through promoter duplication and genetic circuit rewiring into a system in which individual stripes are cued by complexes of preestablished, graded factors (Newman, 1993; Salazar-Ciudad, Solé, and Newman, 2001).

According to the hypothesis outlined in this chapter, physical forces and processes largely irrelevant to producing the forms of individual cells were brought into play when life became multicellular. Initially, the forms assumed by these primitive cell aggregates were predictable, generic outcomes of the physics of excitable soft matter. Much of the history of multicellular life has involved the gradual replacement of physically based self-organization by hierarchically regulated, locally acting molecular interactions as the predominant basis of morphological development. If this hypothesis is correct, the challenge for both developmental and evolutionary biologists will be to experimentally disentangle and conceptually reintegrate these influences in modern organisms and their ancestors.

References

Baldwin JM (1902) Development and Evolution. New York: Macmillan.

Beloussov LV, Dorfman JG, Cherdantzev VG (1975) Mechanical stresses and morphological patterns in amphibian embryos. J Embryol Exp Morphol 34: 559–574.

Boissonade J, Dulos E, DeKepper P (1994) Turing patterns: From myth to reality. In: Chemical Waves and Patterns (Kapral R, Showalter K, eds), 221–268. Boston: Kluwer.

Buss LW (1987) The Evolution of Individuality. Princeton, N.J.: Princeton University Press.

Carroll SB, Gates J, Keys DN, Paddock SW, Panganiban GE, Selegue JE, Williams JA (1994) Pattern formation and eyespot determination in butterfly wings. Science 265: 109–114.

Chuong CM (1993) The making of a feather: Homeoproteins, retinoids and adhesion molecules. BioEssays 15: 513–521.

Chuong CM, Edelman GM (1985a) Expression of cell-adhesion molecules in embryonic induction: 1. Morphogenesis of nestling feathers. J Cell Biol 101: 1009–1026.

Chuong CM, Edelman GM (1985b) Expression of cell-adhesion molecules in embryonic induction: 2. Morphogenesis of adult feathers. J Cell Biol 101: 1027–1043.

Conway Morris S (1989) Burgess shale faunas and the Cambrian explosion. Science 246: 339–346.

Crick FHC (1970) Diffusion in embryogenesis. Nature 225: 420–422.

Crick FHC, Lawrence PA (1975) Compartments and polyclones in insect development. Science 189: 340–347.

de Gennes PG (1992) Soft matter. Science 256: 495–497.

Ekblom P (1992) Renal Development. New York: Raven Press.

Elowitz MB, Leibler S (2000) A synthetic oscillatory network of transcriptional regulators. Nature 403: 335–338.

Evanko SP, Vogel KG (1993) Proteoglycan synthesis in fetal tendon is differentially regulated by cyclic compression in vitro. Arch Biochem Biophys 307: 153–164.

Forgacs G (1995) On the possible role of cytoskeletal filamentous networks in intracellular signaling: An approach based on percolation. J Cell Sci 108: 2131–2143.

Forgacs G, Jaikaria NS, Frisch HL, Newman SA (1989) Wetting, percolation and morphogenesis in a model tissue system. J Theor Biol 140: 417–430.

Forgacs G, Newman SA, Polikova Z, Neumann AW (1994) Critical phenomena in model biological tissues. Colloids and Surfaces B: Biointerfaces 3: 139–146.

Foty RA, Pfleger CM, Forgacs G, Steinberg MS (1996) Surface tensions of embryonic tissues predict their mutual envelopment behavior. Development 122: 1611–1620.

Frenz DA, Akiyama SK, Paulsen DF, Newman SA (1989) Latex beads as probes of cell surface–extracellular matrix interactions during chondrogenesis: Evidence for a role for amino-terminal heparin-binding domain of fibronectin. Dev Biol 136: 87–96.

Frenz DA, Jaikaria NS, Newman SA (1989) The mechanism of precartilage mesenchymal condensation: A major role for interaction of the cell surface with the amino-terminal heparin-binding domain of fibronectin. Dev Biol 136: 97–103.

Garcia-Bellido A (1975) Genetic control of wing disc development in *Drosophila*. Ciba Found Sym 29: 169–178.

Gerhardt M, Schuster H, Tyson JJ (1990) A cellular automation model of excitable media including curvature and dispersion. Science 247: 1563–1566.

Gierer A (1977) Physical aspects of tissue evagination and biological form. Q Rev Biophys 10: 529–593.

Glazier JA, Graner F (1993) A simulation of the differential adhesion-driven rearrangement of biological cells. Phys Rev E 47: 2128–2154.

Goldbeter A (1996) Biochemical Oscillations and Cellular Rhythms: The Molecular Bases of Periodic and Chaotic Behaviour. Cambridge: Cambridge University Press.

Goto T, MacDonald P, Maniatis T (1989) Early and late periodic patterns of even-skipped expression are controlled by distinct regulatory elements that respond to different spatial cues. Cell 57: 413–422.

Graner F (1993) Can surface adhesion drive cell rearrangement? 1. Biological cell-sorting. J Theor Biol 164: 455–476.

Grinnell F (1994) Fibroblasts, myofibroblasts, and wound contraction. J Cell Biol 124: 401–404.

Hall BK, Miyake T (1992) The membranous skeleton: The role of cell condensations in vertebrate skeletogenesis. Anat Embryol Berl 186: 107–124.

Hall BK, Miyake T (2000) All for one and one for all: Condensations and the initiation of skeletal development. BioEssays 22: 138–147.

Halliday NL, Tomasek JJ (1995) Mechanical properties of the extracellular matrix influence fibronectin fibril assembly in vitro. Exp Cell Res 217: 109–117.

Harding K, Hoey T, Warrior R, Levine M (1989) Autoregulatory and gap gene response elements of the even-skipped promoter of *Drosophila*. EMBO J 8: 1205–1212.

Harris AK, Stopak D, Wild P (1981) Fibroblast traction as a mechanism for collagen morphogenesis. Nature 290: 249–251.

Heintzelman KF, Phillips HM, Davis GS (1978) Liquid-tissue behavior and differential cohesiveness during chick limb budding. J Embryol Exp Morphol 47: 1–15.

Herring SW (1993) Formation of the vertebrate face: Epigenetic and functional influences. Am Zool 33: 472–483.

Huan Y, van Adelsberg J (1999) Polycystin-1, the PKD1 gene product, is in a complex containing E-cadherin and the catenins. J Clin Invest 104: 1459–1468.

Ingber DE, Dike L, Hansen L, Karp S, Liley H, Maniotis A, McNamee H, Mooney D, Plopper G, Sims J, et al. (1994) Cellular tensegrity: Exploring how mechanical changes in the cytoskeleton regulate cell growth, migration, and tissue pattern during morphogenesis. Int Rev Cytol 150: 173–224.

Irvine KD, Wieschaus E (1994) Cell intercalation during *Drosophila* germ band extension and its regulation by pair-rule segmentation genes. Development 120: 827–841.

Ish-Horowicz D, Pinchin SM, Ingham PW, Gyurkovics HG (1989) Autocatalytic *ftz* activation and instability induced by ectopic *ftz* expression. Cell 57: 223–232.

Itow T (1986) Inhibitors of DNA synthesis change the differentiation of body segments and increase the segment number in horseshoe crab embryos. Roux's Arch Dev Biol 195: 323–333.

Jablonka E, Lamb MJ (1995) Epigenetic Inheritance and Evolution. Oxford: Oxford University Press.

Jiang T, Jung H, Widelitz RB, Chuong C (1999) Self-organization of periodic patterns by dissociated feather mesenchymal cells and the regulation of size, number, and spacing of primordia. Development 126: 4997–5009.

Kaneko K, Yomo T (1997) Isologous diversification: A theory of cell differentiation. Bull Math Biol 59: 139–196.

Kaneko K, Yomo T (1999) Isologous diversification for robust development of cell society. J Theor Biol 199: 243–256.

Kazmierczak J, Degens ET (1986) Calcium and the early eukaryotes. Mitt Geolog Paläontolog Inst Univ Hamb 61: 1–20.

Koch J (1917) The laws of bone architecture. Am J Anat 21: 177–298.

Kondo S, Asai R (1995) A reaction-diffusion wave on the skin of the marine angelfish *Pomacanthus*. Nature 376: 765–768.

Kornberg L, Earp HS, Parsons JT, Schaller M, Juliano RL (1992) Cell adhesion or integrin clustering increases phosphorylation of a focal adhesion-associated tyrosine kinase. J Biol Chem 267: 23439–23442.

Lambert CA, Soudant EP, Nusgens BV, Lapiere CM (1992) Pretranslational regulation of extracellular matrix macromolecules and collagenase expression in fibroblasts by mechanical forces. Lab Invest 66: 444–451.

Leonard CM, Fuld HM, Frenz DA, Downie SA, Massagué J, Newman SA (1991) Role of transforming growth factor-beta in chondrogenic pattern formation in the embryonic limb: Stimulation of mesenchymal condensation and fibronectin gene expression by exogenous TGF-beta and evidence for endogenous TGF-beta-like activity. Dev Biol 145: 99–109.

Li ML, Aggeler J, Farson DA, Hatier C, Hassell J, Bissell MJ (1987) Influence of a reconstituted basement membrane and its components on casein gene expression and secretion in mouse mammary epithelial cells. Proc Natl Acad Sci USA 84: 136–140.

Lieber T, Wesley CS, Alcamo E, Hassel B, Krane JF, Campos-Ortega JA, Young MW (1992) Single amino acid substitutions in EGF-like elements of Notch and Delta modify *Drosophila* development and affect cell adhesion in vitro. Neuron 9: 847–859.

Lund LR, Romer J, Thomasset N, Solberg H, Pyke C, Bissell MJ, Dano K, Werb Z (1996) Two distinct phases of apoptosis in mammary gland involution: Proteinase-independent and -dependent pathways. Development 122: 181–193.

McGrew MJ, Dale JK, Fraboulet S, Pourquié O (1998) The lunatic fringe gene is a target of the molecular clock linked to somite segmentation in avian embryos. Curr Biol 8: 979–982.

McNamee HP, Ingber DE, Schwartz MA (1993) Adhesion to fibronectin stimulates inositol lipid synthesis and enhances PDGF-induced inositol lipid breakdown. J Cell Biol 121: 673–678.

Micchelli CA, Blair SS (1999) Dorsoventral lineage restriction in wing imaginal discs requires Notch. Nature 401: 473–476.

Mikhailov AS (1990) Foundations of Synergetics. Vol 1. Berlin: Springer.

Mittenthal JE, Mazo RM (1983) A model for shape generation by strain and cell-cell adhesion in the epithelium of an arthropod leg segment. J Theor Biol 100: 443–483.

Miura T, Komori M, Shiota K (2000) A novel method for analysis of the periodicity of chondrogenic patterns in limb bud cell culture: Correlation of in vitro pattern formation with theoretical models. Anat Embryol 201: 419–428.

Miura T, Shiota K (2000a) Extracellular matrix environment influences chondrogenic pattern formation in limb bud micromass culture: Experimental verification of theoretical models. Anat Rec 258: 100–107.

Miura T, Shiota K (2000b) TGFbeta2 acts as an "activator" molecule in reaction-diffusion model and is involved in cell-sorting phenomenon in mouse limb micromass culture. Dev Dyn 217: 241–249.

Mochitate K, Pawelek P, Grinnell F (1991) Stress relaxation of contracted collagen gels: Disruption of actin filament bundles, release of cell surface fibronectin, and down-regulation of DNA and protein synthesis. Exp Cell Res 193: 198–207.

Müller GB, Newman SA (1999) Generation, integration, autonomy: Three steps in the evolution of homology. In: Homology (Bock GR, Cardew G, eds), 65–73. Chichester, England, Wiley.

Müller GB, Streicher J (1989) Ontogeny of the syndesmosis tibiofibularis and the evolution of the bird hind limb: A caenogenetic feature triggers phenotypic novelty. Anat Embryol 179: 327–339.

Newman SA (1992) Generic physical mechanisms of morphogenesis and pattern formation as determinants in the evolution of multicellular organization. In: Principles of Organization in Organisms (Mittenthal J, Baskin A, eds), 241–267. Boston: Addison-Wesley.

Newman SA (1993) Is segmentation generic? BioEssays 15: 277–283.

Newman SA (1994) Generic physical mechanisms of tissue morphogenesis: A common basis for development and evolution. J Evol Biol 7: 467–488.

Newman SA (1996) Sticky fingers: *Hox* genes and cell adhesion in vertebrate limb development. BioEssays 18: 171–174.

Newman SA (1998a) Epithelial morphogenesis: A physico-evolutionary interpretation. In: Molecular Basis of Epithelial Appendage Morphogenesis (Chuong C-M, ed), 341–358. Austin, Tex.: Landes.

Newman SA (1998b) Networks of extracellular fibers and the generation of morphogenetic forces. In: Dynamical Networks in Physics and Biology (Beysens D, Forgacs G, eds), 139–148. Berlin: Springer.

Newman S, Cloître M, Allain C, Forgacs G, Beysens D (1997) Viscosity and elasticity during collagen assembly in vitro: Relevance to matrix-driven translocation. Biopolymers 41: 337–347.

Newman SA, Comper WD (1990) "Generic" physical mechanisms of morphogenesis and pattern formation. Development 110. 1–18.

Newman SA, Frenz DA, Tomasek JJ, Rabuzzi DD (1985) Matrix-driven translocation of cells and nonliving particles. Science 228: 885–889.

Newman SA, Frisch HL (1979) Dynamics of skeletal pattern formation in developing chick limb. Science 205: 662–668.

Newman SA, Müller GB (2000) Epigenetic mechanisms of character origination. J Exp Zool 288: 304–317.

Newman SA, Tomasek JJ (1996) Morphogenesis of connective tissues. In: Extracellular Matrices. Vol 2: Molecular Components and Interactions (Comper WD, ed), 335–369. Reading, Mass.: Harwood Academic.

Nogawa H, Nakanishi Y (1987) Mechanical aspects of the mesenchymal influence on epithelial branching morphogenesis of mouse salivary gland. Development 101: 491–500.

Palmeirim I, Henrique D, Ish-Horowicz D, Pourquié O (1997) Avian *hairy* gene expression identifies a molecular clock linked to vertebrate segmentation and somitogenesis. Cell 91: 639–648.

Patel NH, Kornberg TB, Goodman CS (1989) Expression of *engrailed* during segmentation in grasshopper and crayfish. Development 107: 201–212.

Qian F, Watnick TJ, Onuchic LF, Germino GG (1996) The molecular basis of focal cyst formation in human autosomal dominant polycystic kidney disease type I. Cell 87: 979–987.

Rodriguez-Boulan E, Nelson WJ (1989) Morphogenesis of the polarized epithelial cell phenotype. Science 245: 718–725.

Russell ES (1916) Form and Function. Reprint, Chicago: University of Chicago Press, 1987.

Salazar-Ciudad I, Garcia-Fernandez J, Solé R (2000) Gene networks capable of pattern formation: From induction to reaction-diffusion. J Theor Biol 205: 587–603.

Salazar-Ciudad I, Newman SA, Solé R (2001) Phenotypic and dynamical transitions in model genetic networks: 1. Emergence of patterns and genotype-phenotype relationships. Evol Dev 3: 84–94.

Salazar-Ciudad I, Solé R, Newman SA (2001) Phenotypic and dynamical transitions in model genetic networks: 2. Application to the evolution of segmentation mechanisms. Evol Dev 3: 95–103.

Schaller MD, Parsons JT (1994) Focal adhesion kinase and associated proteins. Curr Opin Cell Biol 6: 705–710.

Seko Y, Takahashi N, Tobe K, Kadowaki T, Yazaki Y (1999) pulsatile stretch activates mitogen-activated protein kinase (MAPK) family members and focal adhesion kinase (p125(FAK)) in cultured rat cardiac myocytes. Biochem Biophys Res Commun 259: 8–14.

Sengel P (1976) Morphogenesis of Skin. Cambridge: Cambridge University Press.

Simpson GG (1953) The Baldwin effect. Evolution 7: 110–117.

Small S, Kraut R, Hoey T, Warrior R, Levine M (1991) Transcriptional regulation of a pair-rule stripe in *Drosophila*. Genes Dev 5: 827–839.

Starmer CF, Biktashev VN, Romashko DN, Stepanov MR, Makarova ON, Krinsky VI (1993) Vulnerability in an excitable medium: Analytical and numerical studies of initiating unidirectional propagation. Biophys J 65: 1775–1787.

Steinberg MS (1998) Goal-directedness in embryonic development. Integ Biol 1: 49–59.

Steinberg MS, Poole TJ (1982) Liquid behavior of embryonic tissues. In: Cell Behavior (Bellairs R, Curtis ASG, eds), 583–607. Cambridge: Cambridge University Press.

Stopak D, Wessells NK, Harris AK (1985) Morphogenetic rearrangement of injected collagen in developing chicken limb buds. Proc Natl Acad Sci USA 82: 2804–2808.

Streicher J, Müller GB (1992) Natural and experimental reduction of the avian fibula: Developmental thresholds and evolutionary constraint. J Morphol 214: 269–285.

Sumida H, Ashcraft RA, Thompson RP (1989) Cytoplasmic stress fibers in the developing heart. Anat Rec 223: 82–89.

Szathmary, E. (2001) Evolution. Developmental circuits rewired. Nature 411: 143–145.

Tomasek JJ, Haaksma CJ, Eddy RJ, Vaughan MB (1992) Fibroblast contraction occurs on release of tension in attached collagen lattices: Dependency on an organized actin cytoskeleton and serum. Anat Rec 232: 359–368.

Unemori EN, Werb Z (1986) Reorganization of polymerized actin: A possible trigger for induction of procollagenase in fibroblasts cultured in and on collagen gels. J Cell Biol 103: 1021–1031.

Vogel KG, Koob TJ (1989) Structural specialization in tendons under compression. Int Rev Cytol 115: 267–293.

Wilkins, AS, 1997. Canalization: A molecular genetic perspective. BioEssays 19: 257–262.

Winfree AT (1994) Persistent tangled vortex rings in generic excitable media. Nature 371: 233–236.

V ORIGINATION AND EVOLVABILITY

The foregoing chapters have provided much evidence that morphological information is not encoded as softwarelike genetic programs. This inevitably raises the question of how, in fact, the interplay between genetic and epigenetic processes originates and alters specific morphological characters in evolution. The answer will amount to a causal understanding of the phenomena of morphological evolution introduced in part II. In addition to the genetic factors focused on by the standard evolutionary paradigm, such a postgenomic consideration of evolution must include epigenetic determinants of form, in particular as they relate to the phenomena of origination and innovation.

Epigenetic is here used in the sense of *epigenesis,* consisting of all conditional, nonprogrammed factors of development that act on the materials of the zygote and its derivatives, including those specified by the genes, to generate three-dimensional biological forms. These comprise physical and other conditional processes active in the molding of tissues at various times in the history of multicellular life, the principal ones having been discussed in parts III and IV. Another use is *epigenetic variation,* variations of the phenotype induced by environmental or behavioral factors acting on genetic variation at the population level. Epigenetic variation can bias phenotypic evolution through the modulation of genetic and developmental reaction norms.

A more restricted use of the term "epigenetic" which has gained recent currency refers to factors that modulate gene expression in development and/or the transmission of genetic information from one generation to the next (*epigenetic inheritance*) by acting on DNA without changing its sequence. These include X-chromosome inactivation, genetic imprinting, and related phenomena of allelic interaction and "paramutation." Although such mechanisms may be the means by which some of the epigenetic effects discussed in this book are mediated at the gene regulatory level, the determinants of epigenesis clearly include a much wider range of cell and tissue level processes, as described in the previous chapters.

Evolvability can be thought of as the potential of a lineage to generate heritable variants that respond productively to external challenges, including those properties of developmental systems that constrain and bias further possibilities for morphological change. It is clear that all the epigenetic determinants of form described above, whether primitive physically based plasticity, divergent reactivity associated with underlying random genetic variabilty, or specifically evolved response capacities, can contribute to evolvability.

Part V explores the consequences of several of the concepts discussed in previous chapters for evolution at the character level in a fashion that goes beyond the canonical notion in which phenotypic evolution is strictly the outcome of evolving genetic programs.

Vidyanand Nanjundiah (chapter 14) discusses how environmentally induced epigenetic variation can be a factor in phenotypic evolution, starting with the observation that the same genotype, in the same environment, can give rise to more than one phenotype and that, conversely, strong phenotypic constancy can exist in a given environment in the face of genetic variability. To explore the possible mechanisms by which such environmentally induced variants can become stabilized genetically, Nanjundiah reviews evidence that environmental or genetic stress can trigger phenotypic variation between individuals of the same, or different, genotypes. He distinguishes between two classes of stabilizing mechanisms: those which depend on genetic uniformity in the initial population (the "Baldwin effect") and those which depend on preexisting genetic variability (the "Waddington effect"), giving rise to genetic assimilation. In presenting a new computational model for genetic assimilation, he argues strongly for a concept in which environmental or genetic stress act as initiating factors for phenotypic innovation by breaking up previously established canalizations of epigenetic reaction norms. Selection in favor of a new phenotype implies selection for novel regulatory gene combinations that now cause the newly "assimilated" phenotype to be again canalized.

Günter Wagner and Chi-hua Chiu (chapter 15) focus on a major evolutionary innovation in vertebrates that may have arisen as a side effect of shifting domains of gene expression: the tetrapod limb, which they link to the evolution of *Hoxa-11* and *Hoxa-13* regulation in developing limb buds. They begin by clearly describing the origination of a developmentally individualized autopodium (hand, foot), whose skeletal elements comprise the mesopodium, the metapodium, and the digits. After analyzing the phylogenetic and developmental settings in which the fin-limb transition took place, they propose a new hypothesis for the origin of the autopodium. From their review of the genetic factors implicated in fin and limb development, they argue that segregation of the primitively overlapping expression domains of *Hoxa-11* and *Hoxa-13* in limb bud development could have effected the morphological transformation of distal fin elements into autopodial elements. In discussing strategies to test their hypothesis, Wagner and Chiu emphasize that any valid explanation of evolutionary novelty must take into account the epigenetic context in which genes act.

Epigenetic context is shown by Georg Striedter (chapter 16) to determine the evolvability of another organ system, in this case, the nervous system. He notes that vertebrate brains do not evolve piecemeal, adding new divisions one by one, but as developmentally and functionally integrated systems. Striedter distinguishes two basic modes in early brain development, the "compartmental mode" and the "dynamic network mode." During early phases of neural development, the neurepithelium is divided, as a result of relatively local cellular and molecular interactions, into numerous distinct compartments that are highly conserved between species. During the second phase of neural development, the formation of axonal connections enables a host of long-range, yet highly specific, developmental interactions that contribute to the resculpting of previously established compartments, the induction of late-appearing divisions, and the establishment of functionally integrated neuronal circuits. Striedter proposes that some of these capacities for epigenetic interaction promote evolutionary change, whereas others constrain it. For instance, because phylogenetic changes affecting the second phase of neural development are likely to generate cascade effects in distant brain regions, they can account for many of the more dramatic species differences in brain organization. The compartmental mode of brain development seems to be more conservative. Striedter argues that brain evolution is the consequence of a balance between change-promoting and conservative mechanisms of neural development.

The rules of epigenesis and their potential influence on evolution become particularly apparent in formal treatments of development. Diego Rasskin-Gutman (chapter 17) observes that morphological organization arises as the product of interactions among spatial entities during ontogeny. Proportions, orientations, connections, and articulations are four levels of morphological organization that can be recognized, separated, and subsequently used for comparative analysis. Using boundary patterns between adjacent elements of vertebrate skulls as an example, Rasskin-Gutman shows that these patterns not only determine the structural scaffolding of skulls but also establish morphological constraints on embryonic development and its evolvability. Taking a graph-theoretical approach, he presents a cellular automata-based program that can be used to model developmental and evolutionary connectivity patterns. It generates sets of boundary patterns that form a "morphospace of connections," which can then be compared with configurations of natural skeletal patterns. Simulated "evolutionary runs" demonstrate constructional epigenetic constraints on the emergence of pattern and novel structures in vertebrate skeletal macroevolution.

14 Phenotypic Plasticity and Evolution by Genetic Assimilation

Vidyanand Nanjundiah

Evolution by natural selection occurs whenever the following three conditions hold: phenotypic variation, differential reproductive success, and heritability. The last of these conditions is important because, in the absence of heritability, there is no (additive) genetic component to the overall phenotypic variance. At the same time, one knows that a given phenotype develops only in the context of a specified environment (see Gilbert, chapter 6, this volume). Moreover, the same genotype can result in different phenotypes in different environments, thereby defining a reaction norm (Suzuki et al., 1986). The concept of reaction norm is related to that of phenotypic plasticity, the capacity of development to lead to the appropriate (i.e., adaptive) phenotype in the appropriate the environment provided that environment falls within the range of experience of the organism's ancestors.

This chapter explores a different aspect of phenotypic plasticity, namely the ability of a genotype to give rise to more than one phenotype even in the same environment. This means that whether the environment is uniform or not, the relationship between genotype and phenotype is one to many and not one to one. We note in passing that multicellular development is a familiar example of cells with the same genome exhibiting extremely diverse phenotypes. However, excepting organisms in which early embryonic development is highly regulative (e.g., mammals), the internal environment in the unfertilized egg is often functionally heterogeneous (Slack, 1985). The implication is that the diversity of cellular phenotypes must just as often be traceable to environmental influences, in this case an environment of maternal origin.

The biological differences between individuals can come about both because of differences in their genetic makeup and because of *epigenetic* factors. The latter include physical factors, secondary modifications in DNA sequence or structure, and variations in patterns of gene expression. Epigenetic factors lie behind the obvious differences between cells of one tissue type and another in an organism. Two individuals that are genetically identical (or virtually so), such as monozygotic twins, can differ in their phenotypes, and therefore differ epigenetically. The differences can be largely or entirely environmental in origin. For example, they can be caused by differences in early postnatal sensory experiences (Hubel, 1988). But often, because we cannot readily ascribe the differences to the environment, we put them down to "developmental noise," a form of random, uncontrolled variation between individuals (Waddington, 1957).

Partly on account of its name, developmental noise tends to be regarded as an unfortunate, but unavoidable, lack of precision in developmental systems, in short, as a nuisance. To the contrary, I argue that developmental noise is yet another manifestation of the organism's being "plastic" in its extended sense of "capable of adapting to varying conditions"

(Merriam-Webster, 1993, 890). The term *phenotypic plasticity* accommodates the possibility that developmental noise can also originate from environmental causes. Besides, the external world, "environment" can include the internal microworld of cells and organisms. Indeed, variations that occur independently of spatial or temporal heterogeneities in the external environment can be exploited for the purpose of evolutionary change. As I hope to show in this chapter, given the right circumstances, phenotypic plasticity can play an important role in evolutionary change. (Figure 14.1 summarizes the essence of my argument.) My use of "plasticity" does not demand that the new phenotype be adaptive, although it is hardly necessary to add that grossly maladaptive phenotypic changes are unlikely to survive in the long run.

Many of these ideas, in particular the notion that developmental noise or phenotypic plasticity could be a significant factor in evolutionary change, were anticipated by Bonner

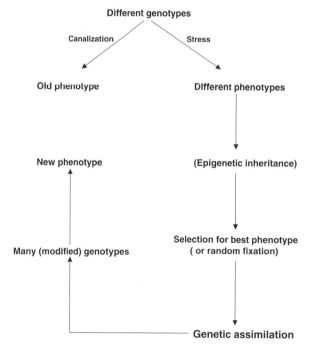

Figure 14.1
Flowchart indicating how genetic assimilation might lead to rapid evolutionary change. A set of genotypes constitutes the "wild type" by virtue of being canalized for an optimal (old) phenotype in a certain environment. Because of genetic or environmental stress, canalization breaks down. This leads to the appearance of a spectrum of phenotypes; selection for the fittest phenotype is followed by its genetic assimilation and canalization. Epigenetic inheritance may be an additional, though not an essential, factor in maintaining the favored phenotype, whose development may also be supported by random fixation of alleles (not discussed in the text).

(1965, 1967) in terms of what he called range variation. One can do no better than quote in detail: "... what apparently is inherited is the ability to vary within certain limits. The variation is therefore not genetic variation in the conventional sense, for the range of variation is genetically determined but the size of any individual within that range is not" (Bonner, 1965, p. 95) [*Note:* "position" or "status" would be less restrictive than "size"]. What advantage might be served in having individual variants that are not genetically determined, in having "this unconventional and relatively haphazard form of inheritance where genetically identical individuals can differ phenotypically over a considerable spectrum"? (Bonner, 1965, p. 95). "The answer is clear"; in the specific case of the cellular slime mold *Dictyostelium discoideum* (of which more below), "if this were not the case it would be immediately impossible to use the variation for the kind of [evolutionarily stable] organizational hierarchy that we are suggesting here" (Bonner, 1965, p. 99). Bonner goes on to discuss the implications of range variation extensively, and the only aspect of his discussion about which one might quibble today is the group-selectionist viewpoint—which may not be wrong in the context of clonal populations of slime molds or ciliates anyway.

The chapter begins by listing examples of phenotypic variation between individuals of the same genotype in the same (uniform) external environment. After outlining plausible means for the origin of such variation, it describes a typical laboratory experiment carried out by Waddington (and repeated by many others), in which he showed that artificial selection, acting on phenotypic variations, can lead to a major morphological change within a very few generations. It provides a conceptual framework for modeling that outcome and speculates on how genetic assimilation may bear on evolutionary changes in nature.

The Same Genotype Can Give Rise to Different Phenotypes in the Same Environment

Ranging over many phyla or divisions, instances of significant phenotypic variation between individuals raised in the same environment abound. In some cases, the individuals are genotypically identical, such as members of a clone. In others, the observed variation clearly appears to be independent of any genotypic differences that might exist.

Instance 1: Bacteria

Spudich and Koshland (1976) demonstrated that clonally related *Escherichia* coli bacteria exhibit substantial differences in their chemotactic behavior under identical experimental conditions. Upon exposure to an attractant, the swimming behavior of cells changes transiently, with the time required to return to the prestimulus pattern of behavior (the response time) varying enormously from cell to cell. The outcome is a cell-to-cell behavioral flexibility that could conceivably prove advantageous to one clone over another. The flexibility is almost certainly not genetic in origin. Although one might object that the external environment may not be exactly the same for all bacteria swimming in a liquid medium, a careful study of the experimental conditions suggests that this is unlikely. A more plausible hypothesis is that the bacteria exhibit different phenotypes on account of statistical fluctuations—"noise"—inherent in the internal biochemistry underlying their chemotactic behavior. This is especially likely if the relevant biochemistry includes autocatalytic reactions (Delbrück, 1940). If the capacity to exhibit such a high degree of phenotypic variation has a genetic basis, clones may possess heritable differences in the extent to which they display a spread in response times. Consequently, the trait may be subject to evolution by natural selection, with selection occurring between clones.

Instance 2: Protozoa

When starved after being raised in a common nutritive environment, genetically identical social amoebas of the species *Dictyostelium discoideum* begin to exhibit striking differences in cell-to-cell properties, differences that culminate in the death of some amoebas and the differentiation of the rest into spores. Nanjundiah and Bhogle (1995) have shown that at least formally, the differences can be ascribed to a stochastic or "coin-tossing" process in which cells acquire different predispositions with different probabilities. Although the predispositions cannot be heritable (because some cells die), one can show that they are correlated with differences that exist in the spore population of the preceding generation, for example, differences in autofluorescence and in the ability to take up certain dyes (Baskar, 1996). The point is that phenotypic differences between genetically identical amoebas, though not environmentally based, are important for the development of *D. discoideum* and may have played a role in the evolution of social behavior in the species.

Instance 3: Vertebrata

Ashoub and colleagues (1958) found that the temperature at which isogenic mice are raised crucially influences their weight at different development stages up to the age of four weeks. In general, the variability of mice raised at an extreme temperature (e.g., 5°C or 28°C) greatly exceeds the variability seen when they are raised at 21°C. Moreover, large litters tend to show greater interindividual variability in birth weights than small litters do. Here again is an instance of significant interindividual variation within a common, though not necessarily identical, genotype.

Thus phenotypic variability can be exhibited in a manner that is, first, epigenetic in origin and, second, made manifest between genetically identical (or very similar) individuals raised in a common environment. What evidence is there that the variations are heritable? In the specific instances just discussed, none. However, there are observations testifying to the fact that interindividual epigenetic differences can indeed be inherited.

Epigenetic Inheritance

The evidence concerning the inheritance of epigenetic traits is extensive, as are discussions of the possible evolutionary implications (Jablonka and Lamb, 1995). McLaren (1962) cites the case of babies born at high altitudes in Peru. These babies have heart defects at birth, presumably on account of anoxia. The fetuses born to mothers with heart defects also tend to have heart defects themselves, showing that the effect can extend over generations. Better-studied examples hint at possible mechanisms. Genetically identical bacteria growing in the same environment can differ in being able to metabolize lactose or not, and can faithfully pass down either characteristic to their progeny (Novick and Wiener, 1957). When seeds of flax, obtained from self-fertilized plants grown in diverse nutrient regimes, are compared in the same environment, subsequent generations show stably altered characteristics. The differences pertain to shape, developmental patterns, height, weight, and nuclear DNA content; the fundamental change appears to occur in DNA sequences coding for the 25S, 18S, and 5S ribosomal RNA (Cullis, 1988). In Mongolian gerbils, intrauterine hormonal interactions between pups in the same litter affect both adult morphology and behavior. Female fetuses that happen to develop between two male fetuses are exposed to high levels of testosterone, an androgenizing agent. Such females exhibit delayed puberty and low lifetime fecundity in comparison with females that, as fetuses, were situated between two other females. Androgenized females produce litters with significantly male-biased sex ratios. In consequence, they tend to have daughters that are themselves androgenized (Clark, Karpluk, and Galef, 1993). Nuclear transplantation experiments in the mouse show altered patterns of gene expression, in particular with respect to genes that encode an olfactory marker protein and components of urinary proteins (Roemer et al., 1997). These patterns persist in subsequent generations. Here the epigenetic change is correlated with changes in the level of DNA methylation in the relevant genes and is stably propagated through heredity.

What principles might underlie epigenetic inheritance? A genetic scheme due to Bussey and Fields (1974) accounts for the inheritance of alternative epigenetic states (figure 14.2). Their model is patterned on the mutual repression exhibited by the *cI* and *cro* genes in bacteriophage lambda. It depends on a network of positive and negative feedbacks between three genes, a regulatory gene (R) and two structural genes ($G1$ and $G2$). The expression of $G1$ by itself characterizes one phenotype and that of $G2$ by itself characterizes a different phenotype. The R product represses $G1$ and $G2$ but activates R; the G1 product activates $G1$ and inhibits $G2$ and R; similarly, the $G2$ product activates $G2$ and inhibits $G1$ and R. To begin with, let us say that the genes encoding $G1$ and $G2$ are repressed. An environmental influence (the inducer) modifies the R product in such a manner that it can no longer regulate its own production or that of G1. Then G1 will continue to be made in the absence of the R product.

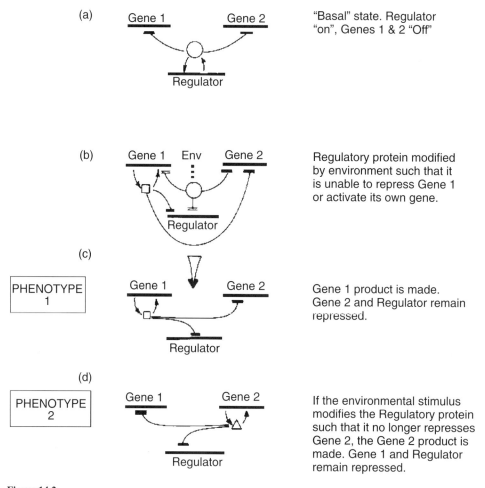

Figure 14.2
Model for hereditary transmission of epigenetically determined phenotypes. (After Bussey and Fields, 1974.)

A parallel series of steps ensures that the production of G2 is triggered and thereafter stably propagated. The explanation is similar to the one alluded to earlier that accounts for the stable transmission of the induced and uninduced states of the *lac* operon in *Escherichia coli* under specified conditions (Novick and Wiener, 1957). Namely, the medium in which the bacteria are grown is a "maintenance" medium. It has just sufficient lactose to ensure that cells with *lac* in an induced state have daughter cells in which *lac* continues to be induced, but not sufficient to cause induction in uninduced cells (or in their daughter cells).

We have seen that alternative phenotypes can be found in the same environment and can be propagated stably. Can such alternatives lead to interesting consequences? Yes, if there are many alternatives and their fitnesses are not the same. The argument rests on two facts. First, a stressful environment (one that departs sufficiently from the one to which the organism is adapted) can trigger phenotypic variation. Second, natural selection acts both on the mean value of a trait and on the extent to which there are variations about the mean. This puts a premium on the degree of reliability with which the optimal phenotype can be specified.

Phenotypic Differences Can Come About as a Result of Stress

Natural selection, in addition to molding the mean value of a trait, tends to improve the extent to which the trait is buffered. In other words, natural selection also sharpens the degree of precision with which the selected phenotype develops. Equivalent terms for *buffering* are *epigenetic stability* and *canalization* (Waddington, 1960). Here we consider the reverse, a breakdown of canalization and why the specification of the phenotype may become imprecise. The stress can be genetic or environmental in origin.

Genetic stress may be an inevitable correlate of developmental complexity. This can be inferred from the observation that organisms have evolved means for overcoming certain unavoidable consequences of possessing a complex internal biochemistry. The best-known example comes from studies of metabolic pathways made up of enzyme-catalyzed reactions, where the flux through a pathway constitutes an important component of the overall organismal phenotype. Kacser and Burns (1981) proved that when the pathway has a large number of intermediates, substrate concentrations are low, and the effects of feedback and nonlinearity are negligible, the flux is automatically buffered with respect to genetic stress. They showed that even when the level of an enzyme decreased by half (as might be caused when an individual was heterozygous for the wild type and a loss-of-function allele), the flux through the pathway changed to a negligible extent. The implication drawn by them was that, notwithstanding a wide range of possible stresses caused by changes in enzyme levels (which in turn could be due to mutations), metabolic fluxes are buffered—apparently automatically.

The assumptions made by Kacser and Burns, specifically, that substrate concentrations be low, feedbacks and nonlinearities unimportant, and oscillatory reactions not possible, turn out to be crucial. If one or more of these assumptions is violated, the system is no longer guaranteed to be insensitive to the effects of mutation, in particular the mutation of regulatory genes (Cornish-Bowden, 1987; Grossniklaus, Madhusudhan, and Nanjundiah, 1996). The resulting genetic stress can impinge significantly on the flux. Therefore, if the observation is that the phenotype, and so fitness, remains unaffected even under stressful conditions, one might reasonably conclude that buffering mechanisms must have been selected for during the course of the organism's evolutionary history. Rutherford and Lindquist (1998) have provided a striking demonstration of how mutations in regulatory genes can destabilize development, conceivably by affecting the flux through one or more biochemical pathways. They found that when mutated, the *Hsp83* gene, which encodes the Hsp90 protein of *Drosophila melanogaster*, causes an enormous range of phenotypes to be manifested. (They went on to demonstrate that it was thereby possible to select for a novel phenotype; the mutant allele could even be dispensed with during the course of the selection.)

When organisms encounter abnormal situations, there may be no escaping the ensuing genetic stress. In normally outbred populations, inbreeding can cause genetic stress. In organisms that are normally inbreeders, such as flax, the *outcrossing* of two inbred strains imposes genetic stress. Inbreeding can lead to phenotypic variation in the progeny that goes well beyond the range found in outbred individuals. A widely accepted belief is that developmental pathways get destabilized by the high levels of homozygosity that follow from inbreeding. Why inbreeding has this effect remains unknown, but an observation made by Biémont, Aouar, and Arnault (1987) may offer a clue. These workers found that after sixty-nine generations of sib mating, an inbred line of *D. melanogaster* showed extensive reshuffling of the mobile genetic element *copia*. It appeared that a specific destabilization of the *copia* element had taken place on account of inbreeding.

The example of *copia* may point to a widespread source of genetic stress. The presence of mobile genetic elements, parasitic entities potentially harmful to their hosts, represents a trade-off between selection acting on the element to favor transposition and selection on the host to favor suppression of transposition. The result can be a stable polymorphism with respect to the distribution of the number of such mobile elements in the genomes of different individuals (Nanjundiah, 1985). Additionally, selection can act on the host so as to suppress transposition and thus improve the stability of its phenotype. But the transposition of mobile genetic elements is commonly replicative (Lewin, 1995) and involves obligatory events that are part of the physiology of the cell, such as transcription, DNA synthesis, and DNA recombination. Therefore, in attempting to suppress the replication of the parasite, the cell risks deleterious side effects that might result from interference with its own functioning.

A way to guard against such side effects might be to make an inhibitor of transposition in just sufficient amounts: one active copy of the gene encoding the inhibitor could block excessive spread of the foreign element, an optimal outcome from the point of view of the host. Two active copies might cause nonspecific deleterious effects; no copies at all would be useless of course. Inbreeding would disrupt this pattern of peaceful coexistence: some inbred lines would lack the inhibitor entirely, whereas others would suffer from a general depression of physiology; in short, the outcome would be dysgenic.

What about the destabilization caused by the *outcrossing* of highly inbred lines, something actually observed when the normal mode of reproduction involves inbreeding? The point to remember is that, irrespective of the nature of the accommodation that has been reached between the transposable DNA parasite and its host, it is unlikely that the same transposable element will have colonized all host lines. When two inbred lines are crossed, because both cellular and chromosomal environments are new for each set of elements, the resident parasitic elements in each genome may be released from the transposition block that existed in their normal hosts. Once again, the outcome can be expected to be dysgenic (Nanjundiah, 1985).

Environmental variations in space, time, or both can constitute an important source of phenotypic variation. Given a sufficiently unpredictable environment, there may simply be no single optimal phenotype (Levins, 1968). Even though environmental *variations* do not occur or are insignificant in the situations we are considering, the environment is stressful all the same. Under such conditions, stress can be a potent factor in eliciting large-scale phenotypic changes (Parsons, 1997). Heat shock, a good example of stress, can cause many cellular proteins to become dysfunctional (on account of abnormal folding, for example), thereby inducing other forms of damage to the cell. The cell tries to protect itself by increasing the rate of production of specialized stress proteins (formerly called "heat shock proteins") to sequester and dispose of the damaged proteins.

Many stress proteins are multifunctional. For example, Hsp90 plays a role in signal transduction, progress through the cell cycle, production of nitric oxide, transcription, and translation (Mayer and Bukau, 1999; Nathan, Vos, and Lindquist, 1997). In order to be able to carry out all these functions under normal conditions, it is evident that the cell must manufacture the stress proteins in sufficient amounts. When it mobilizes them for emergency functions, however, the cell could be compromising one or more of the normal functions mediated by stress proteins. If so, the effect would be a destabilized phenotype (Forsdyke, 1994). A stress-induced depletion of other regulatory molecules could be yet another route to destabilization. The outcome could be that crucial steps in cellular metabolism or gene activation are switched from "on" to "off" or vice versa. It is worth noting that genetic stress can also be thought of as a form of environmental stress, but this time with reference to the internal environment of the cell or organism.

Rapid and Heritable Phenotypic Change Can Occur by Means of Genetic Assimilation

Both Waddington (1953) and, independently, Schmalhausen (see Gilbert, 1994) predicted that if organisms varied genetically in terms of their ability to respond to an environmental stimulus by giving rise to novel phenotypes, natural selection could take advantage of that ability. Their reasoning was based on the assumption that among the phenotypes that resulted as a consequence of the stimulus, some would have a higher reproductive fitness than others. Selection would enrich the population with genotypes that responded to the environment by developing the most advantageous phenotype in a reliable manner. In particular, selection would tend to increase the sensitivity of genotypes to the environment. Those genotypes would be most favored whose threshold of response to the stimulus was extremely low; indeed, so low that the response—the favored phenotype—continued to be expressed in the absence of the stimulus (Waddington, 1953). In this manner, a character originally acquired through exposure to a particular environment would now be stably inherited, or, as Waddington put it, "genetically assimilated."

Waddington went on to demonstrate that genetic assimilation could also work under conditions of artificial selection. Among the experiments carried out by him and repeated by others later, one set involved the transformation of a normal, two-winged stock of *Drosophila melanogaster* into flies with the four wings typical of the extreme Ultrabithorax phenotype (see figure 14.3; Waddington, 1956a,b; Ho et al., 1983; Gibson and Hogness, 1996). Two features made the process especially intriguing. First, it appeared impossible that mutational change could have taken place. Second, the number of generations over which selection needed to be practiced was quite small, of the order of 10–20. Thus genetic assimilation appeared—misleadingly, as Waddington was the first to point out—to resemble an evolutionary transition with Lamarckian overtones. Nevertheless, in his words, "if such a change occurred during phylogenesis it would certainly be accounted a macro-evolutionary phenomenon" (Waddington, 1956b, p. 1).

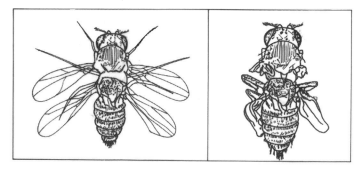

Figure 14.3
Schematic drawings based on originals. (*Left*) Multiply mutant fly with four wings. The anterior pair of wings is usual; the posterior pair is supernumerary and occurs because the tiny dorsal metathoracic appendages, the halteres, have been transformed into almost normal-looking wings. (After Suzuki et al., 1986.) (*Right*) Fly showing effects of genetic assimilation for the Ultrabithorax phenotype. The normal wings have been removed to show more clearly the modified, winglike halteres (dorsal metathorax → dorsal mesothorax transformation). (After Waddington, 1956b.)

The demonstration of genetic assimilation of the Ultrabithorax phenotype went roughly as follows. Eggs aged 2.5–3.5 hours were exposed to ether vapor for about 25 minutes. Some of the resulting adults resembled (not always very closely to begin with) four-winged flies characteristic of the *Ultrabithorax* genotype. Selection could be practiced by breeding from them: after many generations of selection these flies bred true for the new phenotype. Other demonstrations of genetic assimilation were more dramatic (Waddington, 1961), involving sib selection: breeding from the affected flies' brothers and sisters that had not themselves been exposed to the environmental stimulus. In such cases, neither the flies belonging to the assimilated stock nor any of their direct ancestors had ever experienced the stimulus.

In explaining genetic assimilation, Waddington made two assumptions: first, that canalization of the phenotype would have masked the existence of genetic variation in the original wild-type stock; and second, that new combinations of regulatory genes would arise in each generation through recombination. The environmental stimulus would merely unmask, so to speak, the underlying genetic variation (figure 14.4). In the experiment described above, Waddington managed to show by mapping that the selection regime had

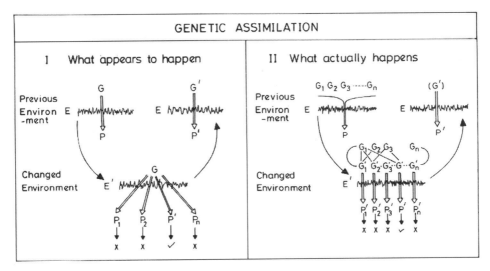

Figure 14.4
Genetic assimilation: appearance versus reality. Panel I assumes a one-to-one relationship between genotype **G** and phenotype **P** in the normal environment **E**, altered to a one-to-many relationship in the stressful environment **E′**. Selection for phenotype **P′** followed by a return to the previous environment results in a new phenotype that breeds true, making it appear that, mysteriously, a (single) new genotype **G′** has appeared. Panel II shows that, because of canalization for the wild-type in the normal environment **E**, many genotypes (G_1, G_2, etc.) can be consistent with the same phenotype **P**. A transfer to **E′** results in a breakdown of canalization; the genotype-phenotype relationship becomes many-to-many. Selection for **P′** implicitly involves selection for a new *set* of genotypes (**G′**). The large arrows symbolize the shift from one environment to the other and back.

indeed given rise to the appearance of at least one new regulatory gene. A recent replication of the experiment (Gibson and Hogness, 1996) showed that four new regulatory gene combinations had appeared in the course of the assimilation of the four-winged phenotype.

A Model for Genetic Assimilation

Waddington's proposal does not seem to have been tested in an explicitly genetic model. The importance of doing so is obvious: if genetic assimilation can result in a rapid, large-scale change in the phenotype, it would constitute a plausible explanation for major evolutionary changes that have taken place relatively rapidly. My colleague Narayan Behera and I (N. Behera and V. Nanjundiah, unpublished) have developed a model for genetic assimilation. The model depends on a computational algorithm we developed for studying the interaction between phenotypic plasticity and regulatory genes (Behera and Nanjundiah, 1997). Genotypes are represented by two randomly generated strings of genes: a *structural string*, which directly influences the phenotype, and a *regulatory string*, which influences the phenotype indirectly by acting on the functioning of the structural genes. The distinction is more symbolic than real. Depending on the context, the same gene can be thought of as structural or regulatory. To use the language of Larsen (chapter 10, this volume), most genes do double duty as both "worker" and "bureaucrat" genes. Conventional thinking, on the other hand, tends to be comfortable with a nomenclature in which the gene encoding the last protein in a biosynthetic pathway is called a "structural gene," whereas a gene encoding any of the proteins that influence the preceding steps (either directly, as substrates or products, or indirectly, as cofactors, activators, or repressors) is called a "regulatory gene." As far as the following discussion goes, I will use "structural" or "regulatory" merely to refer to the more significant aspect of a particular gene in a particular context.

A structural gene can function constitutively, when it is intrinsically "on" or "off," or facultatively, when it has no intrinsically well-defined state but can function *as if* it were "on" or "off," depending on how it is influenced by the set of regulatory genes that act on it. A structural gene locus has three possible allelic states, **1**, **0**, and **X** (figure 14.5), representing the "on," "off," and "either on or off" states of the gene, respectively, with an **X** finally behaving as if it were a **1** or a **0** (as explained below). Once the state of each **X** is unambiguously assigned, the structural gene string has a phenotype associated with it. There is an optimal phenotype associated with each environment and defined by an appropriate string of **1**s and **0**s. The starting environment is denoted by **A**. To begin with, all individuals express the same phenotype, which we may identify with the wild-type phenotype in Waddington's experiments. In every generation, some individuals are transferred, or exposed at a critical stage of their development, to a different environment, denoted by **B**. Operationally, the transfer increases the likelihood that an individual express a (desired) novel phenotype. Let us imagine

Phenotypic Plasticity

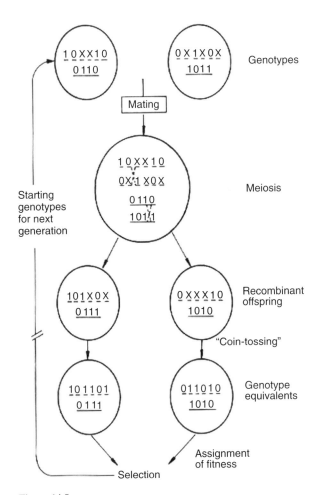

Figure 14.5
Genetic assimilation model, showing two haploid genotypes (small ovals) along with a transient diploid zygote (large oval). Each haploid genotype consists of two strings of genes, a "structural set" (top of each pair, containing **1**, **0**, and **X** alleles) and a "regulatory set" (bottom of each pair, containing only **1** and **0** alleles). Recombinant genotypes are generated via random crossovers occurring during meiosis. By influencing the probabilities of the $X \to 1$ and $X \to 0$ transitions, the regulatory gene set specifies alternative epigenetic states of the same gene **X**. The transitions result in genotype equivalents, affecting the expression of structural genes and helping to specify the phenotype. "Coin tossing" alludes to the probabilistic basis of the transitions.

that the starting phenotype is that of a normal (wild-type) fly and that the desired phenotype is the full Ultrabithorax phenotype with four wings.

The siblings of those individuals that come closest to expressing the altered phenotype after exposure to **B** are chosen for mating. This is merely in order to demonstrate the versatility of the model; obviously, breeding directly from them would work better. We monitor the success of the procedure in terms of a parameter, **H**, which stands for the fraction of the population that expresses the desired phenotype, either after exposure to environment **B** (H_B) or without any such exposure (H_A). In terms of Waddington's experiment, H_A is the fraction of the population that expresses the Ultrabithorax phenotype without any prior exposure to ether and H_B is the fraction that does so after ether exposure. The upshot is that on the whole both H_A and H_B increase over the course of generations. In other words, genetic assimilation occurs. As is true of the experimental observation, in environment **B** (i.e., after exposure to ether) a few genotypes give rise to the assimilated phenotype even in the first generation but far fewer—possibly none—do so in environment **A**. The model does not take advantage of all the restrictions that the experiments allow. It makes no assumptions regarding the nature of the two environments, when the shift occurs or how long it lasts. All it says is that the shift from environment **A** to environment **B** permits a broad spectrum of phenotypes to develop. In contrast, the experimental protocols tend to involve a well-defined environmental stimulus applied within a small time window at a specific stage of development. As Waddington found, this approach can cause a rather narrow range of phenotypes to appear. For example, ether treatment favors the "haltere-to-wing" transition, whereas heat shock affects the development of wing veins and leads to the appearance of cross-veinless phenotypes (Waddington, 1961).

Here, in qualitative terms, is how our model works (the explanation is essentially Waddington's). As I have described the system, selection can act only between alternative phenotypes that are correlated with different structural gene combinations. Regulatory genes would seem to play no role in the specification of phenotypes. However, such a conclusion would be false, because the probability that a particular phenotype is actually attained depends on whether an "**X**" is more likely to behave as a "**1**" than as a "**0**," which in turn is influenced by the regulatory genes. Selection acts on the regulatory genes, albeit indirectly, in our model, favoring combinations of regulatory genes that lead to an increase in the probability that an "**X**" will behave as a "**1**." Thus selection leads to the occurrence, through genetic recombination, of regulatory gene sets that ensure that the phenotype corresponding to maximal fitness is attained with a probability close to 1—irrespective of exposure to the source of environmental stress (exposure to ether in Waddington's experiment). In the beginning, environmental stress played a facilitating role by destabilizing the course of normal development and making the appearance of desired phenotypes more likely than would otherwise be the case. At the end, however, the stress is no longer needed; its role has been made redundant by the canalized action of a new set of genes (figure 14.4).

Evolutionary Possibilities

Phenotypic plasticity can contribute to evolutionary change by two distinct routes, and it is important that we do not conflate them.

In the first route, which I have explored in this article, the original phenotype is canalized in the normal, nonstressful environment, and therefore any underlying genetic variation that exists is cryptic. Stress, whether environmental or genetic in origin, works on the preexisting genetic variation and gives rise to a spectrum of phenotypes, many of them novel, with the optimal phenotype having a higher reproductive success than the others. Sexual reproduction and recombination constantly throw up new arrays of regulatory genes, an obvious prerequisite being that some genetic variation does exist. Selection acts on regulatory gene combinations so as to make the development of the optimal phenotype increasingly probable. To put it differently, selection progressively delinks the appearance of that phenotype from the particular stressful stimulus that potentiated its appearance in the first place. The result is that evolution takes place via genetic assimilation. Epigenetic inheritance is not necessary for this scheme to work; its existence would be an added bonus, as it were. Yet another bonus would be for the optimal phenotype to be produced more or less consistently in the new environment. If that were the case, selection would need to act merely to decrease the variance in the mean level of expression of the optimum.

The second route to evolutionary change, which I have only touched on, could operate in a background of genetic uniformity. Although here, too, phenotypic variation is induced by stress, the cause now lies solely in the manner in which the genotype interacts with the stressful environment. The consequence of the interaction is more than one phenotype. In Bonner's terminology (Bonner, 1965), the consequence is an enlargement in the range variation of the phenotype. Among the phenotypes so generated, the one that leads to the highest reproductive fitness can be rendered constitutive (eventually) via what has been called the "Baldwin effect." In the Baldwin effect, a phenotypic response to a specific environment can become independent of the environment. Baldwin's suggestion was that what started out as a purely physiological adaptation to new conditions could end up, if the right mutational changes took place, as a genetically constitutive outcome (Simpson, 1953). The Baldwin effect may work in situations where the starting population is genetically homogenous and the (stress-induced) variable phenotypes do not differ genetically, at least not in the beginning.

Following a line of reasoning first advanced by West-Eberhard (1986), Stuart Newman (personal communication; also see chapter 13, this volume) has pointed out that it may be useful to broaden the meaning of "genetic assimilation." This could be done by including any set of events whereby a trait which originally depends on an interaction with the environment becomes incorporated into the developmental repertoire of the organism through genetic change. West-Eberhard's (1986) model for major phenotypic change in evolution depends on the ability of more than one phenotype to be consistent with the same genotype or set of genotypes. Her starting assumption is that the same genotype can give rise to distinct but equally well-adapted phenotypes in different individuals belonging to the same population. The alternative phenotypes could persist over generations. Subsequently, perhaps on account of geographical isolation and the demands imposed by a new environment, just one of the alternatives could be favored. The genome would then be "released from the constraints of having to accommodate multiple alternatives," a step that could "facilitate speciation by accentuating divergence from the parent population" (West-Eberhard; 1986, p. 1388).

The less restrictive definition of genetic assimilation may be useful in that it helps us to think in terms of an entire set of phenomena—phenotypic plasticity/epigenetic inheritance/the Baldwin effect/canalization/genetic assimilation—as lying on a conceptual continuum. But by doing so we blur what may be a useful distinction between genetic assimilation and the Baldwin effect; the latter would become just one of the many means through which genetic assimilation could occur. As used here, however, genetic assimilation is quite different from the Baldwin effect. Genetic assimilation requires preexisting genetic variability, whereas the Baldwin effect does not.

In both routes the stressful environment acts as a trigger that permits a whole range of phenotypes to develop. Although any phenotypic trait that thus develops need not be an adaptation to the triggering *condition,* as we have seen, the trait might confer an advantage on its possessor in a quite unrelated environment. In the case of the Waddington-type experiment, the changed environment, **E'** in figure 14.4, is imposed by the experimenter; it is "a peculiar form of predation" (Warburton, 1956, which contains a thought-provoking discussion of this point). Other things being equal, it is reasonable to assume that, for a given intensity of selection, the variance in the mean value of a trait will always be higher when development occurs in a stressful, relatively less tolerant, environment than when it occurs in an environment free of stress. Such an assumption, if true, suggests that selection for a changed phenotype in a stressful environment ipso facto leads to unexpectedly reliable (strongly canalized) development when the original environment is restored (see figure 14.4).

The idea that the extreme phenotypic variation caused by genetic or environmental stress can fuel rapid evolutionary change is not new. Levin (1970) proposed a model for speciation that involved, as a first step, the isolation of a small number of individuals from a larger population. The smaller group could be located at the periphery of the main population, in a region where the environment was suboptimal and, to that extent, stressful. The smaller group would therefore be subject to strong directional selection. That, combined with phenotypic changes resulting from a failure of reliable development (caused by inbreeding), could lead eventually to the stabilization and fixation of a novel phenotype. In Levin's scheme, environmental stress plays an important role in generating developmental instability, but so does a breakdown of canalization because of inbreeding and the concomitant increase in homozygosity.

Both Levin's and West-Eberhard's proposals postulate phenotypic differences within a common gene pool followed by a subsequent step or steps involving natural selection for a genotype which buffers, or canalizes, the newly favored phenotype. These steps must necessarily involve modifier loci. After canalization has been achieved, the modifiers will mask whatever genetic variation there exists within the population. As a result the underlying genetic variation will become cryptic. (The most famous, not to say famously disputed, discussion of the likelihood of such a course of events is Fisher's model for the evolution of dominance; see Nanjundiah, 1993). But the fact that epigenetic states can be maintained and propagated through many generations of reproduction provides, in principle, yet another means for significant phenotypic variations to arise and evolutionary change to occur, this time without any underlying genetic change—at least none to begin with. The inheritance of epigenetic traits, and, more generally, of acquired traits, excites surprise only when it is observed to occur across individual (organismal) generations. The reason behind this is our ingrained belief in the correctness of Weismann's doctrine of the separation between germ line and soma. But in multicellular organisms, the epigenetic traits expressed by differentiated cells are routinely inherited over many cellular generations. Besides, the germ line–soma distinction breaks down in plants and in many invertebrates.

Finally, we need to address an important question. Why do organisms harbor the capacity to exhibit a large degree of phenotypic variation? One possible answer is that selection for complexity—for genetic networks with high levels of connectivity, feedbacks, and nonlinearities—might automatically render the system susceptible to genetic or environmental stresses. In other words, the capacity to exhibit phenotypic variation may be an inescapable consequence of a complex genome and a complex physiological pathways. Alternatively, as we have also seen, the course of developmental canalization may get disrupted by stress. But there is a third possibility. It may be that for many organisms, the native environments are so variable—"on all scales in space and time" according to Bell (1992)—that natural selection has molded their capacity to exhibit a diverse range of phenotypes when called upon to do so.

Acknowledgments

I thank Stuart Newman and Gerd Müller for having motivated me to bring together these ideas. Earlier drafts were read and criticized by Patrick Bateson, Scott Gilbert, Ellen Larsen, Dieter Malchow, Gerd Müller, Stuart Newman, and Mary Jane West-Eberhard, to all of whom I am grateful. This work was partly supported by research grants from the Department of Biotechnology and the Alexander von Humboldt-Stiftung.

References

Ashoub MR, Biggers JD, McLaren A, Michie D (1958) The Effect of the environment on phenotypic variability. Proc R Soc Lond B Biol Sci 148: 192–203.

Baskar R (1996) Early committment of cell types and a morphogenetic role for calcium in *Dictyostelium discoideum*. Ph.D. diss., Indian Institute of Science, Bangalore.

Behera N, Nanjundiah V (1997) *Trans*-gene regulation in adaptive evolution: A genetic algorithm model. J theor Biol 188: 153–162.

Bell G (1992) Five properties of environments. In: Molds, Molecules, and Metazoa (Grant PR, Horn HS, eds), 33–56. Princeton, N.J.: Princeton University Press.

Biémont C, Aouar A, Arnault C (1987) Genome reshuffling of the *copia* element in an inbred line of *Drosophila melanogaster*. Nature 329: 742–744.

Bonner JT (1965) Size and Cycle. Princeton, NJ: Princeton University Press.

Bonner JT (1967) The Cellular Slime Molds (2nd ed). Princeton, NJ: Princeton University Press.

Bussey H, Fields MA (1974) A model for stably inherited, environmentally induced changes in plants. Nature 251: 708–710.

Clark MM, Karpluk P, Galef BG (1993) Hormonally mediated inheritance of acquired characteristics in Mongolian gerbils. Nature 364: 712.

Cornish-Bowden A (1987) Dominance is not inevitable. J theor Biol 125: 333–338.

Cullis CW (1988) Control of variation in higher plants. In: Evolutionary Processes and Metaphors (Ho M-W, Fox SW, eds), 49–61. London: Wiley.

Delbrück M (1940) Statistical fluctuations in autocatalytic reactions. J Chem Phys 8: 120–124.

Forsdyke DR (1994) The heat-shock response and the molecular basis of genetic dominance. J Theor Biol 167: 1–5.

Gibson G, Hogness DS (1996) Effect of polymorphism in the *Drosophila* regulatory gene *Ultrabithorax* on homeotic stability. Science 271: 200–203.

Gilbert SF (1994) Dobzhansky, Waddington, and Schmalhausen: Embryology and the Modern Synthesis. In: The Evolution of Theodosius Dobzhansky (Adams MB, ed), 143–154. Princeton, N.J.: Princeton University Press.

Grossniklaus U, Madhusudhan MS, Nanjundiah V (1996) Nonlinear enzyme kinetics can lead to high metabolic flux control coefficients: Implications for the evolution of dominance. J Theor Biol 182: 299–302.

Ho, M-W, Tucker C, Keeley D, Saunders, PT (1983) Effects of successive generations of ether treatment on penetrance and expression of the *Bithorax* phenocopy in *Drosophila melanogaster*. J Exp Zool 225: 357–368.

Hubel D (1988) Eye, Brain, and Vision. San Francisco: Freeman.

Jablonka E, Lamb MJ (1995) Epigenetic Inheritance and Evolution. Oxford: Oxford University Press.

Kacser H, Burns JA (1981) The molecular basis of dominance. Genetics 97: 639–666.

Levin DA (1970) Developmental instability and evolution in peripheral isolates. Am Nat 104: 343–353.

Levins R (1968) Evolution in Changing Environments. Princeton, N.J.: Princeton University Press.

Lewin B (1995) Genes V. Oxford: Oxford University Press.

Mayer MP, Bukau, B (1999) Molecular chaperones: The busy life of Hsp90. Curr Biol 9: R3222–R3225.

McLaren A (1962). Maternal effects in mammals and their experimental analysis. In: Proceedings of the First International Conference on Congenital Malformations, 211–222. Philadelphia: Lippincott.

Merriam-Webster (1993) Merriam-Webster's Collegiate Dictionary (10th ed). Springfield, Mass.

Nanjundiah V (1985) Transposable element copy number and stable polymorphisms. J Genet 64: 127–134.

Nanjundiah V (1993) Why are most mutations recessive? J Genet 72: 85–97.

Nanjundiah V, Bhogle AS (1995) The precision of regulation in *Dictyostelium discoideum:* Implications for cell type proportioning in the absence of spatial pattern. Indian J Biochem Biophys 32: 404–416.

Nathan DF, Vos MH, Lindquist S (1997) *In vivo* functions of the *Saccharomyces cerevisiae* Hsp90 chaperone. Proc Natl Acad Sci USA 94: 12949–12956.

Novick A, Wiener M (1957) Enzyme induction as an all-or-none phenomenon. Proc Natl Acad Sci USA 43: 553–566

Parsons PA (1997) Evolutionary change: A phenomenon of stressful environments. In: The Web of Life (Padmanaban G, Biswas M, Shaila MS, Vishveshvara S, eds), 25–45. Amsterdam: Harwood Academic.

Roemer I, Reik W, Dean W, Klase J (1997) Epigenetic inheritance in the mouse. Curr Biol 7: 277–280.

Rutherford SL, Lindquist S (1998) Hsp90 as a capacitor for morphological evolution. Nature 396: 336–342.

Simpson GG (1953) The Baldwin effect. Evolution 7: 110–117.

Slack JMW (1985) From Egg to Embryo. Cambridge: Cambridge University Press.

Spudich JL, Koshland DE Jr. (1976) Non-genetic individuality: Chance in the single cell. Nature 262: 467–471.

Suzuki DT, Griffiths AJF, Miller JH, Lewontin RC (1986) An Introduction to Genetic Analysis. New York: Freeman.

Waddington CH (1953) Genetic assimilation of an acquired character. Evolution 7: 118–126.

Waddington CH (1956a) Epigenetics and evolution. Symp Soc Exp Biol 7: 186–199.

Waddington CH (1956b) Genetic assimilation of the *bithorax* phenotype. Evolution 10: 1–13.

Waddington CH (1957) The Strategy of the Genes. London: George Allen & Unwin.

Waddington CH (1960) Experiments in canalizing selection. Genet Res Camb 1: 140–150.

Waddington CH (1961) Genetic assimilation. Adv Genet 10: 257–293.

Warburton FE (1956) Genetic assimilation: Adaptation versus adaptability. Evolution 10: 337–339.

West-Eberhard MJ (1986) Alternative adaptations, speciation, and phylogeny. Proc Natl Acad Sci USA 83: 1388–1392.

15 Genetic and Epigenetic Factors in the Origin of the Tetrapod Limb

Günter P. Wagner and Chi-hua Chiu

This chapter discusses the origin of the tetrapod limb from a morphological and developmental perspective. In accordance with the majority view (see Ahlberg and Milner, 1994; Coates, 1994; Sordino and Duboule, 1996; Capdevila and Izpisúa-Belmonte, 2001), we argue that the origin of the tetrapod limb is coincidental with the origin of the autopodium, that is, with the origin of distinct hands and feet in the paired appendages. In clarifying the notion of an autopodium and proposing an evolutionary-developmental scenario for its origination, our goal is to exemplify the abstract concept of an evolutionary innovation. Our review of the material and the argumentation based on it are thus guided by ideas, our own and others', about what that abstract concept means.

Müller and Wagner (1991; see also Müller, chapter 4, this volume) have argued that the concept of evolutionary innovation is intimately connected to the concept of homology, or to the more neutral concept of character identity (also known as the "biological homology concept"). Following the lead of Shubin and Alberch (1986), we think that the best way to define *character* is to identify the morphogenetic rules underlying its development:

A character is a part of the body that develops according to a coherent (phylogenetically stable) set of morphogenetic rules, which make a distinct range of phenotypic states accessible for this body part, but not for other parts of the organism.

This formulation, though new, expresses essentially the same ideas as the definition of *biological homology* in Wagner, 1989. The only difference is that it does not explicitly invoke developmental constraints or directly refer to developmental processes. By implication, an *evolutionary innovation* can be defined as "a part of the body that follows a set of phylogenetically derived morphogenetic rules, which make a distinct set of morphological states accessible to natural variation." From a developmental perspective, this definition implies that an innovation must in all, or at least in many, cases be tied to a certain morphogenetic field (for the notion of a morphogenetic field, see Gilbert, Opitz, and Raff, 1996). An innovation can be realized by an ancestral field, which has acquired a new set of morphogenetic rules, or by an altogether original morphogenetic field, which executes a new set of morphogenetic rules. Based on this way of thinking, we argue that the key developmental difference between a fin and a tetrapod limb is the existence of a morphogenetic field in limbs that does not exist in fins, namely, the autopodial field. We will review evidence showing that the autopodium actually develops from a morphogenetic field distinct from the proximal parts of the developing limb bud.

The example of the tetrapod autopodium also helps to clarify the integral relationship between genetic and epigenetic factors in the origin of morphological novelties. We agree with Seilacher (1991), Müller and Newman (1999), and Newman and Müller (2001) that the origin of new characters requires epigenetic opportunity for the new morphological states to occur. Genetic factors are required for the heritability and subsequent fixation of new morphological states (compare Müller, chapter 4, this volume). This requirement does not imply, however, that the specific nature of a new character is in any sense determined or explained by the mutations that make the character heritable. Genetic variation also plays a more subtle role, which is often overlooked. Indeed, one can infer that a new structure of the phenotype requires a minimal degree of genetic and developmental autonomy, an inference derived from the expectation that characters are not only observable structures of the phenotype but units or building blocks of phenotypic evolution (Wagner, 1995). To have such autonomy, the new character needs to exhibit heritable variability that is to some degree independent of the rest of the phenotype. Because heritable variation is in most cases caused by genetic variation, independent heritable variation of a character needs to be associated with some degree of developmental, genetic autonomy of the character. We thus argue that the epigenetic opportunity for a new character to arise can only be realized if there is also evolution of developmental autonomy and character individuality (see also Müller, chapter 4, this volume).

The Morphological Pattern of the Fin-Limb Transition

As important new fossil evidence has emerged, the paleontological and anatomical evidence connecting tetrapod limbs with fins has been repeatedly reviewed (see Milner, 1988, 1993; Coates, 1991, 1993, 1994; Vorobyeva, 1991; Carroll, 1992; Ahlberg and Milner, 1994; Shubin, 1995). Our present summary, based on these review papers and on original contributions (Long, 1989; Coates and Clack, 1990; Lebedev and Coates, 1995; Cloutier and Ahlberg, 1996; Coates, 1996; Daeschler and Shubin, 1998; Paton, Smithson, and Clack, 1999; Berman, 2000), is intended to clearly define the morphological transformation we seek to explain.

A number of investigators have proposed that the autopodium is the innovation separating the limb from a fin (see, for example, Ahlberg and Milner, 1994; Coates, 1996; Sordino and Duboule, 1996; Capdevila and Izpisúa-Belmonte, 2001). But what exactly is the autopodium? To answer this question, we have organized the comparative anatomical evidence in the phylogenetic framework of vertebrate evolution. Although the phylogenetic branching patterns of some lineages remain unresolved, there is an emerging consensus on the relationships among taxa (see below) crucial to elucidating the fin-limb transition.

Limbs and Fins: A Basic Taxonomy of Terms

The archetypal limb of a tetrapod consists of three major segments: the stylopodium or upper limb; the zeugopodium or lower limb; and the autopodium or hand/foot (figure 15.1A). The

Origin of the Tetrapod Limb

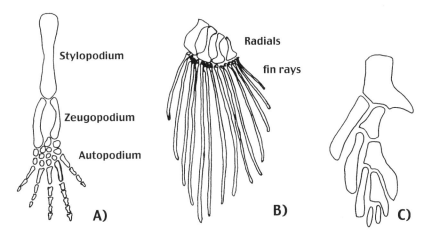

Figure 15.1
Comparison of paired appendage skeletons. (*A*) Typical tetrapod limb with three main segments: the stylopodium, zeugopodium, and autopodium. (*B*) Pectoral fin of a blenny, a perciform fish. The distal elements are fin rays derived from dermal scales and have no counterpart in the tetrapod skeleton. The proximal elements, though endoskeletal, are not homologous to any bone in the tetrapod skeleton. (*C*) Endoskeleton of *Eusthenopteron*, a fossil sarcopterygian relative of tetrapods. Note the branching pattern of skeletal elements, similar to the proximal elements in the tetrapod limb.

stylopodium consists of one long bone: the humerus (attached to the shoulder girdle) in the forelimb; the femur (attached to the pelvic girdle) in the hind limb. The zeugopodium is primarily composed of two long bones: the radius and ulna in the fore limb; the tibia and fibula in the hind limb. The autopodium consists of two segments, a proximal mesopodium and a distal acropodium. The mesopodium, a complex of nodular elements in most tetrapods, is called the "carpus" in the hand and the "tarsus" in the foot. The acropodium is a series of small long bones, the metacarpals in the hand, the metatarsals in the foot, and the digits in both (figure 15.1A).

The typical paired fin of a teleost (e.g., zebrafish) has no specific skeletal elements in common with the tetrapod limb (figure 15.1B). The proximal endoskeletal elements are an anteroposterior series of bones called "radials." Distal to these radials is a row of small cartilages called "distal radials." The most distal skeletal elements are the fin rays, lepidotrichia and actinotrichia, which belong to the dermal skeleton. The tetrapod limb contains no skeletal elements derived from fin rays. The connection between the fin and the limb, however, becomes more evident upon examination of the more complex endoskeletal fin structures of sharks, basal ray-finned fishes such as sturgeons, and the sarcopterygian (lobe-finned) fishes, from which the tetrapods are derived (Janvier, 1996; figure 15.1C).

The tetrapod limb is derived from a posterior part of the fin endoskeleton of elasmobranchs and basal bony fish, the so-called metapterygium, a series of endoskeletal elements that is the first to form in the developing paired fins (Braus, 1906; Shubin, 1995; Mabee, 2000). It arises in close connection to the girdle and, in turn, gives rise to a series of variable elements, usually at its anterior edge. In addition, an independent endoskeletal element called the "protopterygium" develops anterior to the metapterygium in many basal fishes (e.g., bichirs and sturgeons). Whereas teleosts have lost the metapterygium, sarcopterygians have lost the protopterygium and thus develop all their endoskeletal structures, to include the tetrapod limb skeleton, from the metapterygium. The difference between the tetrapod limb and the teleost fin may be explained, then, by a complementary trend in the importance of the metapterygium. These observations limit the usefulness of comparisons between zebrafish fin development and limb development to the most general features, such as the presence of a zone of polarizing activity (ZPA): no specific comparisons are possible between the skeletal elements of these paired appendages.

The Phylogenetic Position of Tetrapods

The hypothesis that tetrapods and sarcopterygian fishes form a clade is widely supported (see Hedges, Moberg, and Maxson, 1990; Schultze and Trueb, 1991; Cloutier and Ahlberg, 1996; Zardoya and Meyer, 1997). The question as to which of the two extant sarcopterygian fish lineages, the lungfish (three extant genera) or the coelacanth *Latimeria,* is closer to the tetrapods remains open (see Rosen et al., 1981; Panchen and Smithson, 1987; Chang, 1991; Schultze, 1991; Hedges and Maxson, 1993; Zardoya, Abouheif, and Meyer, 1996; Zardoya and Meyer, 1997). Because neither of these taxa represents the character state from which the limb is derived (Vorobyeva, 1991; Shubin, 1995), this uncertainty is not relevant to our discussion. There is strong evidence that the panderichthyids are the sister group to the tetrapods and that the osteolepiforms, typified by the well-known *Eusthenopteron,* are the sister group to the panderichthyid-tetrapod clade (Long, 1989; Coates, 1991, 1994; Vorobyeva and Schultze, 1991; Ahlberg and Milner, 1994; Shubin, 1995; Cloutier and Ahlberg, 1996; figure 15.2).

The panderichthyids, a group of Devonian sarcopterygians, share a number of cranial and postcranial characters with the early tetrapods, but not the structure of the distal parts of their paired appendages (Schultze and Arsenault, 1985; Vorobyeva and Schultze, 1991). This group can thus be viewed as tetrapods with paired fins. Like the most basal tetrapods, and much like extant crocodiles, these creatures were shallow water predators (Coates and Clack, 1990; Ahlberg and Milner, 1994).

A third group of lobe-finned fish, the rhizodontids (whose interesting fin structure is discussed below) may be the sister taxon to the ((tetrapod) panderichthyld) osteoleopiform clade, as proposed by Schultze (1987), Long (1989), and Cloutier and Ahlberg (1996), although the phylogenetic position of this group is still a matter of debate. The other sarcopterygians (e.g., lungfish, coelacanths, and porolepiforms) have fin structures that bear little resemblance to the early tetrapods and their immediate relatives.

Origin of the Tetrapod Limb

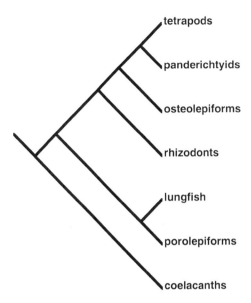

Figure 15.2
Hypothetical phylogenetic branching pattern among sarcopterygian relatives of tetrapods. The phylogeny is simplified from (Cloutier and Ahlberg, 1996, figure 4.)

Phylogeny of Tetrapods

There is strong support for monophyly of all extant and fossil tetrapods (Panchen and Smithson, 1988; Ahlberg and Milner, 1994; Carroll, 1995; Coates, 1996; Janvier, 1996; Laurin, 1998), of amniotes, and of the lissamphibians (frogs, salamanders, and gymnophions; Cannatella and Hillis, 1993; Hedges and Maxson, 1993; Gauthier, Kluge, and Rowe, 1988; Laurin, 1998; figure 15.3). Relationships among the many Carboniferous amphibians and the Lissamphibia, however, remain unresolved (Carroll, 1992, 1995; Milner, 1993; Ahlberg and Clack, 1998; Laurin, 1998; Berman, 2000; Coates, Ruta, and Milner, 2000; Laurin, Gorondot, and Ricqlés, 2000).

Evidence from trace fossils of the middle Upper Devonian, about 370 million years ago, indicates that tetrapod limbs originated in the Devonian (Vorobyeva, 1977; Ahlberg and Milner, 1994). Most of the anatomical evidence about the structure of primitive tetrapod limbs stems from the fossils of three Devonian tetrapods, *Acanthostega, Ichthyostega,* and *Tulerpeton,* found in Late Famennian layers about 362 million years ago (Ahlberg and Milner, 1994). According to cladistic analyses, the Devonian tetrapods are offshoots from the tetrapod stem lineage (Ahlberg and Clack, 1998; Laurin, 1998), thus diverged before the most recent common ancestor of the extant tetrapods (figure 15.3). This phylogenetic

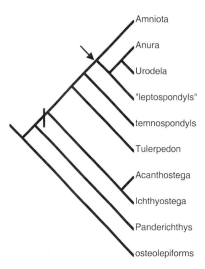

Figure 15.3
Hypothetical phylogenetic branching patterns among major tetrapod taxa, simplified after Ahlberg and Milner, 1994. All recent forms are derived from an ancestor more recent than any of the known Devonian tetrapods. This hypothesis implies that the pentadactyl autopodium arose only once, before the emergence of the most recent common ancestor of living tetrapods (Laurin, 1998).

hypothesis places the most recent common ancestor of extant tetrapods in the Lower Carboniferous period at about 340 million years ago (Paton, Smithson, and Clack, 1999; Laurin, Gorondot, and Ricqlés, 2000).

Stages in the Acquisition of Tetrapod Limb Characters

From the phylogenetic history outlined above, it is clear that the origin of modern tetrapod limbs was not a single event, but instead a series of transformations that continued after the origin of the first unambiguous tetrapod limbs, as shown by the difference between the Devonian and Carboniferous tetrapods. There are at least two major steps to be distinguished: first, the origin of the tetrapod autopodium (which we morphologically define as the "fin-limb transition"); and second, the transformation of the archaic autopodium of Devonian tetrapods into the pentadactyl autopodium of the extant tetrapods (Coates, 1994, 1996).

The three main outgroup taxa of tetrapods, panderichthyids, osteolepiforms, and rhizodontids, have endoskeletal elements corresponding to the stylopodial and zeugopodial elements in a tetrapod limb (Coates, 1991; Vorobyeva, 1991; Ahlberg and Milner, 1994; Shubin, 1995). In addition there are elements that share the position and possibly the developmental derivation of the ulnare and the intermedium. From these observations, most authors have concluded that the stylopodial, zeugopodial, and proximal mesopodial elements have counterparts in the fins of tetrapod ancestors. Moreover, the lineage that leads to the tetrapod ancestor shows a notable tendency to reduce the complexity of the endoskeleton distal to the zeugopodial segment (Coates, 2001). This trend culminated in the pectoral fin of *Panderichthys,* which has only two distal elements, an elongated element corresponding to the intermedium and large bony plate corresponding to the ulnare (Vorobyeva and Schultze, 1991). This pattern suggests that the autopodium did not arise from a transformation of the distal fin skeleton but from largely new elements, with only a few homologues in the fin skeleton.

The neomorphic nature of the autopodium is also reflected in the problems to homologize the digits, which most authors identify as homologues to the radials of a sarcopterygian fin (e.g., Coates, 1994; but see Coates, Ruta, and Milner, 2000). If digits are "segmented radials that do not support fin rays" (Coates, 1994), the question remains, to which radials do they correspond? Given that in most recent tetrapods the digits derive from the digital arch in a sequence from posterior to anterior (at least digits DIV, DIII, and DII, but not DV; Burke and Alberch, 1985; and perhaps not even DI; see original data on the alligator in Müller and Alberch, 1990, figures 4 and 7), it is tempting to assume that the digits correspond to postaxial radials (Ahlberg and Milner, 1994). Postaxial radials, however, are not described among the close outgroups of tetrapods. Osteolepiforms tend to only have anterior radials and rhizodontids have terminal radials, much like digits (e.g., *Barameda;* Long, 1989; *Sauripterus;* Daeschler and Shubin, 1998). Postaxial radials are present in lungfish, the coelacanth *Latimeria,* and the shark *Xenacanthus* (Braus, 1906; Shubin, 1995) but these lineages are not directly ancestral to tetrapods. Finally, cell fate mapping in bird limb buds does not show a "bending" of posterior growth axis in the autopodium (Vargesson et al., 1997). Hence it is not obvious that digits and the digital arch can be understood as a bent metapterygial axis as proposed by (Shubin and Alberch, 1986). Below we argue that the digital arch may have evolved during the stabilization of the pentadactyl autopodium, rather then during the fin-limb transition itself.

The only consistent differences between sarcopterygian fins and the variety of primitive tetrapod limbs are the mesopodial-acropodial pattern of skeletal elements in the autopodium and the absence of fin rays. This implies the coincident origin of digits and the mesopodium, carpus and tarsus. The typical mesopodium of extant tetrapods consists of a complex array of three kinds of nodular elements: the proximal tarsals (ulnare/fibulare, intermedium, and radiale/tibiale), the central carpals/tarsals, and the distal carpals/tarsals supporting the metapodial elements. Among the recent tetrapods, there are a few examples in which tarsal or carpal elements are secondarily elongated (Blanco, Misof, and Wagner, 1998). We will discuss these exceptions below.

Not all mesopodial elements of crown group tetrapods are found in the most basal stem tetrapods. The elaboration of the mesopodium occurred after the origin of the digits

(Smithson et al., 1993; Coates, 1996). The carpus of *Acanthostega* is not known except for the presence of an elongated fibulare (Coates and Clack, 1990). The tarsus of *Acanthostega*, however, consists of elements corresponding to the proximal tarsals (tibiale, intermedium and fibulare) and four distal tarsals supporting digits DII, DIII, DIV and DV, although the preservation of the specimen suggests that some additional elements may have been lost (Coates, 1996). The hind limb of *Ichthyostega* also has the proximal tarsals and two distal tarsals, as well as one central element wedged between the intermedium and the two distal carpals (Coates and Clack, 1990). Even among stem amniotes, the tarsus is still more primitive than in crown amniotes (Smithson et al., 1993). Hence the mesopodium consists of plesiomorphic elements that were integrated and transformed into the mesopodium (the ulnare/fibulare and the intermedium) and of new elements, many arising after the origin of the autopodium. Clearly, not all mesopodial elements of Devonian forms are nodular (e.g., the carpal intermedium of *Acanthostega*) and many elements are added later. But all tetrapod limbs have some nodular elements inserted between the zeugopodium and the digits (in the structure we refer to as the "mesopodium"), regardless of the degree of elaboration.

We conclude that a developmental scenario for the origin of the autopodium has to account for the origin of a zeugopodial-mesopodial transition but not necessarily for the completely elaborated mesopodium seen in modern tetrapods. This transition corresponds to a marked difference in skeletogenetic mode, from the development of large elongated elements to smaller and most often nodular elements, that occurs in all tetrapod limbs but not in any sarcopterygian fin.

After the establishment of the meso-acropodial pattern in the Devonian, the tetrapod limb continued to evolve. The Devonian forms have an autopodium that is structurally distinct from all the limbs of extant tetrapods and from all known limbs of Carboniferous forms (Coates, 1991, 1994). These forms are all polydactylous, ranging from eight digits in *Acanthostega* (Coates and Clack, 1990) to six digits in *Tulerpeton* (Lebedev and Coates, 1995). Furthermore, the *Ichthyostega* foot (the hand is not known) is heterodactylous, which means that the digits are heterogeneous in size trends and cross section (Coates and Clack, 1990). Finally, the number of mesopodial elements is smaller, as discussed above. Similarly, the urodeles have a radically different mode of hand/foot development than all other extant tetrapods (Braus, 1906). From this it is clear that the pentadactyl tetrapod limb morphology stabilized after the actual fin-limb transition (Ahlberg and Milner, 1994; Coates, 1994, 1996; Laurin, 1998; Paton, Smithson, and Clack, 1999; Laurin, Gorondot, and Ricqlés, 2000). The question thus arises, how does the development of extant tetrapod limbs relate to the morphology of stem tetrapods, with their polydactylous limbs and primitive mesopodial (Wagner, Chiu, and Laublichler, 2000)?

Recent tetrapods differ in the mode of digit development. There are at least four modes for deployment of digits (figure 15.4). In the most common mode, the digital arch grows from the ulnare/fibulare in a posterior to anterior direction and digits sprout from the postaxial side from the digital arch (figure 15.4a). Digit-forming condensations without connection to the digital arch have also been observed; these give rise to the most posterior digits in the amniotes, DV, and sometimes also to the most anterior digit, DI (see, for example, Burke and Alberch, 1985; Müller and Alberch, 1990; Burke and Feduccia, 1997). A single digit can develop from the radiale and tibiale each, the prehallux and the prepollex (figure 15.4b). The digits I and II in urodeles are developmentally derived from the intermedium (Schmalhausen, 1910; Hinchliffe, 1991; Blanco and Alberch, 1992; Vorobyeva and Hinchliffe, 1996; Hinchliffe and Vorobyeva, 1999; figure 15.4c). Hence the digital arch is certainly not the only mode for deployment of digits in limb development, and there is variation in the "digitogenic pathways" among recent tetrapods. We thus suggest that, in the limb buds of Devonian tetrapods, several digitogenic pathways might have been used simultaneously, which may account for the higher digit number compared to extant tetrapods. This hypothesis may also account for heterodactly in *Ichthyostega* (Coates, 1991), assuming that digits with different morphologies are derived from different digitogenic pathways. These suggestions are testable with loss-of-function mutations (see below). The stabilization of the autopodial morphology may have then resulted from suppression of some the digitogenic pathways, such as that from the radiale/tibiale in amniotes, and the expansion of the digital arch. Consequently, the extent of the digital arch found in amniotes and frogs may be a derived developmental character of extant eutetrapods. We therefore

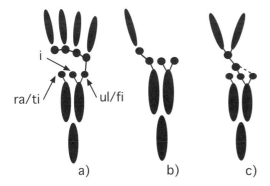

Figure 15.4
Modes of digit development, corresponding to the three proximal mesopodial elements in the tetrapod limb, which are the radiale/tibiale, the intermedium, and the ulnare/fibulare. (*a*) In almost all recent tetrapods most digits are derived from the digital arch which is emanating from the ulnare/fibulare. (*b*) One digit also can arise from the radiale or tibiale and is called the "prepollex" or "prehallux," respectively. A predigit is a common feature of anuran feet, and is an occasional natural variant in newts (J. Rienesl and G.P. Wagner) and some lizards. (*c*) The third mode of digit development is connected to the intermedium and is seen in many extant urodele species (Schmalhausen, 1910; Blanco and Alberch, 1992; Vorobyeva and Hinchliffe, 1996; Vorobyeva, Ol'shevskaya, and Hinchliffe, 1997).

conclude that the digitogenic pathway of most recent tetrapods may not be a guide to the developmental mechanisms for the transformation of fins to limbs. In particular, the digital arch may not be a defining feature of the autopodium.

Definition of the Autopodium

According to our hypothesis, the autopodium can be defined as the distal segment of a vertebrate paired appendage that consists of mesopodial elements, which are mostly nodular, and acropodial elements, which are an anteroposterior series of small long bones (metacarpals, metatarsals, and phalanges). The acropodial elements are characteristically separated from the zeugopodium by one or more rows of mesopodial elements: this is the only consistent morphological difference between fins and limbs, whether one interprets the acropodial elements as radials or not.

There are two recent tetrapod groups in which the proximal mesopodial elements have been transformed into two long bones resembling zeugopodial elements: the anurans and the crocodilians, with elongated tarsal elements (Blanco, Misof, and Wagner, 1998) and carpal elements (O.C. Rieppel, unpublished), respectively. In both cases, the proximal tarsals/carpals are true long bones with a bony collar and cartilaginous distal and proximal ends. These elements ossify together with the other long bones rather than with the other mesopodial elements. There is evidence that the transformation of the anuran tarsal elements represents a distal shift in the zeugo-autopodial border (Blanco, Misof, and Wagner, 1998). Among the primates, elongated bones develop in the tarsus of the galago *Otolemur* and the tarsier *Tarsius,* but there is no conclusive evidence as to the mode of ossification (G.P. Wagner and C.-h. Chiu, unpublished). These transformations are complementary to the "mesopodialization," that is, proximal shift of the zeugo-autopodial border, observed in aquatic reptiles (Caldwell, 1997). These exceptions are likely due to evolutionary variation in the zeugo-autopodial border, as suggested by Blanco and collaborators (Blanco, Misof, and Wagner, 1998), and are not in contradiction with the definition of the autopodium as a configuration of mesopodial and acropodial segments.

Based on this definition of an autopodium, the two critical questions regarding the origin of the autopodium are the following. What are the genetic and developmental mechanisms that establish the zeugo-mesopodial boundary? Is the origin of these mechanisms also involved in the origin of the autopodium?

Development of the Autopodium

A *morphological novelty* can be defined as a character derived for a clade (i.e., autapomorphic), hence, not present in the ancestor of a more inclusive clade (Müller and Wagner, 1991). It follows that a morphological novelty is correlated with changes to an existing developmental program or creation of a new developmental pathway. Elucidating how the zeugopodial-mesopodial transition, the innovation in tetrapod limb evolution, arose requires a brief review of the genetic factors involved in limb development.

Tetrapod limbs originate from groups of cells in the lateral plate mesoderm and develop into mesenchymal buds surrounded by ectoderm (Searls and Janners, 1971). As growth continues distally, part of the ectoderm thickens, forming the apical ectodermal ridge (AER). In amniotes the AER is essential for continued cell proliferation during limb bud growth. If the AER is removed, cell proliferation is reduced, leading to truncated limbs (Saunders, 1948). Not all tetrapods have an AER, however; for example, urodeles (Karczmar and Berg, 1951) and the directly developing frog *Eleuterodactlyus coqui* (Richardson et al., 1998) do not. Most strikingly, removal of the distal ectodermal cup of the limb buds of urodeles and *E. coqui* does not lead to an arrest of limb bud growth (Lauthier, 1985; Richardson et al., 1998). Although it is still a matter of debate whether fish (e.g., zebrafish) have an "AER-like" structure (Geraudie, 1978; Grandel and Schulte-Merker, 1998), recent genetic evidence suggests that the fin fold has AER activity (Neumann et al., 1999). A group of mesenchymal cells located in the posterior margin of the developing limb bud forms the zone of polarizing activity (ZPA), which controls patterning along the anteroposterior axis mediated by *sonic hedgehog* (*shh*; Saunders and Gasseling, 1968; Riddle et al., 1993). The mesenchymal cells at the distal end of the developing limb bud form the progress zone (PZ), where cell proliferation is maintained by signaling from the AER (Summerbell, Lewis, and Wolpert, 1973), which in turn is maintained by the ZPA as development proceeds. This signaling mechanism provides cells of the PZ positional clues as they later develop most of the endoskeletal elements of the limb in a proximal to distal sequence (Summerbell, Lewis, and Wolpert, 1973).

Several signaling molecules (e.g., *shh, engrailed,* retinoic acid, *wnt,* fibroblast growth factors or FGFs), and transforming growth factors or TGFs that are involved in developmental patterning and growth of the limb have been identified (reviewed in Schwabe, Rodriguez-Esteban, and Izpisúa-Belmonte, 1998). In this chapter, we focus on evidence for the developmental autonomy of the autopodium and on the current evidence about the developmental origin of the zeugopodial-autopodial transition.

Evidence for the Developmental Autonomy of the Autopodium

Three distinct phases of expression of the *Abd-B*-like HoxA and HoxD group 9–13 genes in developing chick and mouse limb buds have been described (Nelson et al., 1996). In the first phase, group 9 (*Hoxa-9, Hoxd-9*) and group 10 (*Hoxa-10, Hoxd-10*) genes are expressed uniformly in the mesoderm. Group 11 through 13 genes are not expressed in phase one. In phase two, the *Hoxd-9* through *Hoxd-13* genes are sequentially activated at the posterior-distal edge of the limb bud. With the exception of *Hoxa-13,* which is expressed only during phase three, the *Hoxa* genes are expressed uniformly in phase two.

During phase three, which corresponds to the stage in development when autopodium skeletal elements are formed, *Hoxd-13* through *Hoxd-10* are sequentially activated in reverse order, breaking the "temporal" and "spatial" colinearity rule (Nelson et al., 1996). Transgenic experiments have shown that, whereas expression of *Hoxd* genes in the early phases is regulated by several enhancer elements of each locus (Beckers, Gérard, and Duboule, 1996; Hoeven, Zákány, and Duboule, 1996) in the third phase, expression of all *Hoxd* genes is controlled by a single "global" enhancer (Hérault et al., 1999). The phase-three expression of *Hoxa-13* depends on fibroblast growth factors (FGFs) secreted from the apical ectodermal ridge in the chick limb, but FGF-4-soaked beads cannot activate *Hoxa-13* expression in phase two (Vargesson et al., 2001). This indicates that *Hoxa-13* is under different control in phase three than in phase two. Interestingly, phase-three expression of *Hoxa-13* appears earlier than that of *Hoxd-13,* suggesting that genes on the HoxA cluster may be "upstream" to genes on the HoxD cluster and that paralogous genes of the A and D clusters are under different regulatory controls during autopodial development (Nelson et al., 1996). This is supported by analysis of mice mutant for posterior genes of the HoxD or HoxA clusters where loss of function alleles lead to polydactyly (HoxD) or to loss of digits (HoxA; Zákány et al., 1997).

Hoxa-11/Hoxd-11 double-knockout mice have relatively normal upper limbs and hands, whereas the long bones of the lower arm are reduced to nodular elements (Davis et al., 1995). In contrast, *Hoxa-13/Hoxd-13* double-knockout mice have relatively normal upper and lower limbs, whereas their hands/feet are severely abnormal (Fromental-Ramain et al., 1996). Overexpression of *Hoxa-13* in the chick wing leads to a loss of the long bone character of the ulna and radius (randomization of the orientation of mitosis of chondrocytes) coupled with the development of several small ectopic cartilages, reminiscent of mesopodial elements (Yokouchi et al., 1995).

There is also evidence that chondrification of the mesenchymal condensations in the proximal limb bud or in the autopodial anlage is caused by different molecular mechanisms. Activin A is a member of the TGF-β superfamily of growth factors (Stern et al., 1995), which is antagonized by follistatin (DeWinter et al., 1996). Activin A plays a role in chondrogenesis during digit formation (Merino et al., 1999). This activity can be inhibited by follistatin treatment. Interestingly, activin A is not able to induce ectopic chondrogenesis in early stages of limb development. In addition, follistatin inhibits cartilage formation in the autopodium but not in the proximal regions of the limb bud (Merino et al., 1999). This indicates that chondrogenesis is induced through different molecular pathways in the autopodium and in the proximal parts of the limb.

Development of the Zeugopodial-Autopodial Transition

The expression domains of *Hoxa-11* and *Hoxa-13* in the mouse (Haack and Gruss, 1993) and the chick (Yokouchi, Sasaki, and Kuroiwa, 1991) are mutually exclusive: *Hoxa-11* is restricted to the zeugopodium, whereas *Hoxa-13* is expressed only in the autopodium proper. The restriction of *Hoxa-11* to the zeugopodial-autopodial boundary has also been shown for *Xenopus* (Blanco, Misof, and Wagner, 1998), supporting the hypothesis that this expression dynamic was already present in the most recent common ancestor of extant tetrapods. Distal displacement of the *Hoxa-11* expression domain in chick limbs leads to a loss of the zeugopodial-autopodial transition (Mercanter et al., 1999). These findings suggest that *Hoxa-11* and *Hoxa-13* are involved in determining the hand/foot field, which is to say, the limit between the developing zeugopodium and the developing autopodium.

In striking contrast to tetrapods, the expression domains of *Hoxa-11* and *Hoxa-13* orthologues in the paired fin development in the teleost zebrafish are overlapping (Sordino, Hoeven, and Duboule, 1995; Sordino and Duboule, 1996). However, the situation is complicated by the recent discovery that zebrafish possess two HoxA clusters (a and b), each containing a group 11 gene (*Hoxa-11a*, *Hoxa-11b*) and a group 13 gene (*Hoxa-13a*, *Hoxa-13b*; Amores et al., 1998). In addition, *Hoxa-11* is also expressed in the cells that enter the fin fold (C.-h. Chiu and C. Pazmandi, unpublished), which is a cell population not found in limb buds. As we shall see, in the following section, evolution of *Hoxa-11* and *Hoxa-13* regulation may have been a key step in the fin-limb transition.

Genetic Hypotheses for the Origin of the Autopodium

Two specific hypotheses have been put forth to explain the origin of the autopodium by a genetic mechanism. One is related to the maintenance of the progress zone and its associated interactions between the zone of polarizing activity (ZPA) and the apical ectodermal ridge (AER); Thorogood, 1991; Sordino and Duboule, 1996, whereas the other focuses on Hox gene regulation in the autopodial anlage (Gerard, Duboule, and Zákány, 1993; Hoeven, Zákány, and Duboule, 1996).

The development of the distal parts of the tetrapod limb depends on the sustained activity of the progress zone, which in turn depends on the activity of the AER (at least in most tetrapods; see above). The AER in turn is dependent on a sustained interaction with the ZPA (see above). Geraudie (1978) and Thorogood (1991) have proposed that the absence of distal endoskeletal structures in the actinopterygian fin is due to the premature cessation of AER-like activity because the ectoderm folds onto itself to become the fin fold. In support of this hypothesis, Sordino and Duboule (Sordino, Hoeven, and Duboule, 1995) reported that the expression dynamics of *sonic hedgehog* (*Shh*), a genetic marker of the zone of polarizing activity, differ in developing fin and limb buds. In zebrafish, *Shh* expression remains in a proximal location, which is consistent with the idea that fish lack a "distal" phase of ZPA-AER-like interaction. This finding could explain the absence of progress-zone-mediated growth and the formation of distal endoskeletal structures. In tetrapods, in contrast, *Shh* expression moves distally as the limb bud grows. From these observations, it has been hypothesized that the origin of the autopodium is due to a distalization of the ZPA.

The second hypothesis is based on the surprising discovery that the inverted colinearity of HoxD gene expression in the autopodium (Nelson et al., 1996) is caused by a single enhancer element (Gerard, Duboule, and Zákány, 1993; Hoeven, Zákány, and Duboule, 1996). It is thus easy to imagine that the autopodial expression pattern of Hox genes resulted from a few mutations. Indeed, the acquisition of this enhancer element may have been a key step in the origin of the autopodium and may be responsible for the postero-anterior development of the digital arch. Below we evaluate these two hypotheses and propose a third, which is complementary, rather than alternative, to at least one of the proposals reviewed above.

The paleontological evidence reviewed in "The Morphological Pattern of the Fin-Limb Transition" above indicates that the tetrapod limb is derived from the paired fins of sarcopterygian fishes, and that the closest known relative of tetrapods, the panderichthyids, only possess two distal endoskeletal elements, whereas all the other outgroups have many more (e.g., *Eusthenopteron* and *Sauripterus*). From this observation, we conclude that the origin of the autopodium is not coincidental with the first appearance of additional endoskeletal elements distal to the putative zeugopodial homologue in *Panderichthys*. Because the tetrapod limb is not derived from the actinopterygian fin, the Thorogood-Sordino-Duboule hypothesis may account for the stunted development of the actinopterygian fin, as exemplified by zebrafish, but cannot account for the origin of the autopodium. The appendages ancestral to the tetrapod limb possessed endoskeletal elements distal to the zeugopodium. Therefore the lack of distal skeletal elements does not in itself account for the difference between a sarcopterygian fin and a tetrapod limb: the distal skeletal elements in the sarcopterygian fins do not form an autopodial configuration and are not obviously individualized from the proximal parts, as is the autopodium of tetrapods.

From the above reasoning, it is possible that the origin of the global enhancer element in the HoxD cluster may have caused the origin of the digital arch and other osteological features specific to the autopodium. Interestingly, knockout phenotypes of *AbdB*-like genes from the A and D cluster of the mouse show that the deletion of D cluster genes leads to a polydactylous phenotype with fully formed but shortened digits, whereas the deletion of A cluster genes leads to digit loss (Zákány et al., 1997). Zákány and colleagues have suggested that the autopodial enhancer acts downstream of *Hoxa-13*, which determines the distal part of the limb bud to become an autopodium and that the HoxD cluster gene function is phylogenetically derived relative to the functional role of *Hoxa-13* in autopodium development. We therefore propose that the critical developmental change underlying the morphological innovation (zeugopodial-mesopodial transition) is the origin of the genetic mechanism responsible for determining the autopodial field.

The developmental genetic evidence reviewed above indicates that the spatially exclusive expression of *Hoxa-11* and *Hoxa-13* is involved in the determination of the autopodial field. We therefore hypothesize that the evolution of the *Hoxa-11/Hoxa-13* expression pattern may be causally involved in the origin of the autopodium. Although generally consistent with the hypothesis that the third phase of Hox gene expression is involved in the origin of the tetrapod limb (Zákány and Duboule, 1999) our hypothesis differs from it in several specific respects. In particular, it is clear that function of the 5′ HoxD genes is not necessary for digit development (Zákány et al., 1997). Further, we assume that the *Hoxa-13* function is an apomorphic character of tetrapods rather than plesiomorphic, as suggested by Zákány and colleagues (1997).

The expression patterns of *Hoxa-11* and *Hoxa-13* in forms basal to the tetrapod lineage are not known. In the zebrafish pectoral fin bud, these genes have overlapping expression domains (Sordino, Hoeven, and Duboule, 1995; see discussion above). Interestingly, *Hoxa-11* (the b paralogue; C.-h. Chiu) and *Hoxa-13* (it is not yet clear which paralogue) overlap completely in the distal part of the fin bud, but not proximally (Neumann et al., 1999). Hence there is already a proximodistal difference in the *Hoxa-11/Hoxa-13* expression, with *Hoxa-13* being expressed only distally. But there is no local exclusivity of *Hoxa-11* and *Hoxa-13*. There is also some indirect evidence that *Hoxa-13* is regulated by *Shh* and hence by the zone of polarizing activity of the fin bud. In the zebrafish mutant *sonic you* (*syu*), a loss-of-function mutation of the zebrafish orthologue of *Shh* (Schauerte et al., 1998), expression of *Hoxa-13* is lost in the fin bud (Neumann et al., 1999). Moreover, in *syu* mutants, *Hoxa-11* is only expressed proximally, but not in the region where it overlaps with *Hoxa-13* in the wild type. This finding could mean that there is already a ZPA-dependent expression of *Hoxa-13* in fish, just as in the tetrapod autopodium (Vargesson et al., 2001), but no distal suppression of *Hoxa-11*. One has to note, however, that the regulation of *Hoxa-13* by *Shh* has not been demonstrated beyond reasonable doubt: the lack of expression of *Hoxa-13* in *syu* mutants could also be due to the stunted fin bud development typical for these mutant phenotypes. Regardless of whether *Hoxa-13* is directly regulated by *Shh* (Neumann et al., 1999) or by the FGFs from the AER (Vargesson et al., 2001), the zebrafish expression patterns suggest that the main genetic change necessary to establish a tetrapod-like expression pattern may be *cis*-regulatory mutations at the *Hoxa-11* locus, leading to the derived status of distal repression of *Hoxa-11*.

The situation in zebrafish and other teleosts is more complex, however, because they possess at least two copies (paralogues) of *Hoxa* genes of which there is only a single counterpart (orthologue) in tetrapods (Amores et al., 1998). Awareness of this finding has not yet penetrated the developmental literature: much of the expression data was reported before the discovery of the additional gene copies in zebrafish; even some recent developmental papers do not take notice of this fact.

Another caveat in interpreting the zebrafish results is that it is not clear which cell types express *Hoxa-11* in the fin bud. Antibody staining by C.-h. Chiu and C. Pazmandi, unpublished, clearly shows that *Hoxa-11* is also expressed in the cells that enter the fin fold and presumably contribute to the development of the fin rays. In tetrapods, there is no corresponding cell population, which makes the comparison of "expression patterns" between teleosts and tetrapods difficult: such patterns may reflect expression in nonhomologous cell populations.

All these are compelling reasons to move from zebrafish to nontetrapod species that represent more closely the character state ancestral for tetrapods. Prime candidates are the extant sarcopterygians, lungfish, and coelacanth, as well as basal ray-finned fishes such as bichirs and sturgeons. The closest relatives of tetrapods among extant taxa are the lungfish, in which a study of gene expression, though difficult, is technically possible. On the other hand, the lungfish paired fin skeleton is quite different from that of osteolepiforms. Indeed, among extant forms, the metapterygium most similar to that known from osteolepiforms is found in sturgeons and the paddlefish *Polyodon* (Mabee, 2000). Hence the lungfish may not represent the character state most similar to that ancestral for tetrapods, which implies that developmental data from both extant lungfish as well as several basal ray-finned fish are needed to infer the ancestral genetic regulatory network. The situation is complicated, by Cloutier and Ahlberg's phylogenetic analysis (1996), however, which implies that lungfish fins may represent the ancestral fin skeleton for the Tetrapodomorpha (see Coates, 2001). Thus the similarity between the paddlefish metapterygium and that of the osteolepiforms might not reflect inheritance of an ancestral character state.

Conclusion

The development of the tetrapod limb highlights distinct differences in the mode of pattern formation between the phylogenetically old, proximal parts of the limb and the recent (apomorphic) distal autopodium. In the autopodium, the skeletal condensations are small and display different developmental polarities with a tendency for development to progress in the anteroposterior direction. These differences are mediated through changes in the cell-cell and cell-matrix interactions (Newman, 1996; see also Bissell et al., Larsen, Steinberg, and Newman, chapters 7, 8, 9, and 13, this volume) and thus represent the epigenetic basis for the evolutionary origin of the autopodium as a morphological unit distinct from any fin structure. In addition, this new mode of skeletogenesis occurs in an apomorphic developmental field that represents a different mode of gene regulation. We conclude that the evolutionary origin of the autopodium consists of both a change in the mode of pattern formation and the origin of a new developmental field possessing distinct modes of gene regulation that can accommodate the development of novel structures.

Acknowledgments

The work of this chapter was supported, in part, by National Science Foundation grants IBN-9507046 and IBN-9905403 (to Günter Wagner) and by a National Science Foundation–Sloan Postdoctoral Research Fellowship in Molecular Evolution (to Chi-hua Chiu). We thank Michael Coates and Neil Shubin for clarifying the paleontological data, and we thank the members of the Ruddle and Wagner laboratories for extensive and stimulating discussions of the topic of this chapter.

References

Ahlberg PE, Clack JA (1998) Lower jaws, lower tetrapods: A review based on the Devonian genus *Acanthostega*. Trans R Soc Edin Earth Sci 89: 11–46.

Ahlberg PE, Milner AR (1994) The origin and early diversification of tetrapods. Nature 368: 507–514.

Amores A, Force A, Yan Y-L, Joly L, Amemiya C, et al. (1998) Zebrafish Hox clusters and vertebrate evolution. Science 282: 1711–1714.

Beckers J, Gérard M, Duboule D (1996) Transgenic analysis of a potential *Hoxd-11* limb regulatory element present in tetrapods and fish. Dev Biol 180: 543–553.

Berman DS (2000) Origin and early evolution of the amniote occiput. J Paleontol 74: 938–956.

Blanco MJ, Alberch P (1992) Caenogenesis, development variability, and evolution in the carpus and tarsus of the marbled newt *Triturus marmoratus*. Evolution 46: 677–687.

Blanco MJ, Misof BY, Wagner GP (1998) Heterochronic differences of *Hoxa-11* expression in *Xenopus* fore- and hind limb development: Evidence for a lower limb identity of the anuran ankle bones. Dev Genes Evol 208: 175–187.

Braus H (1906) Die Entwicklung der Form der Extremitäten und des Extremitätenskeletts. In: Handbuch der vergleichenden und experimentellen Entwicklungslehre der Wirbeltiere (Hertwig O, ed), 167–338. Jena: Gustav Fisher.

Burke AC, Alberch P (1985) The development and homology of the chelonian carpus and tarsus. J Morphol 186: 119–131.

Burke AC, Feduccia A (1997) Developmental patterns and the identification of homologies in the avian hand. Science 278: 666–668.

Caldwell MW (1997) Modified perichondral ossification and the evolution of paddle-like limbs in ichthyosaurs and plesiosaurs. J Vert Paleontol 17: 534–547.

Cannatella DC, Hillis DM (1993) Amphibian relationships: Phylogenetic analysis of morphology and molecules. Herpetol Monog 7: 1–7.

Capdevila J, Izpisúa-Belmonte JC (2001) Perspectives on the evolutionary origin of tetrapod limbs. In: The Character Concept in Evolutionary Biology (Wagner GP, ed), 531–558. San Diego, Calif.: Academic Press.

Carroll RL (1992) The primary radiation of terrestrial vertebrates. Annu Rev Earth Planet Sci 20: 45–84.

Carroll RL (1995) Problems of the phylogenetic analysis of paleozoic choanates. Bull Mus Natl Hist Nat, Paris 4 sér. 17: 389–445.

Chang M-M (1991) "Rhipidistans," dipnoans, and tetrapods. In: Origins of the Higher Groups of Tetrapods (Schultze H-P, Trueb L, eds), 3–38. Ithaca, N.Y.: Comstock.

Cloutier R, Ahlberg PE (1996) Morphology, characters and the interrelationships of basal sarcopterygians. In: Interrelationships of Fishes (Stiassny MLJ, Parenti LR, Johnson GD, eds), 445–480. San Diego, Calif.: Academic Press.

Coates M (1991) New paleontological contributions to limb ontogeny and phylogeny. In: Developmental Patterning of the Vertebrate Limb (Hinchliffe RJ, ed), 325–337. New York: Plenum Press.

Coates MI (1993) Ancestors and homology: The origin of the tetrapod limb. Acta Biotheor 41: 411–424.

Coates MI (1994) The origin of vertebrate limbs. Development suppl.: 169–180.

Coates MI (1996) The Devonian tetrapod *Acanthostega gunnari* Jarvik: Postcranial anatomy, basal tetrapod relationships and patterns of skeletal evolution. Trans R Soc Edin Earth Sci 87: 363–427.

Coates MI (2001) Fins to limbs: What the fossils say. Paper presented to Society for Integrative and Comparative Biology, Chicago, January.

Coates MI, Clack JA (1990) Polydactyly in the earliest known tetrapod limbs. Nature 347: 66–69.

Coates MI, Ruta M, Milner AR (2000) Early tetrapod evolution. TREE 15: 327–328.

Daeschler EB, Shubin N (1998) Fish with fingers? Nature 391: 133.

Davis AP, Witte DP, Hsieh-Li HM, Potter SS, Capecchi MR (1995) Absence of radius and ulna in mice lacking *Hoxa-11* and *Hoxd-11*. Nature 375: 791–795.

DeWinter JP, Dijke PT, deVries CJM, Achterberg TAE, Sugino H, et al. (1996) Follistatins neutralize activin bioactivity by inhibition of activin binding to its type II receptors. Mol Cell Endocrinol 116: 105–114.

Fromental-Ramain C, Warot X, Messadecq N, LeMeur M, Dollé P, et al. (1996) *Hoxa-13* and *Hoxd-13* play a crucial role in the patterning of the limb autopod. Development 122: 2997–3011.

Gauthier JA, Kluge AG, Rowe T (1988) The early evolution of Amniota. In: The Phylogeny and Classification of the Tetrapods (Benton MJ, ed), 103–155. Oxford: Clarendon Press.

Gérard M, Duboule D, Zákány J (1993) Structure and activity of regulatory elements involved in the activation of the *Hoxd-11* gene during late gastrulation. EMBO J 12: 3539–3550.

Geraudie J (1978) The fine structure of the early pelvic fin bud of the trouts *Salmo gaidneri* and *S. trutta fario*. Acta Zool 59: 85–96.

Gilbert SF, Opitz JM, Raff RA (1996) Resynthesizing evolutionary and developmental biology. Dev Biol 173: 357–372.

Grandel H, Schulte-Merker S (1998) The development of the paired fins the zebrafish (*Danio rerio*). Mech Dev 79: 99–120.

Haack H, Gruss P (1993) The establishment of murine *Hox-1* expression domains during patterning of the limb. Dev Biol 157: 410–422.

Hedges SB, Hass CA, Maxson LR (1993) Relations of fish and tetrapods. Nature 363: 501–502.

Hedges SB, Maxson LR (1993) A molecular perspective on lissamphibian phylogeny. Herpetol Monog 7: 27–42.

Hedges SB, Moberg KD, Maxson LR (1990) Tetrapod phylogeny inferred from 18S and 28S ribosomal RNA sequences and a review of the evidence for amniote relationships. Mol Biol Evol 7: 607–633.

Hérault Y, Beckers J, Gérard M, Duboule D (1999) Hox gene expression in limbs: Colinearity by opposite regulatory controls. Dev Biol 208: 157–165.

Hinchliffe R (1991) Developmental approaches to the problem of transformation of limb structure in evolution. In: Developmental Patterning of the Vertebrate Limb (Hinchliffe JR, ed), 313–323. New York: Plenum Press.

Hinchliffe JR, Vorobyeva EI (1999) Developmental basis of limb homology in urodeles: Heterochronic evidence from the primitive hynobiid family. In: Homology (Bock GR, Cardew G, eds), 95–105. Chichester, England: Wiley.

Hoeven Fvd, Zákány J, Duboule D (1996) Gene transposition in the HoxD complex reveal a hierarchy of regulatory controls. Cell 85: 1025–1035.

Janvier P (1996) Early Vertebrates. Oxford: Clarendon Press.

Karczmar AG, Berg GG (1951) Alkaline phosphatase during limb development and regeneration of *Amblystoma opacum* and *Amblystoma punctatum*. J Exp Zool 117: 139–163.

Laurin M (1998) The importance of global parsimony and historical bias in understanding tetrapod evolution: 1. Systematics, middle ear evolution and jaw suspension. Ann Sci Nat Zool 19: 1–42.

Laurin M (1998) A reevaluation of the origin of pentadactyly. Evolution 52: 1476–1482.

Laurin M, Gorondot M, Ricqlés Ad (2000) Early tetrapod evolution. Trends Ecol Evol 15: 118–123.

Lauthier M (1985) Morphogenetic role of epidermal and mesodermal components of the fore- and hind limb buds of the newt *Pleurodeles waltlii* (Urodela, Amphibia). Arch Biol Bruss 96: 23–43.

Lebedev OA, Coates MI (1995) The postcranial skeleton of the Devonian tetrapod *Tulerpeton curtum*. Zool J Linn Soc 113: 307–348.

Long JA (1989) A new rhizodontiform fish from the early Carboniferous of Victoria, Australia, with remarks on the phylogenetic position of the group. J Vert Paleont 9: 1–17.

Mabee PM (2000) Developmental data and phylogenetic systematics: Evolution of the vertebrate limb. Am Zool 40: 789–800.

Mercanter N, Leonardo E, Azpiazu N, Serrano A, Morata G, et al. (1999) Conserved regulation of proximodistal limb axis development by *Meis1/Hth*. Nature 402: 425–429.

Merino R, Macias D, Gañan Y, Rodriges-Leon J, Economides AN, et al. (1999) Control of digit formation by activin signalling. Development 126: 2161–2170.

Milner AR (1988) The relationships and origin of living amphibians. In: The Phylogeny and Classification of the Tetrapods (Benton MJ, ed), 59–102. Oxford: Clarendon Press.

Milner AR (1993) The paleozoic relatives of lissamphibians. Herpetol Monog 7: 8–27.

Müller GB, Alberch P (1990) Ontogeny of the limb skeleton in *Alligator mississippiensis:* Developmental invariance and change in the evolution of archosaur limbs. J Morphol 203: 151–164.

Müller GB, Newman SA (1999) Generation, integration, autonomy: Three steps in the evolution of homology. In: Homology (Bock GR, Cardew G, eds), 65–73. Chichester, England: Wiley.

Müller GB, Wagner GP (1991) Novelty in evolution: Restructuring the concept. Annu Rev Ecol Syst 22: 229–256.

Nelson CE, Morgan BA, Burke AC, Laufer E, DiMambro E, et al. (1996) Analysis of Hox gene expression in the chick limb bud. Development 122: 1449–1466.

Neumann CJ, Grandel H, Gaffield W, Schulte-Merker F, Nüsslein-Volhard C (1999) Transient establishment of anterior-posterior polarity in the zebrafish pectoral fin bud in the absence of *sonic hedgehog* activity. Development 126: 4817–4826.

Newman SA (1996) Sticky fingers: Hox genes and cell adhesion in vertebrate limb development. BioEssays 18: 171–174.

Newman SA, Müller GB (2001) Epigenetic mechanisms of character origination. In: The Character Concept in Evolutionary Biology (Wagner GP, ed), 559–579. San Diego, Calif.: Academic Press.

Panchen A, Smithson T (1987) Character diagnosis, fossils and the origin of tetrapods. Biol Rev Camb Philos Soc 62: 341–438.

Panchen AL, Smithson TR (1988) The relationships of early tetrapods. In: The Phylogeny and Classification of the Tetrapods (Benton MJ, ed), 1–32. Oxford: Clarendon Press.

Paton RL, Smithson TR, Clack JA (1999) An amniote-like skeleton from the early Carboniferous of Scotland. Nature 398: 508–513.

Richardson MK, Carl TF, Hanken J, Elinson RP, Cope C, et al. (1998) Limb development and evolution: A frog embryo with no apical ectodermal ridge (AER). J Anat 192: 379–390.

Riddle RD, Johnson RL, Laufer E, Tabin C (1993) *Sonic hedgehog* mediates the polarizing activity of the ZPA. Cell 75: 1401–1416.

Rosen D, Forey P, Gardiner B, Patterson C (1981) Lungfish, tetrapods, paleontology and pleisiomorphy. Bull Am Mus Nat Hist 167: 163–274.

Saunders JWJ (1948) The proximodistal sequence of the origin of the parts of the chick wing and the role of the ectoderm. J Exp Zool 108: 363–404.

Saunders JWJ, Gasseling MT (1968) Ectoderm-mesenchymal interaction in the origins of wing symmetry. In: Epithelial-Mesenchymal Interactions (Fleischmajer R, Billingham RE, eds), 78–97. Baltimore: Williams and Wilkins.

Schauerte HE, Eeden FJMv, Fricke C, Odenthal J, Strähle U, et al. (1998) *Sonic hedgehog* is not required for the induction of medial floor plate cells in the zebrafish. Development 125: 2983–2993.

Schmalhausen JJ (1910) Die Entwicklung des Extremitätenskeletts von *Salamandrella kayserlingii*. Anat Anz 37: 431–446.

Schultze H-P (1987) Dipnoans as sarcopterygians. J Morphol suppl 1: 39–74.

Schultze H-P (1991) Controversial hypotheses on the origin of tetrapods. In: Origins of the Higher Groups of Tetrapods (Schultze H-P, Trueb L, eds), 29–67. Ithaca, N.Y.: Comstock.

Schultze H-P, Arsenault M (1985) The panderichthyid fish *Elpistostege:* A close relative of tetrapods? Paleontology 28: 293–310.

Schultze H-P, Trueb L (eds) (1991) Origins of the Higher Groups of Tetrapods. Ithaca, N.Y.: Comstock.

Schwabe JWR, Rodriguez-Esteban C, Izpisúa-Belmonte J-C (1998) Limbs are moving: Where are they going? Trends Genet 14: 229–235.

Searls RL, Janners MY (1971) The initiation of limb bud outgrowth in the embryonic chick. Dev Biol 24: 198–213.

Seilacher R (1991) Self-organizing mechanisms in morphogenesis and evolution. In: Constructional Morphology and Evolution (Schmidt-Kittler N, Vogel K, eds), 251–271. Berlin: Springer.

Shubin N (1995) The evolution of paired fins and the origin of tetrapod limbs. Evol Biol 28: 39–86.

Shubin NH, Alberch P (1986) A morphogenetic approach to the origin and basic organization of the tetrapod limb. Evol Biol 20: 319–387.

Smithson TR, Carroll RL, Panchen AL, Andrews SM (1993) *Westlothiana lizziae* from the Viséan of East Kirkton, West Lothian, Scotland, and the amniote stem. Trans R Soc Edin Earth Sci 84: 383–412.

Sordino P, Duboule D (1996) A molecular approach to the evolution of vertebrate paired appendages. TREE 11: 114–119.

Sordino P, Hoeven Fvd, Duboule D (1995) Hox gene expression in teleost fins and the origin of vertebrate digits. Nature 375: 678–681.

Stern CD, Yu RT, Kakizuka A, Kintner CR, Mathews LS, et al. (1995) Activin and its receptors during gastrulation and later phases of mesoderm development in the chick embryo. Dev Biol 172: 192–205.

Summerbell D, Lewis JH, Wolpert L (1973) Positional information in chick limb morphologenesis. Nature 244: 492–496.

Thorogood P (1991) The development of the teleost fin and implications for our understanding of tetrapod limb evolution. In: Developmental Patterning of the Vertebrate Limb (Hinchliffe RJ, ed), 347–354. New York: Plenum Press.

Vargesson N, Clarke JD, Vincent K, Coles C, Wolpert L, et al. (1997) Cell fate in the chick limb bud and relationship to gene expression. Development 124: 1909–1918.

Vargesson N, Kostakopoulou K, Drossopoulou G, Papageorgiou S, Tickel C (2001) Characterization of *Hoxa* gene expression in the chick limb bud in response to FGF. Dev Dyn 220: 87–90.

Vorobyeva EI (1977) Morphology and nature of the evolution of crossopterygian fishes. Trudy paleont Inst 163: 1–240.

Vorobyeva EI (1991) The fin-limb transformation: paleonotological and embryological evidence. In: Developmental Patterning of the Vertebrate Limb (Hinchliffe JR, ed), 339–345. New York: Plenum Press.

Vorobyeva EI, Hinchliffe JR (1996) Developmental pattern and morphology of *Salamandrella keyserlingii* limbs (Amphibia, Hynobiidae) including some evolutionary aspects. Russ J Herpetol 3: 68–81.

Vorobyeva EI, Ol'shevskaya OP, Hinchliffe JR (1997) Specific features of development of the paired limbs in *Ranodon sibiricus* Kessler (Hynobiidae, Caudata). Russ J Dev Biol 28: 150–158.

Vorobyeva E, Schultze H-P (1991) Description and systematics of panderichthyid fishes with comments on their relationship to tetrapods. In: Origins of the Higher Groups of Tetrapods (Schultze H-P, Trueb L, eds), 68–109. Ithaca, N.Y.: Comstock.

Wagner GP (1989) The biological homology concept. Annu Rev Ecol Syst 20: 51–69.

Wagner GP (1995) The biological role of homologues: A building block hypothesis. N Jb Geol Paläont Abh 195: 279–288.

Wagner GP, Chiu C-H, Laubichler M (2000) Developmental evolution as a mechanistic science: The inference from developmental mechanisms to evolutionary processes. Am Zool 40: 819–831.

Yokouchi Y, Nakazato S, Yamamoto M, Goto Y, Kameda T, et al. (1995) Misexpression of *Hoxa-13* induces cartilage homeotic transformation and changes cell adhesiveness in chick limb buds. Genes Dev 9: 2509–2522.

Yokouchi Y, Sasaki H, Kuroiwa A (1991) Homeobox gene expression correlated with the bifurcation process of limb cartilage development. Nature 353: 443–445.

Zákány J, Duboule D (1999) Hox genes in digit development and evolution. Cell Tissue Res 296: 19–25.

Zákány J, Fromental-Ramain C, Warot X, Duboule D (1997) Regulation of number and size of digits by posterior Hox genes: A dose-dependent mechanism with potential evolutionary implications. PNAS 94: 13695–13700.

Zardoya R, Abouheif E, Meyer A (1996) Evolutionary analyses of *hedgehog* and *Hoxd-10* genes in fish species closely related to the zebrafish. PNAS 93: 13036–13041.

Zardoya R, Meyer A (1997) The complete DNA sequence of the mitochondrial genome of a "living fossil," the coelacanth (*Latimeria chalumnae*). Genetics 146: 995–1010.

16 Epigenesis and Evolution of Brains: From Embryonic Divisions to Functional Systems

Georg F. Striedter

Once upon a time, I hoped that many genes would be uniquely expressed in very specific brain regions and thus serve as reliable "markers" of homology. Unfortunately, we now know that most genes are expressed in several different locations and that many homologies based on the expression patterns of single genes have turned out to be controversial, to say the least (e.g., Janies and DeSalle, 1999; and Conway Morris, chapter 2, Willmer, chapter 3, Müller, chapter 4, Wagner and Chiu, chapter 15, this volume). I now tend to think of brain regions as being defined or "specified" by some sort of combinatorial molecular code (e.g., Redies, 1997). If this is true, then it should be possible to infer homologies from the expression patterns of several different genes.

Although this argument is appealing in its simplicity, many genes are expressed only during specific developmental stages, which implies that any combinatorial "code" would have to be decoded sequentially rather than simultaneously. In practice, this means that one must consider a structure's developmental history before one can know its "code" and find its homologue in other species. But once a structure's developmental history becomes part of the argument about its homology, troublesome questions arise (Striedter, 1998). For example, can homologous structures derive from nonhomologous precursors, and are all the adult derivatives of homologous precursor regions homologous to one another (see Müller, chapter 4, this volume)?

In grappling with these questions, I began to think of morphological structures as the valleys in epigenetic landscapes of the kind originally proposed by Waddington (see figure 16.1). Specifically, I proposed that the valleys in an epigenetic landscape be thought of as attractors (analogous to attractor states in other dynamical systems) that may appear and disappear during individual development, but recur reliably across generations (Striedter, 1998). Because phylogeny can be conceptualized as a succession of epigenetic landscapes, homologues can be defined as corresponding valleys (i.e., attractors) that have recurred reliably since their origin in a single ancestral population. According to the "epigenetic homology" concept, homologues are distinct, identifiable units in both ontogeny and phylogeny.

This way of thinking about morphological homology resolves some conundrums that have long plagued evolutionary developmental biologists. For instance, the epigenetic homology concept is consistent with the view that homologous structures may derive from nonhomologous precursors, because it is possible (though perhaps rare) for cells and tissues to converge onto corresponding attractors from various prior states. Similarly, because attractors can be robust to minor changes in mechanism, the epigenetic homology concept is consistent with the view that the mechanisms underlying a structure's development may change over evolutionary time (Striedter, 1998). Most important, the epigenetic approach

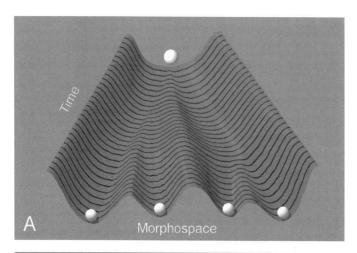
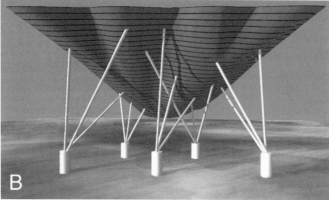

Figure 16.1
Epigenetic landscapes (*A*) were used by Waddington (1957) to illustrate how an initially homogeneous morphological system develops into a number of distinct subsystems along a diverging series of buffered pathways or valleys. The branching topology of epigenetic landscapes nicely captures the progressive compartmentalization seen during the early phases of brain development. The shape of an epigenetic landscape is determined by a complex web of gene, protein, and environmental interactions (*B*). Some of these interactions may be axon-mediated, particularly during the later phases of brain development, making it possible for one brain region to influence the ontogenetic trajectory of a distant brain region. Such dynamic network effects are likely to promote functional integration and facilitate brain evolution. (Modified with permission from Striedter, 1998.)

to homology suggests that the mechanisms and rules of morphological development should be closely related to the mechanisms and rules of morphological evolution.

Pursuing the latter idea, this chapter reviews what is currently known about the mechanisms of brain development and then examines how these mechanisms might constrain or facilitate brain evolution, or both. It argues that there are two fundamentally different modes of brain development. The first, compartmental mode is characterized by a progressive compartmentalization of the embryonic brain and is largely complete by the time most neurons become functionally interconnected. The second, dynamic network mode of brain development is characterized by axon-mediated developmental interactions and involves both trophic and activity-dependent mechanisms. It concludes that, certain exceptions notwithstanding, the compartmental mode of brain development acts as a conservative force in brain evolution, whereas the dynamic network mode provides for functional integration of disparate brain regions and, thereby, promotes evolutionary changes in brain organization.

The Compartmental Mode of Brain Development

The notion that brain morphogenesis involves the formation of distinct developmental compartments has a venerable history. Back in the nineteenth century, embryologists first noted that the brains of all vertebrates, at a stage shortly after neural tube closure, exhibit a number of transversely oriented ring-shaped bulges, or neuromeres. Although neuromeres are most evident in the hindbrain, where they are called "rhombomeres," many researchers (e.g., Bergquist and Källén, 1954) have described neuromeres in the midbrain and forebrain as well, at least during some stages of early development. The most widely accepted interpretation of these neuromeres was that they constitute transverse rings of increased proliferative activity, separated by narrow zones of decreased proliferation. Subsequent developmental processes were thought to further subdivide these neuromeric compartments, eventually generating a checkerboard pattern of distinct developmental compartments, from which young neurons migrated out and formed the adult cell groups (see Nieuwenhuys, 1998).

These classical theories about neuromeres were largely forgotten until modern cellular and molecular methods enabled researchers to show that (1) rhombomeres are indeed spatially distinct domains of increased proliferative activity (Guthrie, Butcher, and Lumsden, 1991); (2) cells rarely migrate across rhombomere boundaries (Fraser, Keynes, and Lumsden, 1990); and (3) most rhombomere boundaries are in register with the expression boundaries of vertebrate Hox genes and other transcription factors (Wilkinson and Krumlauf, 1990). Since then, many of these genes have been demonstrated to be essential for normal rhombomere development (Carpenter et al., 1993; Bell, Wingate, and Lumsden, 1999), and gene expression and lineage restriction domains that correspond to postulated neuromeres have now been described also in the diencephalon and telencephalon (Bulfone et al., 1993;

Figdor and Stern, 1993). Although the existence and boundaries of some neuromeres remain controversial, particularly in the telencephalon (Alvarez-Bolado, Rosenfeld, and Swanson, 1995), the classical neuromeric theories have, in most respects, been well supported by the modern cellular and molecular data (see Puelles and Rubenstein, 1993).

The initial formation of neuromeres is due in large part to the spatial segregation of cells with different affinities for one another. In analogy to how body segments form in *Drosophila,* it is thought that morphogenetic gradients within the early neural tube provide individual cells with specific "genetic addresses," which then cause cells in different spatial positions to express different kinds and quantities of various cell adhesion molecules (Lawrence and Struhl, 1996). Because cells with similar affinities tend to self-aggregate, whereas cells with different affinities tend to segregate, differential cell affinity could in and of itself lead to the formation of developmental compartments that are defined by lineage restriction. Evidence in favor of this hypothesis derives primarily from rhombomere transplantation and in vitro cell-mixing experiments (Guthrie, Prince, and Lumsden, 1993; Stoykova et al., 1997; Wizenmann and Lumsden, 1997). The alternative hypothesis, that neuromeres form because specialized "border cells" physically obstruct cell mixing, is unlikely to be correct: experimental elimination of the border cells does not significantly disturb neuromere formation (Nittenberg et al., 1997). It remains possible, however, that some neuromere boundaries, particularly in the diencephalon, form at least in part because fiber tracts, many of which course along neuromere boundaries, physically obstruct cell movements (Wilson, Placzek, and Furley, 1993).

Soon after the neuromeres have formed, they largely disappear, making it tempting to dismiss neuromeres as "merely" ephemeral structures with little bearing on the structural or functional organization of the adult brain (Herrick, 1933). The adult hindbrain, for example, is clearly organized into longitudinal cell columns, not into the transverse rings or wedges that would be expected if the rhombomeric pattern of organization were retained into adulthood. Moreover, it is not evident what, if anything, the adult derivatives of any given rhombomere have in common with one another, either structurally or functionally. For example, although it has been postulated that the neurons derived from rhombomeres 7 and 8 share electrical membrane properties that make them especially suited for the generation of rhythmic activity patterns (Bass and Baker, 1997), it is likely that cells in other rhombomeres are also able to generate rhythmic activity (Fortin et al., 1995). Similarly, although forebrain neuromeres probably exist, it is difficult to see what structural or functional features are shared by the adult derivatives of any given forebrain neuromere. Therefore, the suggestion that neuromeres are transient embryonic features, of little relevance to adult brain organization, should be taken seriously.

The question of neuromere transience was recently examined by means of detailed cell-fate-mapping studies (Marín and Puelles, 1995; Wingate and Lumsden, 1996; Díaz et al.,

1998). These studies showed that embryonic rhombomeres do in fact give rise to transverse wedges within the adult brain stem and that the boundaries of these wedges are often aligned with the rostral or caudal boundaries of adult cell groups. That some wedge boundaries appear rather fuzzy can in most cases be explained by the relatively late migration of cells (and entire cell groups) rostrally or caudally away from their original embryonic location. Ultimately, because many cell groups derive from multiple rhombomeres and therefore fuse across rhombomere boundaries, the rhombomere boundaries do disappear in many locations, although some may remain as cytoarchitecturally subtle (and as yet unknown) boundaries between different subdivisions of larger, rhombomere-spanning cell groups (Marín and Puelles, 1995). But, even though the available cell-fate-mapping data support the hypothesis that the embryonic rhombomeres are causally related to the structural organization of the adult hindbrain, some rostrocaudal cell group boundaries in the adult brain stem do not form in accordance with rhombomere boundaries (Díaz et al., 1998). Therefore, factors other than rhombomeric origin must be involved in generating at least some rostrocaudal cell group boundaries.

Significant progress has also been made in understanding dorsoventral patterning in the brain. Within the hindbrain, several genes, including some required for normal development, are expressed in clearly longitudinal domains (Graham, Maden, and Krumlauf, 1991; Davenne et al., 1999). Whether these longitudinal gene expression domains are characterized by lineage restriction and whether their boundaries correspond to adult cell group boundaries remain to be seen, but it is already known that some cell groups in the embryonic brain stem migrate dorsoventrally for considerable distances to reach their adult locations (Tan and Le Douarin, 1991). In the forebrain, numerous gene products are also localized into domains that parallel the brain's long axis (Bulfone et al., 1993; Shimamura et al., 1997). Some of these longitudinal domains may be lineage restriction domains but, as in the hindbrain, some forebrain cells are known to migrate far away from their original embryonic locations (Anderson et al., 1997). In addition, the dynamics of gene expression and morphogenetic distortion in the telencephalon make it difficult to determine for any given gene expression domain whether it is transverse or longitudinal. Nonetheless, it is reasonable to conclude from all these data that the structural organization of the entire adult brain, including the telencephalon, develops in large measure by the formation of successively smaller developmental compartments, which ultimately give rise to the adult neuronal cell groups.

Once the brain is divided into many cell groups, these must be interconnected by axons. In this process, cell adhesion molecules are likely to play a significant role: functionally interconnected cell groups and their axons frequently express the same cell adhesion molecules (Redies and Takeichi, 1996). Because many cell adhesion molecules bind homophilically, it is postulated that axons adhere preferentially to other neurons that express similar concentrations and combinations of cell adhesion molecules (Redies, 1997). Given the enormous variety of cell adhesion molecules known to be expressed in the nervous system and the rapid rate at which more are being discovered (Kohmura et al., 1998), such a combinatorial adhesive code could account for much of the brain's connectional specificity. Complementing this adhesive chemoaffinity model (Sperry, 1963) is the hypothesis that growing axons are guided into the proximity of their ultimate target by both attractive and repulsive molecular signals they encounter along the way (Braisted, Tuttle, and O'Leary, 1999).

In general, then, brain development can be envisioned as a process in which the growing axons play a relatively passive role, responding to the information that was previously established by the process of compartmentalization. On the other hand, as we shall see, there is also considerable evidence that the axons themselves can actively influence the brain's own morphogenesis.

The Dynamic Network Mode of Brain Development

The brain is unique among organs in that many of its cells are interconnected across long distances by highly specialized processes, namely, axons (but see Ramirez-Weber and Kornberg, 1999). These interconnections endow the brain with much of its information-processing capacity; they can also play an active role in the brain's own construction. As understood here, axon-mediated developmental interactions include trophic and activity-dependent interactions that can influence the formation and survival of both neurons and their connections (Purves, 1988; Katz and Shatz, 1996). A critical feature of these interactions is that developmental alterations in one part of the brain can effect changes in relatively distant, axonally interconnected parts of the brain.

The best-studied type of axon-mediated developmental interaction is the modulation of neuronal cell death by target-derived trophic signals. It has long been known that the amount of normally occurring cell death among motor neurons can be increased by removing the target muscles, and reduced by providing additional target muscles (Holliday and Hamburger, 1976; Oppenheim, 1981). Trophic dependencies between neurons and their targets have now been observed also in the central nervous system (e.g., Hughes and Lavelle, 1975) and are known to be mediated by a variety of retrograde neurotrophic signals (see Johnson, 1999). In addition, neuron survival may depend on afferent innervation (see Linden, 1994). Collectively, these trophic interactions ensure that each muscle or neuron receives the appropriate amount of innervation, and such "population-matching" interactions could, at least in theory, cascade throughout a neuronal circuit (Katz and Lasek, 1978). It has been pointed out, however, that trophic cascades of cell group size are likely to be buffered out whenever neural circuits converge or diverge, that is, whenever neurons can derive trophic support from multiple sources and whenever decreases in afferents from one source lead to compensatory increases in afferents from other sources (Finlay, Wikler, and Sengelaub, 1987). The extent and precise form of trophic cascades are thus depend on how the individual neurons are interconnected and how they respond to changes in other parts of the circuit.

Major changes in the size of one brain area can affect not only the size but also the connections of other cell groups. Neonatal destruction of the cochlea on one side, for example, leads to the degeneration of the (now-denervated) ipsilateral ventral cochlear nucleus (VCN). It also leads to compensatory changes in the connections of the ipsilateral VCN, which sprouts connections to brain stem auditory nuclei that normally receive inputs only from the ipsilateral VCN (Kitzes et al., 1995; Tierney, Russell, and Moore, 1997). Similarly, neonatal removal of one eye leads to compensatory innervation from the other eye to the dorsal lateral geniculate nucleus (LGNd) and changes the pattern of interhemispheric connections in the visual cortex (Sengelaub and Finlay, 1981; Guillery et al., 1985; Olavarria, Malach, and Van Sluyters, 1987). Although the induced projections in these examples, were of the same modality as the eliminated connections, compensatory innervation may derive also from other modalities. Thus early removal of both eyes leads to the induction of novel projections from a brain stem somatosensory cell group to the LGNd (Asanuma and Stanfield, 1990). These cross-modal compensatory projections are relatively small, but can be enlarged if eye removal is accomplished very early on in development and if the normal target of the ascending somatosensory projections is experimentally reduced in size. These data show that compensatory innervation can lead to functionally significant changes in neural circuitry.

The best-studied examples of experimentally induced changes in neural circuits involve the rerouting of retinal projections in hamsters and ferrets to thalamic nuclei that normally process other sensory modalities (Schneider, 1973). In these studies, neonatal lesions were used to reduce the size of the normal retinal targets (the LGNd or the superior colliculus) and to deafferent thalamic nuclei belonging to other sensory modalities (Frost, 1981; Sur, Pallas, and Roe, 1990; Angelucci et al., 1997). These manipulations induced the development of compensatory projections from the retina to the denervated auditory and somatosensory thalamic nuclei (as well as to visual thalamic nuclei not normally receiving retinal inputs). These induced projections, though sometimes due to the retention of normally transient axon collaterals, also involved the formation of at least some normally nonexistent connections (Bhide and Frost, 1992; Pallas, Halm, and Sur, 1994). The experimental manipulations also induced a projection from a normally visual thalamic nucleus (the lateral posterior nucleus) to the primary "auditory cortex," without altering many other aspects of thalamo-cortical interconnectivity (Pallas, Roe, and Sur, 1990). Intriguingly, the "auditory cortex" of the "rewired" animals processed visual instead of auditory information and did so in a manner reminiscent of primary visual cortex (Sur, Garraghty, and Roe, 1988). These similarities are generally attributed to fundamental similarities in the intrinsic circuitry of all sensory cortices, rather than to any dramatic intracortical reorganization. However, some of the horizontal connections intrinsic to the rewired "auditory" cortex were altered in such a way that they came to resemble those of primary visual cortex (Gao and Pallas, 1999).

Cross-modal rerouting of sensory inputs to the thalamus may also have occurred naturally during the evolution of blind mole rats. Specifically, the LGNd of blind mole rats, whose retinas are highly degenerate, was reported to receive compensatory innervation from the auditory midbrain and then to send this auditory information to the cytoarchitecturally identified primary "visual" cortex (Heil et al., 1991; Doron and Wollberg, 1994). Other authors, however, have argued that the LGNd in blind mole rats is actually quite small (e.g., Leder, 1975), that it projects specifically (albeit nontopographically) to the primary visual cortex (Cooper, Herbin, and Nevo, 1993a), and that the region identified by some as LGNd actually comprises portions of the medial geniculate and posterior thalamic nuclei, both of which process auditory information also in sighted rodents (Rehkämper, Necker, and Nevo, 1994). According to this alternative explanation, no rerouting of afferents has taken place: blind mole rats have simply reduced their thalamocortical visual system, although the reduction in the visual system may have been accompanied by an increase in the size of the somatosensory system (Necker, Rchkämper, and Nevo, 1992). This interpretation is supported by the observation that sensory afferents in hamsters and ferrets (see above) appear to be rerouted only if their normal targets have also been eliminated, which is not the case in the blind mole rats. However, further studies on the cortical representations of the various sensory systems in blind mole rats will be needed to resolve this issue to everyone's satisfaction.

In addition to modifying the size and connections of other brain regions, axon-mediated interactions can also play an important role in the induction and histological differentiation of other brain areas, particularly the neocortex. This is perhaps most evident in the somatosensory system, where cortical differentiation coincides with the arrival of the thalamic afferents (Shlaggar and O'Leary, 1994) and peripheral changes consistently lead to matching changes in cortical representation (Woolsey, 1990; Catania and Kaas, 1997). Moreover, embryonic cortical tissue that normally gives rise to visual cortex can differentiate into apparently normal somatosensory cortex (with its unique vibrissa-related cytoarchitectural organization) when transplanted into a location where it receives somatosensory thalamic afferents (Schlaggar and O'Leary, 1991). Thalamic afferents appear to be important also for the differentiation of visual cortex: early enucleation in primates leads to a reduction in the size of the primary visual cortex and can, under some circumstances, lead to the appearance of a novel visual cortical area (Dehay et al., 1996). Although it is clear, therefore, that thalamic afferents play an important role in the areal differentiation of neocortex, some aspects of cortical regionalization precede thalamic innervation and are not dependent on it for their normal development (Gitton, Cohen-Tannoudji, and Wassef, 1999; Miyashita-Lin et al., 1999). It should also be pointed out that heterotopic cortical transplants do develop some features according to their original embryonic location (Ebrahimi-Gaillard and Roger, 1996; Frappé, Roger, and Gaillard, 1999). It is thus most likely that molecular signals intrinsic to the embryonic neocortex interact with signals derived from the incoming thalamic afferents to sculpt the adult pattern of areal differentiation (Levitt, Barbe, and Eagleson, 1997).

Many of the axon-mediated interactions described above probably arise as a result of competition among neurons for trophic support; some can occur even when the neurons are electrically silent (Chiaia et al., 1992). On the other hand, electrical activity can also influence some aspects of neuronal structure (Katz and Shatz, 1996; Crair, 1999). For instance, intracerebral injections of tetrodotoxin, which eliminate both spontaneous and environmentally elicited brain activity, prevent the refinement of retinotopic maps in the midbrain, the formation of ocular dominance columns in the primary visual cortex, and the normal segregation of retinogeniculate afferents into eye-specific laminae (Meyer, 1983; Stryker and Harris, 1986; Shatz and Stryker, 1988). In addition, neonatal eyelid suture or strabismus, which alters stimulus-driven but not spontaneous activity, prevents the normal development of horizontal connections within the primary visual cortex and of callosal connections between the two visual cortices (Callaway and Katz, 1991; Schmidt et al., 1997; Zufferey et al., 1999).

Interestingly, in all these cases, the manipulations of neural activity alter the spatial distribution of axonal arbors within their normal target areas, without inducing gross cytoarchitectural changes (other than changes in cell group size) or projections to novel targets. One possible exception to this rule is that neural activity can interfere with some aspects of thalamocortical axon targeting (Catalano and Shatz, 1998), and aberrant thalamocortical projections could conceivably induce aberrant cortical architecture (see above). In general, however, the activity-dependent developmental processes studied thus far seem to have more in common with the mechanisms underlying plasticity in adult animals (Merzenich et al., 1988) than they do with the activity-independent axon-mediated interactions that help to construct the brain's cytoarchitectural entities and fundamental circuits during embryonic development (Purves et al., 1994).

Dichotomizing Development

The compartmental network mode of brain development involves the brain's progressive compartmentalization into smaller units, which then become interconnected by means of a molecular affinity code, whereas the dynamic network mode involves axon-mediated influences on neural differentiation and connectivity. Such a distinction has been implicit in the previous literature, with most authors focusing on one or the other of the two modes in their reviews of brain development. Support for this distinction derives from the fact that true axons are unique to the nervous system and thus axon-mediated developmental interactions distinguish brain development from the development of other organ systems. In addition, the compartmental and dynamic network modes of brain development are largely separate in time: most axonal connections form after most of the brain's fundamental divisions have already been established. This temporal separation is hardly precise, however. Many early axons grow along early compartmental boundaries (Easter et al., 1994) and are probably guided to their general target areas by molecular signals emanating from or inherent in the brain's developmental compartments (Tessier-Lavigne and Goodman, 1996; Braisted, Tuttle, and O'Leary, 1999).

Further blurring the distinction between these two modes is their ability to interact with one another: changes in the relative size of one or more developmental compartments can, for example, alter the outcome of later competitive interactions between axons and yield very different adult circuitry (Deacon, 1990). Moreover, some of the region-specific patterns of cadherin expression in the developing brain may be induced by the ingrowth of axons rather than serving as a preexisting molecular code for where the ingrowing axons should terminate (Huntley and Benson, 1999). Thus axon-mediated interactions may themselves influence brain compartmentalization. Finally, it is misleading to argue that the dynamic network mode of brain development is somehow "more epigenetic" than the compartmental mode, in the sense that it is less dependent on the genome (Katz, 1982): both modes depend on responses of the genome to other molecules, and even sensory stimuli ultimately exert their morphogenetic effects through genomic responses. These considerations suggest that, although the compartmental and dynamic network modes of brain development are logically (and to some extent temporally) distinct from one another, one should avoid dichotomizing them excessively.

Functional Integration and Evolvability

Perhaps the most important reason for distinguishing between the compartmental and dynamic network modes of brain development is that the two have very different implications for the process of brain evolution. Specifically, the compartmental mode of brain development acts as a conservative force in brain evolution, whereas the dynamic network mode facilitates evolutionary change, primarily by promoting functional integration.

Because morphological structures, to work, must be functionally integrated with the rest of the organism, evolutionary changes are rarely restricted to single structures. But how can the coordinated changes required for functional integration be achieved if evolution is based only on chance mutations? In an early attempt to deal with this problem, Wilhelm Roux (1895) proposed that there are specific developmental mechanisms that promote functional integration between different body parts. According to Roux and others (e.g., Alberch, 1982), these integration-promoting mechanisms, most occurring relatively late in development, depend on interactions between body parts and between the body and its environment. Thus they resemble or partly overlap with the axon-mediated interactions that characterize the dynamic network mode of brain development described in this chapter.

Many previous authors have noted that, to be functional, evolutionary changes in one brain region must be associated with evolutionary changes in other brain regions, and that brain evolution would be arduous, to say the least, if the brain developed in a purely compartmental manner, with all connections specified according to an explicit molecular code (e.g., Stent, 1981). Functional integration is readily achieved in the dynamic network mode of brain development, however: all neurons receive at least some innervation and the balance of inputs to any given neuron reflects the size relationships of its various input structures (Katz, 1982; Finlay, Wikler, and Sengelaub, 1987; Deacon, 1990).

In contrast, the compartmental mode of brain development tends to oppose evolutionary change because it does not, by itself, promote the functional integration required for survival and speciation. Furthermore, it is likely that the evolutionary stability of many brain compartments has been enhanced over time by the evolution of mutually reinforcing developmental mechanisms (epigenetic networks) that ensure the emergence of these compartments even in the face of genomic change (Newman, 1993). Such epigenetic canalization (Waddington, 1957) would be expected particularly for early embryonic compartments and may, for example, explain why the hindbrain's rhombomeric organization is so highly conserved between taxa (Gilland and Baker, 1993).

On the other hand, the compartmental mode of brain development need not always be a brake on evolutionary change: major changes in brain compartmentalization, when viable, are likely to trigger a wealth of changes in neural connections and open up many new directions for evolutionary change (Raff, 1996). Conversely, the dynamic network mode does not facilitate just any kind of evolutionary change. Changes in one part of the developing nervous system generally elicit highly specific and predictable changes in other parts of the system (Katz et al., 1981). For example, similar neonatal brain lesions in hamsters and ferrets cause very similar patterns of compensatory innervation (Frost, 1981; Pallas, Roe, and Sur, 1990). Indeed, if evolution were ever to create hamsters and ferrets with severely reduced superior and inferior colliculi, one would expect both new species to exhibit similar patterns of retinal projections to normally nonvisual thalamic targets. The existence of developmental "constraints" may help to explain some such cases of parallelism in brain evolution (Wilczynski, 1984).

The dynamic network mode of brain development may constrain brain evolution in another sense: its adult products may not be optimal in terms of physiological or behavioral function. There is no a priori reason, for example, to expect that rerouting retinal afferents to the medial geniculate nucleus should be functionally more advantageous than rerouting them to some other dorsal thalamic nucleus. Compensatory innervation may therefore lead to functional integration without necessarily achieving optimal function.

Finally, that the compartmental and dynamic network modes of brain development differ in their propensity to generate cascades or avalanches of developmental change (Katz et al., 1981) implies a difference in their tendency to support evolutionary change. In the compartmental mode, because the mechanisms that lead to compartment formation tend to be local, changes in one brain region tend to remain localized to that brain region. Even diffusible morphogens, which are known to play an important role in the formation of some developmental compartments, are unlikely to diffuse over great distances and may be unable to diffuse across compartmental boundaries (Lawrence and Struhl, 1996). Axon-mediated interactions, on the other hand, can span relatively long distances and can, at least theoretically, be chained together to generate complex cascades of developmental change. Although axon-mediated cascades tend to be dampened when they involve neurons that can obtain trophic support from multiple sources (Finlay, Wikler, and Sengelaub, 1987), they can be significant in systems that are characterized by a low degree of connectivity and a high dependence on trophic support. The visual system of blind mole rats, for example, has become vestigial in most respects but retains a relatively large pathway from the retina to the suprachiasmatic nucleus (Cooper, Herbin, and Nevo, 1993b), which is involved in the control of circadian rhythms and unlikely to depend on axonal connections for trophic support (Lehman et al., 1995).

Conclusion

In very general terms, then, brain development can be seen as a compromise between phylogenetically conservative mechanisms, associated primarily (but not exclusively) with the compartmental mode of brain development, and radical, change-promoting mechanisms, associated primarily (but again not exclusively) with the dynamic network mode. This balance of opposing forces is unlikely to be an accident of nature, for an excess of conservative mechanisms would tend to obstruct evolutionary change, whereas an excess of change-promoting mechanisms would lead to developmental instability even within a species. By analogy to other complex dynamic systems, one might see brain development as being poised at a "critical" state, where evolutionary changes in development are possible but not catastrophic (Bak, 1996).

In contrast to the most frequently studied nonbiological complex systems, however, the brain is not homogeneous with respect to its developmental mechanisms. Thus the compartmental mode dominates development in some brain regions, particularly in the hindbrain, whereas the dynamic network mode dominates development in the forebrain, particularly in the thalamocortical system. These regional differences in the brain's developmental mechanisms correlate, to a considerable degree, with regional differences in brain function. Thus aspects of hindbrain organization involved in the control of physiologically vital functions, such as respiration and feeding, are highly conserved across species. On the other hand, the forebrain, which is more plastic both ontogenetically and phylogenetically, functions primarily in behaviors that vary greatly from species to species, adapting each to its niche.

As this chapter has shown, an understanding of how brains develop can enrich our understanding of how brains evolved and why they function the way they do. This kind of integration between developmental, evolutionary, and functional data is, in my opinion, the ultimate goal of all brain research.

References

Alberch P (1982) The generative and regulatory roles of development in evolution. In: Environmental Adaptation and Evolution (Massakowski D, Roth D, eds), 19–36. New York: Fischer.

Alvarez-Bolado G, Rosenfeld MG, Swanson LW (1995) Model of forebrain regionalization based on spatiotemporal patterns of POU-III homeobox gene expression, birthdates, and morphological features. J Comp Neurol 355: 237–295.

Anderson SA, Eisenstat DD, Shi L, Rubenstein JLR (1997) Interneuron migration from basal forebrain to neocortex: Dependence on *Dlx* genes. Science 278: 474–476.

Angelucci A, Clascá F, Bricolo E, Cramer KS, Sur M (1997) Experimentally induced retinal projections to the ferret auditory thalamus: Development of clustered eye-specific patterns in a novel target. J Neurosci 17: 2040–2055.

Asanuma C, Stanfield BB (1990) Induction of somatic sensory inputs to the lateral geniculate nucleus in congenitally blind mice and in phenotypically normal mice. Neurosci 39: 533–545.

Bak P (1996) How Nature Works: The Science of Self-Organized Criticality. New York: Springer.

Bass AH, Baker R (1997) Phenotypic specification of hindbrain rhombomeres and the origins of rhythmic circuits in vertebrates. Brain Behav Evol 50, suppl 1: 3–16.

Bell E, Wingate RJT, Lumsden A (1999) Homeotic transformation of rhombomere identity after localized *Hoxb1* misexpression. Science 284: 2168–2171.

Bergquist H, Källén B (1954) Notes on the early histogenesis and morphogenesis of the central nervous system in vertebrates. J Comp Neurol 100: 627–659.

Bhide PG, Frost DO (1992) Axon substitution in the reorganization of developing neural connections. Proc Natl Acad Sci USA 89: 11847–11851.

Braisted JE, Tuttle R, O'Leary DDM (1999) Thalamocortical axons are influenced by chemorepellent and chemoattractant activities localized to decision points along their path Dev Biol 208: 430–440.

Bulfone A, Puelles L, Porteus MH, Frohman MA, Martin GR, Rubenstein JL (1993) Spatially restricted expression of *Dlx-1, Dlx-2* (*Tes-1*), *Gbx-2,* and *Wnt-3* in the embryonic day 12.5 mouse forebrain defines potential transverse and longitudinal boundaries. J Neurosci 13: 3155–3172.

Callaway EM, Katz LC (1991) Effects of binocular deprivation on the development of clustered horizontal connections in cat striate cortex. Proc Natl Acad Sci USA 88: 745–749.

Carpenter EM, Goddard JM, Chisaka O, Manley NR, Capecchi MR (1993) Loss of *Hox-A1* (*Hox-1.6*) function results in the reorganization of the murine hindbrain. Development 118: 1063–1075.

Catalano SM, Shatz CJ (1998) Activity-dependent cortical target selection by thalamic axons. Science 281: 559–562.

Catania KC, Kaas JH (1997) The mole nose instructs the brain. Somatosens Mot Res 14: 56–58.

Chiaia NL, Fish SE, Bauer WR, Bennet-Clarke CA, Rhoades RW (1992) Postnatal blockade of cortical activity by tetrodotoxin does not disrupt the formation of vibrissa-related patterns in the rat's somatosensory cortex. Dev Brain Res 66: 244–250.

Cooper HM, Herbin M, Nevo E (1993a) Visual system of a naturally micropthalmic mammal: The blind mole rat, *Spalax ehrenbergi*. J Comp Neurol 328: 313–350.

Cooper HM, Herbin M, Nevo E (1993b) Ocular regression conceals adaptive progression of the visual system in a blind subterranean mammal. Nature 361: 156–159.

Crair MC (1999) Neuronal activity during development: Permissive or instructive? Curr Opin Neurobiol 9: 88–93.

Davenne M, Maconochie MK, Neun R, Pattyn A, Chambon P, Krumlauf R, Rijli FM (1999) *Hoxa2* and *Hoxb2* control dorsoventral patterns of neuronal development in the rostral hindbrain. Neuron 22: 677–691.

Deacon TW (1990) Rethinking mammalian brain evolution. Am Zool 30: 629–705.

Dehay C, Gourd P, Berland M, Killackey H, Kennedy H (1996) Contribution of thalamic input to the specification of cytoarchitectonic cortical fields in the primate: Effects of bilateral enucleation in the fetal monkey on the boundaries, dimensions, and gyrification of striate and extrastriate cortex. J Comp Neurol 367: 70–89.

Díaz C, Puelles L, Marín F, Glover JC (1998) The relationship between rhombomeres and vestibular neuron populations as assessed in quail-chicken chimeras. Dev Biol 202: 14–28.

Doron N, Wollberg Z (1994) Cross-modal neuroplasticity in the blind mole rat *Spalax ehrenbergi*: A WGA-HRP tracing study. NeuroReport 5: 2697–2701.

Easter SSJ, Burrill J, Marcus RC, Ross LS, Taylor JSH, Wilson SW (1994) Initial tract formation in the vertebrate brain. Prog Brain Res 102: 79–93.

Ebrahimi-Gaillard A, Roger M (1996) Development of spinal cord projections from neocortical transplants heterotopically placed in the neocortex of newborn hosts is highly dependent on the embryonic locus of origin of the graft. J Comp Neurol 365: 129–140.

Figdor MC, Stern CD (1993) Segmental organization of embryonic diencephalon. Nature 363: 630–634.

Finlay BL, Wikler KC, Sengelaub DR (1987) Regressive events in brain development and scenarios for vertebrate evolution. Brain Behav Evol 30: 102–117.

Fortin G, Kato F, Lumsden A, Champagnat J (1995) Rhythm generation in the segmented hindbrain of chick embryos. J Physiol 486: 735–744.

Frappé I, Roger M, Gaillard A (1999) Transplants of fetal frontal cortex grafted into the occipital cortex of newborn rats receive a substantial thalamic input from nuclei normally projecting to the frontal cortex. Neurosci 89: 409–421.

Fraser S, Keynes R, Lumsden A (1990) Segmentation in the chick embryo hindbrain is defined by cell lineage restrictions. Nature 344: 431–435.

Frost DO (1981) Orderly anomalous retinal projections to the medial geniculate, ventrobasal and lateral posterior nuclei of the hamster. J Comp Neurol 203: 227–256.

Gao W-J, Pallas SL (1999) Cross-modal reorganization of horizontal connectivity in auditory cortex without altering thalamocortical projections. J Neurosci 19: 7940–7950.

Gilland E, Baker R (1993) Conservation of neuroepithelial and mesodermal segments in the embryonic vertebrate head. Acta Anat 148: 110–123.

Gitton Y, Cohen-Tannoudji M, Wassef M (1999) Specification of somatosensory area identity in cortical explants. J Neurosci 19: 4889–4898.

Graham A, Maden M, Krumlauf R (1991) The murine *Hox-2* genes display dynamic D-V patterns of expression during central nervous system development. Development 112: 255–264.

Guillery RW, LaMantia AS, Robson JA, Huang K (1985) The influence of retinal afferents upon the development of layers in the dorsal lateral geniculate nucleus of mustelids. J Neurosci 5: 1370–1379.

Guthrie S, Butcher M, Lumsden A (1991) Patterns of cell division and interkinetic nuclear migration in the chick embryo hindbrain. J Neurobiol 22: 742–754.

Guthrie S, Prince V, Lumsden A (1993) Selective dispersal of avian rhombomere cells in orthotopic and heterotopic grafts. Development 118: 527–538.

Heil P, Bronchti G, Wollberg Z, Scheich H (1991) Invasion of visual cortex by the auditory system in the naturally blind mole rat. NeuroReport 2: 735–738.

Herrick CJ (1933) Morphogenesis of the brain. J Morphol 54: 233–258.

Holliday M, Hamburger V (1976) Reduction of naturally occurring motoneuron loss by enlargement of the periphery. J Comp Neurol 170: 311–320.

Hughes WF, Lavelle A (1975) The effects of early tectal lesions on development in the retinal ganglion cell layer of chick embryos. J Comp Neurol 162: 265–284.

Huntley GW, Benson DL (1999) Neural (N)-cadherin at developing thalamocortical synapses provides an adhesion mechanism for the formation of somatotopically organized connections. J Comp Neurol 407: 453–471.

Janies D, DeSalle R (1999) Development, evolution, and corroboration. Anat Rec 257: 6–14.

Johnson JE (1999) Neurotrophic factors. In: Fundamental Neuroscience (Zigmond MJ, Bloom FE, Landis SC, Roberts JL, Squire LR, eds), 611–635. San Diego, Calif.: Academic Press.

Katz LC, Shatz CJ (1996) Synaptic activity and the construction of cortical circuits. Science 274: 1133–1138.

Katz MJ (1982) Ontogenetic mechanisms: The middle ground of evolution. In: Evolution and Development (Bonner JT, ed), 207–212. Berlin: Springer.

Katz MJ, Lasek RJ (1978) Evolution of the nervous system: Role of ontogenetic buffer mechanisms in the evolution of matching populations. Proc Natl Acad Sci USA 75: 1349–1352.

Katz MJ, Lasek RJ, Kaiserman-Abramof IR (1981) Ontophyletics of the nervous system: Eyeless mutants illustrate how ontogenetic buffer mechanisms channel evolution. Proc Natl Acad Sci USA 78: 397–401.

Kitzes LM, Kageyama GH, Semple MN, Kil J (1995) Development of ectopic projections from the ventral cochlear nucleus to the superior olivary complex induced by neonatal ablation of the contralateral cochlea. J Comp Neurol 353: 341–363.

Kohmura N, Senzaki K, Hamada S, Kai N, Yasuda R, Watanabe M, Ishii H, Yasuda M, Mishina M, Yagi T (1998) Diversity revealed by a novel family of cadherins expressed in neurons at a synaptic complex. Neuron 20: 1137–1151.

Lawrence PA, Struhl G (1996) Morphogens, compartments, and pattern: Lessons from *Drosophila?* Cell 85: 951–961.

Leder M-L (1975) Zur mikroskopischen Anatomie des Gehirns eines blinden Nagers (*Spalax leucodon*, Nordmann, 1840) mit besonderer Berücksichtigung des visuellen Systems: 2. Das visuelle System. Zool Jb Anat 94: 74–100.

Lehman MN, Lesauter J, Kim C, Berriman SJ, Tresco PA, Silver R (1995) How do fetal grafts of the suprachiasmatic nucleus communicate with the host brains? Cell Transplant 4: 75–81.

Levitt P, Barbe MF, Eagleson KL (1997) Patterning and specification of the cerebral cortex. Annu Rev Neurosci 20: 1–24.

Linden R (1994) The survival of developing neurons: A review of afferent control. Neurosci 58: 671–682.

Marín F, Puelles L (1995) Morphological fate of rhombomeres in quail/chick chimeras: A segmental analysis of hindbrain nuclei. Eur J Neurosci 7: 1714–1738.

Merzenich MM, Recanzone G, Jenkins WM, Allard TT, Nudo RJ (1988) Cortical representational plasticity. In: Neurobiology of Neocortex (Rakic P, Singer W, eds), 41–67. New York: Wiley.

Meyer RL (1983) Tetrodotoxin inhibits the formation of refined retinotopography in goldfish. Dev Brain Res 6: 293–298.

Miyashita-Lin EM, Hevner R, Wassarman KM, Martinez S, Rubenstein JLR (1999) Early neocortical regionalization in the absence of thalamic innervation. Science 285: 906–909.

Necker R, Rehkämper G, Nevo E (1992) Electrophysiological mapping of body representation in the cortex of the blind mole rat. NeuroReport 3: 505–508.

Newman SA (1993) Is segmentation generic? BioEssays 15: 277–283.

Nieuwenhuys R (1998) Morphogenesis and general structure. In: The Central Nervous System of Vertebrates, vol. 1 (Nieuwenhuys R, Ten Donkelaar HJ, Nicholson C, eds), 159–229. Berlin: Springer.

Nittenberg R, Patel K, Joshi Y, Krumlauf R, Wilkinson DG, Bricknell PM, Tickle C, Clarke JDW (1997) Cell movements, neuronal organisation and gene expression in hindbrains lacking morphological boundaries. Development 124: 2297–2306.

Olavarria J, Malach R, Van Sluyters RC (1987) Development of visual callosal connections in neonatally enucleated rats. J Comp Neurol 260: 321–348.

Oppenheim RW (1981) Neuronal cell death and some related regressive phenomena during neurogenesis: A selective historical review and progress report. In: Studies in Developmental Neurobiology: Essays in Honor of Victor Hamburger (Cowan WM, ed), 74–133. New York: Oxford University Press.

Pallas SL, Hahm J, Sur M (1994) Morphology of retinal axons induced to arborize in a novel target, the medial geniculate nucleus: 1. Comparison with arbors in normal targets. J Comp Neurol 349: 343–362.

Pallas SL, Roe AW, Sur M (1990) Visual projections induced into the auditory pathway of ferrets: 1. Novel inputs to primary auditory cortex (AI) from the LP/pulvinar complex and the topography of the MGN AI projection. J Comp Neurol 298. 50–68.

Puelles L, Rubenstein JLR (1993) Expression patterns of homeobox and other putative regulatory genes in the embryonic mouse forebrain suggest a neuromeric organization. Trends Neurosci 16: 472–479.

Purves D (1988) Body and Brain: A Trophic Theory of Neural Connections. Cambridge, Mass.: Harvard University Press.

Purves D, Riddle DR, White LE, Gutierrez-Ospina G, LaMantia A-S (1994) Categories of cortical structure. Prog Brain Res 102: 343–355.

Raff RA (1996) The Shape of Life: Genes, Development, and the Evolution of Animal Form. Chicago: University of Chicago Press.

Ramirez-Weber F-A, Kornberg TB (1999) Cytonemes: Cellular processes that project to the principal signaling center in *Drosophila* imaginal discs. Cell 97: 599–607.

Redies C (1997) Cadherins and the formation of neural circuitry in the vertebrate CNS. Cell Tissue Res 290: 405–413.

Redies C, Takeichi M (1996) Cadherins in the developing central nervous system: An adhesive code for segmental and functional subdivisions. Dev Biol 180: 413–423.

Rehkämper G, Necker R, Nevo E (1994) Functional anatomy of the thalamus in the blind mole rat *Spalax ehrenbergi:* An architectonic and electrophysiologically controlled tracing study. J Comp Neurol 347: 570–584.

Roux W (1895) Gesammelte Abhandlungen über Entwicklungsmechanik der Organismen. Leipzig: Engelmann.

Schlaggar BL, O'Leary DDM (1991) Potential of visual cortex to develop arrays of functional units unique to somatosensory cortex. Science 252: 1556–1560.

Schlaggar BL, O'Leary DDM (1994) Early development of the somatotopic map and barrel patterning in rat somatosensory cortex. J Comp Neurol 346: 80–96.

Schmidt KE, Kim DS, Singer W, Bonhoeffer T, Löwel S (1997) Functional specificity of long-range intrinsic and interhemispheric connections in the visual cortex of strabismic cats. J Neurosci 17: 5480–5492.

Schneider GE (1973) Early lesions of superior colliculus: Factors affecting the formation of abnormal retinal connections. Brain Behav Evol 8: 73–109.

Sengelaub DR, Finlay BL (1981) Early removal of one eye reduces normally occurring cell death in the remaining eye. Science 213: 573–574.

Shatz CJ, Stryker MP (1988) Prenatal tetrodotoxin infusion blocks segregation of retinogeniculate afferents. Science 242: 87–89.

Shimamura K, Martinez S, Puelles L, Rubenstein JLR (1997) Patterns of gene expression in the neural plate and neural tube subdivide the embryonic forebrain into transverse and longitudinal domains. Dev Neurosci 19: 88–96.

Sperry RW (1963) Chemoaffinity in the orderly growth of nerve fiber patterns and connections. Proc Natl Acad Sci USA 50: 703–710.

Stent GS (1981) Strength and weakness of the genetic approach to the development of the nervous system. Annu Rev Neurosci 4: 163–193.

Stoykova A, Götz M, Gruss P, Price J (1997) Pax6-dependent regulation of adhesive patterning, R-cadherin expression and boundary formation in developing forebrain. Development 124: 3765–3777.

Striedter GF (1998) Stepping into the same river twice: Homologues as recurring attractors in epigenetic landscapes. Brain Behav Evol 52: 218–231.

Stryker MP, Harris WA (1986) Binocular impulse blockade prevents the formation of ocular dominance columns in cat visual cortex. J Neurosci 6: 2117–2133.

Sur M, Garraghty PE, Roe AW (1988) Experimentally induced visual projections into auditory thalamus and cortex. Science 242: 1437–1441.

Sur M, Pallas SL, Roe AW (1990) Cross-modal plasticity in cortical development: Differentiation and specification of sensory neocortex. Trends Neurosci 13: 227–233.

Tan K, Le Douarin NM (1991) Development of the nuclei and cell migration in the medulla oblongata. Anat Embryol 183: 321–343.

Tessier-Lavigne M, Goodman CS (1996) The molecular biology of axon guidance. Science 274: 1123–1132.

Tierney TS, Russell FA, Moore DR (1997) Susceptibility of developing cochlear nucleus neurons to deafferentation-induced death abruptly ends just before the onset of hearing. J Comp Neurol 378: 295–306.

Waddington CH (1957) The Strategy of the Genes. London: Allen and Unwin.

Wilczynski W (1984) Central neural systems subserving a homoplasous periphery. Am Zool 24: 755–763.

Wilkinson DG, Krumlauf R (1990) Molecular approaches to the segmentation of the hindbrain. Trends Neurosci 13: 335–339.

Wilson SW, Placzek M, Furley AJ (1993) Border disputes: Do boundaries play a role in growth-cone guidance? Trends Neurosci 16: 316–323.

Wingate RJT, Lumsden A (1996) Persistence of rhombomeric organisation in the postsegmental hindbrain. Development 122: 2143–2152.

Wizenmann A, Lumsden A (1997) Segregation of rhombomeres by differential chemoaffinity. Mol Cell Neurosci 9: 448–459.

Woolsey TA (1990) Peripheral alteration and somatosensory development. In: Development of Sensory Systems in Mammals (Coleman JR, ed), 461–516. New York: Wiley.

Zufferey PD, Jin F, Nakamura H, Tettoni L, Innocenti GM (1999) The role of pattern vision in the development of cortico-cortical connections. Eur J Neurosci 11: 2669–2688.

17 Boundary Constraints for the Emergence of Form

Diego Rasskin-Gutman

In chapter 13 of *On the Origin of Species,* Darwin portrays morphology as "the most interesting department of natural history." Lacking knowledge of genetics and biochemistry, he was nevertheless able to bring together a whole theory of evolution, which, although refined, remains essentially unchanged through today. In spite of such a promising beginning, after almost 150 years, the mainstream of evolutionary biology has abandoned the analysis of form and finds itself confined to the analysis of molecular data. On many occasions, evolution is defined purely at the level of changes in allele frequencies in a population leading to speciation (textbooks are good examples of this reductionistic trend; see, for example, Beck, Liem, and Simpson, 1991), rather than using a more integrative approach where the genetic, morphological, and population levels are taken into account. Thus morphology has been gradually left out of the evolutionary picture, employed only when needed to make claims about adaptation and functionality.

To make matters worse for morphology, the research program for molecular evolution is conceptually simple, involving the comparative analysis of well-defined molecular units (i.e., bases, amino acids). All that is needed for comparison are some assumptions about the tempo in which the units mutate. Disregarding the true nature of the homology between equivalent positions among each one of these units (see, for example, Hall, 1994; Müller, chapter 4, this volume), the comparison of, say, the amino acids of a protein such as hemoglobin between a cow and a pig is straightforward.

Morphological information, on the other hand, is not that simple. Shape is an elusive concept, intimately related to human perception and our scale of observation. On many occasions, shape is loosely defined as "external appearance," with no suitable units that can account for, say, the formation of a crest and the length of a bone at the same time. The shape of bone trabeculae are elements that cannot be used in a comparative analysis of, for example, the visible features of skull bones of theropod dinosaurs (for the same reason the genetic code has hardly anything to say about these same features). In contrast with the neat, linear information encapsulated in the sequence of molecular units, an array of cells is involved in the generation of a three-dimensional entity that exhibits shape, size, and relations to neighboring tissues. Although, to be sure, nucleic acids and proteins also present three-dimensional features, except for RNA secondary structure—the only instance in which genetic material and shape are directly linked (see, for example, Fontana and Schuster, 1998)—these features are simply not used in comparative analysis.

All of which raises the question, should we abandon any attempt to use morphological information as an indicator of evolution? The answer is, of course not. Disciplines such as developmental evolutionary biology and paleontology (to mention two of the most obvious ones) would not benefit at all from such a decision: form data are needed, not only to account for the processes that generate morphological patterns, but also to elaborate phylogenetic hypotheses. On a more fundamental level, molecular evolution and morphological evolution are two quite different processes; the end products of both are linked and expressed by the nonlinear events that occur during the development of an organism. One could entertain the idea that morphological evolution may be undergoing different paces and is subject to law-like changes independent from the genome. This would imply that we should explore new avenues regarding the generation and transformation of form to elucidate and study an evolutionary story separate from that of the genetic code. In contrast with genes, where changes are restricted to the accumulation of mutations, morphology is a product of developmental pathways that are part of an epigenetic interplay between physical forces and geometrical arrangements.

This chapter explores a level of morphological organization identified as "connectivity" for tetrapod skeletons, but easily extendable to other organisms. It introduces the physical correlates of connectivity relations or "boundaries" in biological systems, briefly outlining a system for the efficient description of organic form. It then analyzes patterns of form in the framework of a morphospace of connections, highlighting the importance of boundary patterns with two kinds of evidence: empirical data on bones of tetrapod skulls and computer simulations from a theoretical approach to generate possible connectivity relations that may correlate with those present in organisms. Theoretical computer simulations generate several morphospaces of connections made under different assumptions as well as "evolutionary runs" that trace changes of connections throughout generations. These theoretical constructions, in turn, permit the exploration of the possible sets of boundary patterns that may emerge under different connectivity constraints.

Levels of Morphological Organization

A system of description that could suitably separate different types of morphological information would allow us to efficiently assess patterns of form. Derived from comparative information, these patterns would in turn suggest the existence of or the need for certain kinds of processes to account for them. Such a system might be configured like the one outlined in figure 17.1 (see Rasskin-Gutman and Buscalioni, 2001), where each level of description provides its own type of information that can be used in comparative analyses. Different analytical tools have to be used in each case, with a proper formalization and use of morphospace representations according to their respective level of description. For the analysis of the external appearance of adult organisms, shape and size are the most detailed

ORGANIZATIONAL LEVEL	DESCRIPTOR	FORMALIZATION	MORPHOSPACE
PROPORTIONS	ELEMENT	CHARACTER MATRIX	HYPERSPACE
ORIENTATIONS	COMPOUND	ANGLES, POSITIONS	DISPOSPACE
CONNECTIONS		BOUNDARY PATTERNS	CONNECTIVITY SPACE
ARTICULATIONS	MECHANISM	ANGLES; DISTANCES	CONFIGURATION SPACE

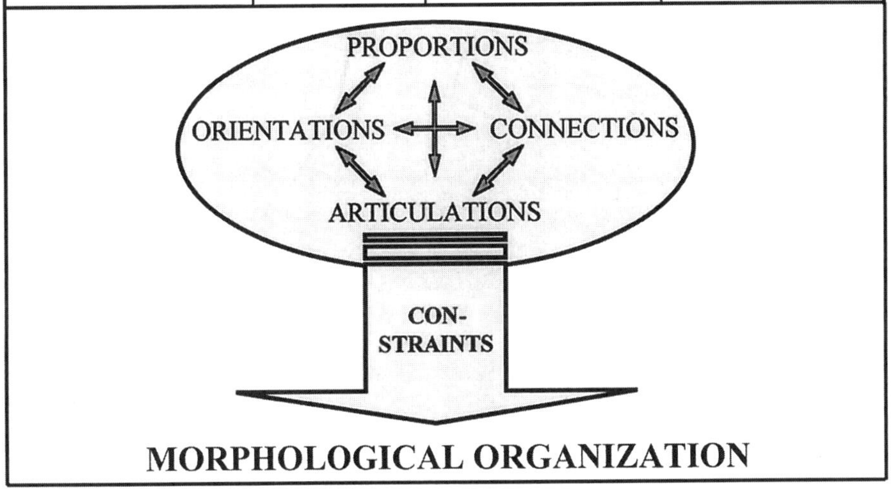

Figure 17.1
(*Above*) System of analysis of morphological organization. (*Below*) Each organizational level acts on the others in a nonhierarchical way. Constraints arise as a result of these interactions.

properties, whereas orientations, connections, and articulations characterize higher levels of description.

Each of these four levels is present at different times and, most important, at different scales, during embryonic development. Connectivity appears early in ontogeny as the most fundamental relation among embryonic cells. Later, a cascade of differentiations and secondary inductions takes place, starting to shape tissues and organs. Proportions and orientations along the three spatial axes of the embryo take over, giving rise to a new level of connections, this time between tissues. Articulations, which allow movement among elements such as in hard skeletal parts, appear much later, and can be said to be a by-product of connections.

The formation of vertebrate limbs (see, for example, Hinchliffe and Johnson, 1980) provides a suitable instance of the interplay among morphological levels of organization. The initial state of a developing limb bud is a proliferation of mesenchymal cells. The shape and size of the individual cells determine relations of connectivity among them, which are prominent at this stage. The resulting mass of cells, packed together, forms a bud. Thus the connectivity properties of the individual cells generate a higher level of organization, the limb bud, which exhibits, on its own, properties of proportion (size and shape) far removed from the proportions exhibited by the individual cells. In turn, the proportions of the limb bud determine the number and position of cell condensations that appear in the mesenchyme of precartilage areas, forming the primordia of future bones of the limb, which start to assume identities of their own. Later, each condensed precartilage center shows a preferential orientation as well as connectivity relations at a new organizational level, where the future bones are the new elements, and the individual cells are no longer suitable to describe the system.

These four levels of morphological organization constrain each other during development. Shape constrains packing, which constrains connectivity, which provokes inductions, which generates new elements that exhibit shape and connection properties, and so on. There is no hierarchy among these levels, but rather a nested succession of events in which one property level originates structures that are then subjected to the influence of another property level. This is the reason why any system of morphological description fails to accommodate all levels at once.

The information provided by each level requires a formalization that is given by data matrices in specific ways, leading to the construction of diverse morphospaces with varied levels of complexity. Morphospaces provide frameworks in which both the variation of morphological organization and morphological generation can be assessed. But they do this in an extended way, allowing the visualization of possible forms of morphological organization that encompass and go beyond those known in nature (see McGhee, 1999; Chapman and Rasskin-Gutman, 2000; Rasskin-Gutman, 2000; Rasskin-Gutman and Buscalioni, 2001). Thus morphospaces are heuristic tools in which the generation of morphological organization can be efficiently addressed. If thoroughly modeled, they can predict what sorts of processes are the most likely ones to account for patterns found in nature. Boundary patterns, defined as the formalization and physical correlates of the connectivity level of morphological organization, can be generated and combined to model varied morphospaces in which naturally occurring patterns can be analyzed.

Boundary Patterns

Some Early Attempts

Connections have been identified in the past as one of the most fundamental aids to recognize homology: among others, Cuvier, Geoffroy Saint Hilaire, and Remane used the principle of connections to recognize structural identities throughout organisms (see, for example, Riedl, 1978; Rieppel, 1988). However, there have been only a few attempts to provide a comparative framework that would recognize boundary patterns as a level of morphological organization in their own right. Several ideas were developed in that direction, mainly by Woodger, Rashvesky, and Riedl.

A first attempt to use boundary information of vertebrate skeletal parts was made by Woodger (1945), who tried to codify relationships in the tetrapod limb with three descriptors, namely, "being distal to," "being postaxial to," and "to be articulated with." The data were then put together in a table where each bone was unequivocally characterized by its relational position. But the method has never been applied, mostly due to its level of generality, as immediately recognized by Woodger.

Rashevsky (1954, pp. 343–344) attempted to work out connectivity from the point of view of physiological relations among functional modules (the principle of "biotopological equivalence"). Regarding morphology, he stated the following:

It may be possible to find a topological principle which relates the biological function graph to a particular three-dimensional bounded topological complex.... Considering an animal as a "system of linked levers"... we notice that the (linear) topology or structure of the lever system is closely related to the structure of the group of motions it can perform. Vice versa, the group of motions may well determine the arrangement and structure of the lever system.

Although Riedl (1978) diagrammed the connectivity relations between bones to depict the "mammalian morphotype," no attempts were made to use such a descriptive approach in a comparative framework. More recently, Young (1993) looked into connectivity relations and cellular components as a proxy for a new, modern notion of archetype.

Of all these attempts to codify connectivity information, Rashevsky came closest to offering an efficient descriptive tool in terms of graphs and bounded topologies that could be used for comparative purposes. Rather than follow Rashevsky's emphasis on physiological relations, however, this chapter will describe vertebrate skeletal parts, coupling graph theory with modeling strategies based on cellular automata.

Boundaries, Connections, and Modules

Boundaries originate automatically when an entity takes shape and, more important for the central theme of this chapter, when two or more entities make contact and form a new one at a higher level of organization. Thus atoms form molecules by establishing covalent bonds; layers of heterogeneous tissues arise out of the same precursor tissue; bones follow an arrangement that makes them part of a network of connections, which, as we have seen, can be identified as a lever system. The common phenomenon in all these cases is the emergence of a new compound out of simpler elements. In the case of a molecule such as water, new physical properties arise out of this association that were not present in the isolated atoms of hydrogen and oxygen. The same can be said of organismal units. Cells of mesenchymal tissue must be in contact during secondary induction processes in order to generate new, differentiated tissues, such as epidermal glands (Bard, 1990). Cells that are adjacent can communicate via membrane gap junctions or by morphogens that elicit specific gene expression patterns. As a result, one cell can induce its neighbors to follow a path of differentiation that, separately, they would never have taken. The attachment of the hind limbs to the pelvic girdles' socketlike structure (acetabulum) provides an example for the need of connections to form anatomical parts at the level of the skeleton. The acetabulum originates as a result of the connection among the three bones that form the pelvis (ilium, ischium, and pubis). Note that each bone, as it stands on its own, has no functional capacities; the capacities to, say, transmit force and allow femoral movement, arise only after the bones are put in contact as a connected network. The organization (in this case at the level of connectivity) of the whole system is what confers functional capacities onto it. Thus morphological organization is logically prior to any functional capacities. Geoffroy's old motto "Function follows form" is here absolutely vindicated.

This chapter uses topological relations of the skeletal parts as frameworks onto which other properties can be mapped. *Boundary* or *connectivity patterns* are defined as "the morphological arrangements that arise when two or more elements are in physical contact." The degree of connection is an abstract specification of each connected system; each element has a "degree" of connectivity defined as the number of other elements to which it connects. This quantitative measure can be used to evaluate general properties that apply to any kind of connected system. *Boundary modules,* an abstraction of the physical contact between elements, are defined as "coherent groups of the smallest number of elements into which a connected system can be separated." Modules can be linear (two elements connected by a single edge); triangular (three elements mutually connected); or of higher order with more than three elements connected in a ringlike fashion (quadrangular, pentagonal, hexagonal, etc.).

Morphospaces are collections of morphological systems that bear some resemblance to biological systems. As noted above, they form broad frameworks in which natural occurrences can be compared to forms that never appeared in nature. Generative morphospaces are built by establishing a generative rule that forms a set of different patterns. In this context, we can define a morphospace of connections of **n** elements for the probability space **S**

Boundary Constraints

and the generative rule **G**, as the collection of all configurations generated by **G** that meet the constraint set by **S**, where **S** is a probability space of connectivity degrees, and **G** is a set of rules to form boundary patterns as described below.

Material and Methods

The analysis of boundary patterns needs (1) an efficient way to code and quantify the notion of connectivity; and (2) a generative tool to form patterns given a set of conditions. Quantification and generation allow us to thoroughly explore the space of possible patterns, while, at the same time, making operative comparisons between the theoretical patterns and those patterns found in nature.

Quantifying the Morphospace of Connections

For the descriptive task, I have used graph theory, a well-defined mathematical discipline that provides efficient ways to describe the general properties of boundary patterns (see, for example, Harary, 1972, for a general introduction; and Rashevsky, 1955; Trucco, 1956a,b, for some pioneer ideas about the information contents of graphs). Here a bone is represented as the vertex of a graph and the boundaries formed by two physically adjacent bones as the edges that join two vertices (figure 17.2). Each skeletal pattern is then abstracted as

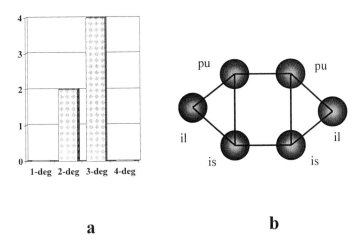

Figure 17.2
(*a*) Frequencies of appearance of connectivity degrees. Ilia are 2-degree bones, pubes and ischia are 3-degree bones. (*b*) Graph representation of a skeletal part. A pelvic girdle of a theropod dinosaur (ventral view). Note the quadrangular pattern between the pubes and the ischia (region of ventral symphyses) and both triangular patterns where the acetabula form, allowing articulation with the femoral heads. il, ilium; pu, pubis; is, ischium.

a graph and characterized by a table of frequencies for the degree of connectivity of each bone. We can use these frequencies to set a constraint to generate connectivity patterns in a morphospace of connections, so that only connections that meet these conditions are considered. For a number **n** of bones, we can use a frequency table to specify a space of probability, **S**, where

$$\mathbf{S} = \{p(0), p(1), p(2), \ldots, p(k)\}$$

$$\Sigma\, p(i) = 1; \quad \text{for } i = 0 \text{ to } k,$$

and where $p(0)$ indicates the probability of having degree 0 (to be isolated), $p(1)$ is the probability of having degree 1 (to be connected only to one node), and so on. An example is given in figure 17.3, along with one interpretation of a possible shape that the boundary patterns may take, always preserving the connectivity relations. In this example, the morphospace of connections is formed by all nonisomorphic configurations of $\mathbf{n} = 10$ bones for the space of probabilities $P = \{p(0) = 0; p(1) = 1/10; p(2) = 5/10; p(3) = 2/10; p(4) = 1/10; p(5) = 1/10\}$. An interpretation of the proportions of the vertices such as in figure 17.3 is called a "virtual skeleton." Although the number of graphs that meet each space of probabilities is finite, and the upper bound of this number can be computed with the Polya formula (Harary, 1972), a complex mathematical analysis is needed to find out the exact number of nonisomorphic graphs that meet a given space of probabilities.

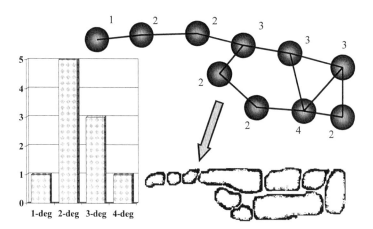

Figure 17.3
(*Left*) A set of frequencies for connectivity degrees defines a set of constraints for a finite number of configurations. (*Above*) One of the graphs that meets these constraints. (*Below*) Interpretation of the graph as a "virtual skeleton." Proportions are irrelevant as long as the connections are satisfied. Note the unavoidable opening inside the pentagonal boundary pattern.

A sample with nine skull roofs of tetrapods was used to look for real patterns and to make a preliminary estimate of the frequencies of connections. The skulls belong to seven diapsids, one anapsid, and one synapsid, taken from diagrams after Goodrich, 1930, figures 336, 351, 352, 356, 369; Rieppel, 1993, figures 7.7A, 7.7B, and 7.7E; and Stahl, 1974, figure 7.31. The connectivity information of the boundary patterns used refers to (1) the total number of bones to which a given bone is connected (degree of a bone); and (2) the kinds of boundary modules they form (figure 17.4). Although a bigger sample is needed to have enough confidence in these frequencies, the estimates can be used as parameters to feed the cellular automata as explained in the next section, and will probably not differ too much as long as they are taken from skull roofs.

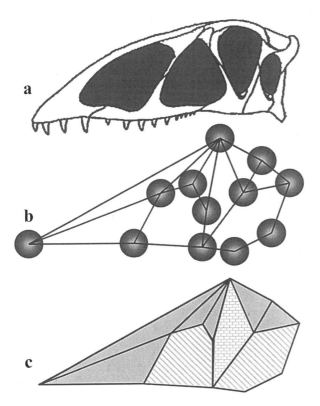

Figure 17.4
(*a*) Diagram of the lateral view of the pterosaur *Dimorphodon* (Reptilia, Diapsida). (*b*) Graph of connections. (*c*) Diagram of the boundary modules. Note that for any three-dimensional structure, the connective net of bones can be reduced to a two-dimensional graph with or without crossings.

Generating and Exploring the Morphospace of Connections

A stand-alone simulation program designed and programmed in Turbo C++ for a DOS environment, FRONTERA, was developed to generate and explore boundary patterns. The program generates graphs taking into account conditions relative to the connectivity degrees of each node and the final number of nodes sought. These conditions can be changed by the user to suit special purposes. (FRONTERA can be downloaded from the KLI homepage *www.kli.ac.at*). The architecture of FRONTERA is based on cellular automata models, consisting of a layout of cells that can be turned on and off according to a set of local rules, which refer to the internal status of the cell and the status of the immediate neighbors. In a cellular automaton following the "Moore neighborhood," each cell is evaluated in relation to its nine neighbors (see, for example, Rietman, 1989). At each cycle, the "universe" (the whole set of cells) changes globally after every cell has been evaluated, according to probabilistic production rules that relate the on/off status of neighbors. As cycles continue evolving, cells that are turned on form a boundary pattern in which connection is defined by adjacency between cells. Thus each switched-on cell is treated as the vertex of an evolving graph.

The following is an example of a set of production rules for the automata. These rules are evaluated by the program for each cell at every new cycle:

1. If the cell is *off* and up to four neighbors are *on,* then with $P(0.01)$ turn the cell *on* (a rule to add a cell to the pattern in relation to the neighbors).

2. If the cell is *on,* then continue being *on* (a rule for a cell already in the pattern to continue being part of the pattern, in order to preserve the evolving configuration).

An additional rule controls the deletion of cells that are part of the pattern in order to prevent stagnation at a configuration from which the pattern cannot go further:

3. If the cell is *on,* then with $P(0.001)$ turn the cell *off,* then test connectivity.

Note that the probability of deletion is much smaller than the probability for additions. This is a way to simulate "history," that is, to prevent a situation of permanent addition and deletion where no pattern would be preserved. Also, after one cell has been randomly selected to disappear, a check is performed in order to be sure that the pattern preserves connectivity; if not, the cell remains turned on. Before starting a new cycle, the algorithm checks out two stop rules. One is used to see if the desired number of vertices has been achieved and the other to check that the pattern of connection degree for each vertex (the probability space) has been attained.

The program operates by adding and deleting vertices from a user-defined pattern (the initial condition) according to the local rules of interaction outlined above. From the initial configuration, the pattern keeps "evolving" until the stop rule is satisfied. This stop rule, which can be changed by the user, can be based on the space of probabilities derived from the empirical data. In this way, the program tries to arrive at a pattern that has the same number of vertices (bones) and the same pattern of edges (connections) as seen in real configurations. By specifying ranges of the possible number of vertices for a given degree (as in the example below), the possibilities are enlarged, and a broader number of real patterns can be approximated. The program can be used to generate morphospaces of connections for a particular set of constraints (as explained above) or to analyze evolutionary runs, in which the whole sequence from the initial configuration to the final pattern is studied (as explained below). Analysis of evolutionary runs allows us to observe the generation of boundary patterns as a sequence of events that can be interpreted as macroevolutionary dynamics.

Results

Empirical Data

The frequency distribution for the degree of connection of individual bones shows that bones connected three and four times are the most abundant (see figure 17.5). Based on these results, a theoretical assessment of the kinds of skulls that one would expect to find in nature is shown in figure 17.6. On the other hand, the boundary modules derived from the connectivity data show that around 75 percent of them are triangular, whereas the remaining ones are hexagonal, pentagonal, quadrangular, and linear, with no preference among these.

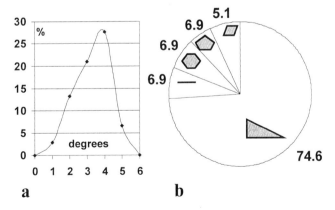

Figure 17.5
(*a*) Frequency of degree of connections for a sample of ten tetrapod skulls. (*b*) Distribution of boundary patterns for the same sample.

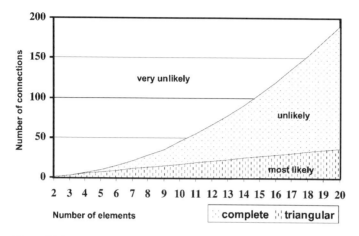

Figure 17.6
Diagram illustrating a rough expectancy for tetrapod skulls in nature. Skulls will tend to triangulate spaces, usually around several fenestrae.

Simulations

Patterns resulting from a simulation run with the program FRONTERA are shown in figure 17.7. An initial hexagonal pattern of 6 vertices was set up in a universe of 8 × 6 cells. The stop rule required a final pattern with 13 vertices. The specified degrees were 1 or 2 vertices of degree 1; 2 to 5 vertices of degree 2; 3 to 5 vertices of degree 3; 3 to 6 vertices of degree 4; and 1 or 2 vertices of degree 5, approximating the results of the empirical data.

The program ran through 635 cycles before finding a pattern that met the requirements of the stop rules. During the process, 22 stable boundary patterns were generated. The final pattern configuration had 13 vertices and 16 connections: 2 vertices of degree 1; 4 vertices of degree 2; 3 vertices of degree 3; 3 vertices of degree 4; and 1 vertex of degree 5. The evolution curve (figure 17.8), defined as the number of additions or deletions of vertices to the initial pattern, showed a rapid change toward a plateau (a period of stasis) up to cycle 67. In this cycle, the pattern already had 13 vertices and the subsequent cycles were attempts by the algorithm to find a configuration that met the stop rules. This was achieved in the remaining 568 cycles. The first pattern after the stasis period was a reversion to the previous pattern (pattern 41). Another reversion event occurred in pattern 240, which again duplicated pattern 41. After that, a "novelty" event took place that made the whole system able to evolve and find a final configuration. Figure 17.8 shows the evolution curve as well as the evolution of the number of edges and the "compactedness" of the graph, defined as the ratio between the number of actual edges and the number of possible edges for a given

Boundary Constraints

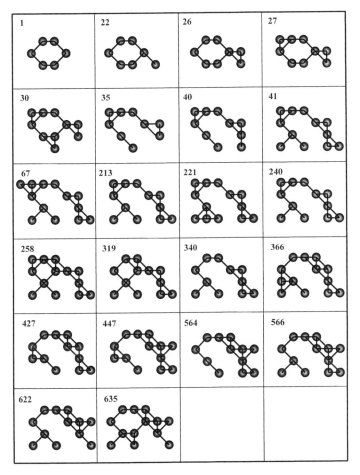

Figure 17.7
Patterns of a simulation run with the program FRONTERA. Numbers indicate the cycle at which each pattern was generated.

number of vertices. Compactedness can be interpreted as a possible measure of the complexity of the configuration.

Discussion

Multiple boundary-related processes take place during embryogenesis, forming the appearance of the organism as it grows. Some of these processes are under genetic control and

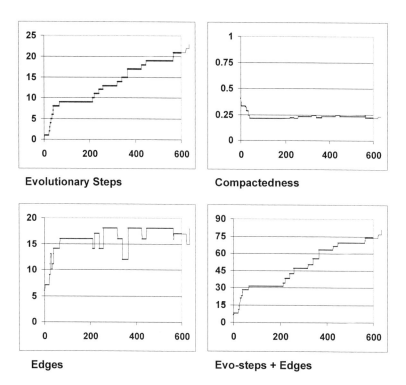

Figure 17.8
(*Upper left*) Curve illustrating changes in evolutionary steps as defined by the addition or deletion of elements. (*Upper right*) Curve of changes in compactedness. (*Lower left*) Changes in number of edges. (*Lower right*) Evolutionary steps as defined by changes in number of elements plus absolute changes in number of edges.

some are mechanically induced. One of the large-scale boundary phenomena in vertebrate organisms is the establishment of boundaries between adjacent bones. Can this information be useful for comparative analysis at macroevolutionary scales? Thus far, the morphological data from bones in the literature are mostly related to shape and size, used to analyze the proportions of detailed structures in each separate bone. Except for the references commented on above, connections have been rarely used, or used only to derive functional properties of skeletal parts, such as the analysis of the "force lines" of dermal skull bones in fishes (Thomson, 1995) or the analysis of the system of joints in dinosaur skulls (Weishampel, 1993).

Can such different features as the diameter of a notch and the angle at which bones articulate, or the curvature of a ridge and the length of a whole bone, be meaningfully combined? This chapter argues that each level of morphological organization conveys different information and that, to understand processes, each should be treated separately. Although, when the aim is to infer patterns of relationships among taxa, as in cladistic analysis, mixing information would seem more suitable, even if patterns can be discerned, the processes that produce them simply cannot be followed when features of different levels of morphological organization are mixed together. The shape of emerging elements such as bones varies amply in every lineage and for each structure, although this variation is constrained by the boundary patterns established during growth. Conversely, a growing bone may establish a boundary with another bone by a modification in proportions. Identifying and separating these processes are crucial to our understanding of the design of the body plans of organisms.

The preliminary data from the sample identify a set of boundary patterns that integrates a skull. This set includes linear, triangular, quadrangular, pentagonal, and hexagonal patterns, which can be thought of as the "construction modules" that make up a skull. The triangular modules are the most conspicuous ones, which may reflect the simple constructional constraint for covering a surface with many elements, where every element touches at least three other adjacent elements. Also, the data suggest that the design of a skull always consists of an orbital pentagonal or hexagonal module surrounded by triangular ones, and may sometimes consist of other, quadrangular or higher-order modules in the postorbital region of the skull.

Every high-order pattern implies the existence of an opening. The orbital fenestra of these reptilian skulls is the opening around which every other module is situated. The postorbital region is highly variable. This variation coincides with the opening of other fenestrae in this area, an event that has occurred in the synapsid and diapsid lineages. In contrast, the rostral area seems to be more conservative. The opening of skull fenestrae is a process that occurs in all lineages. The type of boundary modules and their distribution around skull openings suggest that boundaries constrain their construction.

Specifically, by looking at the types of modules where fenestrae open, one can identify two types of fenestrae, corresponding to two different processes: active fenestrae, which appear within a triangular module; and passive fenestrae, which appear within a higher-order module. Active fenestrae occur when the presence of another tissue impedes the closing of the boundary area (e.g., a nerve), or when bone is resorbed later in the course of development (e.g., the nares). In contrast, passive fenestrae occur within higher-order boundary patterns when openings simply cannot close unless the patterns become triangular by the growth of one of the bones involved. A reasonable expectation of a model of fenestra types is that passive fenestrae will remain passive throughout a given lineage unless a specific growth of one of the bones provokes contact with another bone, which could provoke the closure of the opening. In the opposite case, when the common ancestor of a lineage has an active fenestra, this can only be closed up if the tissue involved in preventing the closure disappears. Thus closure of a passive fenestra requires heterochrony (relative growth of one bone), whereas closure of an active fenestra, which can be explained by the action of an external tissue, does not.

The simulation run with FRONTERA showed that the program is a suitable tool to evaluate macroevolutionary dynamics at the level of boundary patterns. Its modeling of processes such as reversion, convergence, stasis, and novelty highlighted the problem of homology. During quite a large number of cycles, the evolving pattern remained in a period of stasis, finding the same pattern as if it had been a repeated process of convergence. Then a "novelty" occurred that allowed the pattern to reach the final stop rules.

Using this program, the assessment of homology relations can be tackled easily provided there is confidence in the whole sequence of changes from the initial pattern to the final one, as shown in figure 17.9, where the initial hexagon can be misidentified if compared only with patterns 258 and 319. When the whole sequence is present, the homologous areas can be detected trivially.

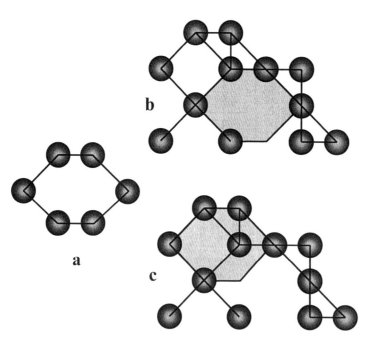

Figure 17.9
Illustration of the problem of homology. Initial hexagon (*a*) cannot be assessed correctly if compared only with patterns 258 (*b*) and 319 (*c*) (shaded hexagons). Knowing the whole sequence makes the homology test trivial; the correct answer is pattern (*c*).

Conclusions

Two ways to analyze boundary patterns have been indicated. The first is by using a proper morphological framework and an efficient way to code the information of this organizational level. The second is by using a simulation program based on cellular automata to generate evolutionary runs. Graph theory has much to offer for the first task, and many structural properties of the boundary patterns can be analyzed that may lead to a better understanding of skeletal design. As used here, the simulation approach has certain problems, the most important being that the layout of the automata universe already sets up a constraint on the production of patterns. Thus some real boundary patterns found in skulls cannot be reproduced in a rectangular universe with a Moore neighborhood. This problem can be surmounted by using a different layout, by allowing many cells to be part of the same abstracted bone, or by using a different modeling approach, one not based on cellular automata.

To be sure, the level of abstraction in the analysis of boundary patterns is a long way from explaining all constraints involved in an evolutionary process. However, by abstracting modules of different order, some constraints have been highlighted, such as the possible existence of two different ways to generate fenestrae, with explicit and testable predictions about evolutionary expectations in vertebrate lineages.

Can we derive lawlike principles at the boundary level? How are the boundary patterns related to each other? Are they constraining the appearance of other boundary patterns? Can we derive the subset of possible articulations out of the whole set of boundaries, and if so, how? These are questions that could be tackled within the framework of the morphospace of connections by analyzing the dynamics involved in the establishment of boundaries during embryonic development and by looking at macroevolutionary boundary patterns. This is a task that almost certainly would have intrigued Darwin, just as he would have been fascinated with the discovery of the genetic code.

References

Bard JBL (1990) Morphogenesis: The Cellular and Molecular Processes of Developmental Anatomy. Cambridge: Cambridge University Press.

Beck WS, Liem KF, Simpson GG (1991) Life: An Introduction to Biology (3d ed). New York: HarperCollins.

Chapman RE, Rasskin-Gutman D (2000) Quantifying morphology. In: Palaeobiology, vol 2 (Briggs D, Crowther P, eds), 489–492. Oxford: Blackwell.

Fontana W, Schuster P (1998) Continuity in evolution: On the nature of transitions. Science 280: 1451–1455.

Goodrich ES (1930) Studies on the Structure and Development of Vertebrates. Vol 1. London: Macmillan. Reprint, Mineola, N.Y.: Dover, 1958.

Hall BK (1994) Homology: The Hierarchical Basis of Comparative Biology. San Diego, Calif.: Academic Press.

Harary F (1972) Graph Theory. Reading, Mass.: Addison-Wesley.

Hinchliffe JR, Johnson DR (1980) The Development of the Vertebrate Limb: An Approach through Experiment, Genetics, and Evolution. Oxford: Clarendon Press.

McGhee GR (1999) Theoretical Morphology: The Concept and its Applications. New York: Cambridge University Press.

Rashevsky N (1954) Topology and life: In search of general mathematical principles in biology and sociology. Bull Math Biophys 16: 317–348.

Rashevsky N (1955) Life, information theory, and topology. Bull Math Biophys 17: 229–235.

Rasskin-Gutman D (2000) The operational strategies of theoretical morphology. Histor Biol 14: 305–308.

Rasskin-Gutman D, Buscalioni AD (2001) Theoretical morphology of the archosaur (Reptilia: Diapsida) pelvic girdle. Paleobiology 27: 59–78.

Riedl R (1978) Order in Living Organisms: A Systems Analysis of Evolution. London: Wiley.

Rieppel OC (1988) Fundamentals of Comparative Biology. Basel: Birkhäuser.

Rieppel OC (1993) Patterns of diversity in the reptilian skull. In: The Vertebrate Skull, vol 2 (Hanken J, Hall BK, eds), 344–390. Chicago: University of Chicago Press.

Rietman E (1989) Exploring the Geometry of Nature. Blue Ridge Summit, Pa.: Windcrest.

Stahl BJ (1974) Vertebrate History: Problems in Evolution. New York: McGraw-Hill. Revised edition. Mineola, N.Y.: Dover, 1985.

Thomson KS (1995) Graphical analysis of dermal skull roof patterns. In: Functional Morphology in Vertebrate Paleontology (Thomason IJ, ed), 193–204. Cambridge: Cambridge University Press.

Trucco E (1956a) A note on the information content of graphs. Bull Math Biophys 18: 129–135.

Trucco E (1956b) On the information content of graphs: Compound symbols; different states for each point. Bull Math Biophys 18: 237–253.

Weishampel D (1993) Beams and machines: Modeling approaches to analyses of skull form and function. In: The Vertebrate Skull, vol 3 (Hanken J, Hall BK, eds), 304–344. Chicago: University of Chicago Press.

Woodger JH (1945) On Biological Transformations: Essays on Growth and Form Presented to D'Arcy Wentworth Thompson (Le Gros Clark WE, Medawar PB, eds), 95–120. Oxford: Oxford University Press.

Young BA (1993) On the necessity of an archetypal concept in morphology: With special reference to the concepts of "Structure" and "Homology." Biol Philos 8: 225–248.

Contributors

Mina J. Bissell
Lawrence Berkeley National Laboratory
Berkeley, California

Roy J. Britten
California Institute of Technology
Corona del Mar, California

Chi-hua Chiu
Department of Ecology and Evolutionary Biology
Yale University
New Haven, Connecticut

Simon Conway Morris
Department of Earth Sciences
University of Cambridge
Cambridge, England

Scott F. Gilbert
Martin Biology Research Laboratories
Swarthmore College
Swarthmore, Pennsylvania

Kunihiko Kaneko
Department of Pure and Applied Sciences
College of Arts and Sciences
University of Tokyo
Tokyo, Japan

Ellen Larsen
Department of Zoology
University of Toronto
Toronto, Canada

I. Saira Mian
Lawrence Berkeley National Laboratory
Berkeley, California

Gerd B. Müller
Institute of Zoology
University of Vienna
Vienna, Austria

Vidyanand Nanjundiah
Developmental Biology and Genetics Laboratory
Indian Institute of Science
Bangalore, India

Stuart A. Newman
Department of Cell Biology and Anatomy
New York Medical College
Valhalla, New York

H. Frederik Nijhout
Department of Biology
Duke University
Durham, North Carolina

Olivier Pourquié
Stowers Institute for Medical Research
Kansas City, Missouri

Derek Radisky
Lawrence Berkeley National Laboratory
Berkeley, California

Diego Rasskin-Gutman
Integrative Morphology Group
Institute of Anatomy
University of Vienna
Vienna, Austria

Malcolm Steinberg
Department of Molecular Biology
Princeton University
Princeton, New Jersey

Georg F. Striedter
Department of Psychobiology
University of California at Irvine
Irvine, California

Eva Turley
Lawrence Berkeley National Laboratory
Berkeley, California

Günter P. Wagner
Department of Ecology and Evolutionary Biology
Yale University
New Haven, Connecticut

Pat Willmer
School of Biology
University of Saint Andrews
Fife, Scotland

Index

Abiotic conditions, 96–97
Acetylation, 107
Action potential, 170
Activin A, 276
Adhesion
  and brain development, 290, 291–292
  differential, 147, 149, 224–226
  differential hypothesis, 152–158
  in embryos, 154–155
  gradients, 142
  heterotypic, 148
  intercellular, 152
  as key principle, 60, 77–79, 145–149
  and limbs, 40
  and lumen formation, 226–227
  nonadhesive surfaces, 143–144
  relative strengths, 147–149
  and segmentation, 229
  and surface tension, 149–152
  timing hypotheses, 145–147
Altitude, 249
Amoebas, 248
Amphibians
  in abiotic conditions, 96–97
  digits, 125
  egg surfaces, 143
  embryos, 137, 138–139, 154
  global body pattern, 207
  predator-induced polyphenism, 95
Amphioxus, 138
Androgenization, 249
Apical ectodermal ridge, 275, 277
Apoptosis, 125, 227, 292–293
Arthropods, 26, 38, 123
Articulations, 307–308, 310, 318
Attractor states, 287
Autocatalysis, 198. *See also* Positive feedback
Autopodium
  definition, 274
  description, 267
  development, 274–277
  and digits, 271
  genetics, 275–280
  origin, 272, 274, 277–280
  as transition, 266
  zeugopodial boundary, 276–277
Axons, 291–292, 295, 296, 298

Bacteria
  E. coli, 212, 247–248, 250
  and epigenesis inheritability, 249
  multicellular, 15
  phenotypic plasticity, 247–248
  and polyphenisms, 95–96
  and sex determination, 93

Baldwin effect, 234, 242, 259, 260
B/cdc2 complex, 185
B-cells, 96
Behavior, 27, 55
bFGF, 107, 108
Bicoid protein, 175, 176, 178
Blind mole rats, 293–294, 298
Blood cells, 206
Body plans, 26, 45, 59
Bones, 97, 315, 319. *See also* Skeleton; Skulls
Border cells, 290
Boundaries, 306, 310. *See also* Connectivity
Brachiopods, 41
Brain
  development of, 289–296, 298–299
  epigenetic homology, 287–289
  evolution of, 296–298
  functional integration, 296–297
  neocortex, 294–295
  neural connections, 80
  species differences, 243
Breast cancer, 108
Brownian motion, 209, 224–225
Bryozoans, 41
Bureaucrat genes, 121, 125, 256
Burrowing, 22
Butterflies, 170–172, 231

Cadherins, 152–159, 226–227
Calcium, 153, 183–184, 198
Cambrian explosion, 20–27
Camouflage, 93
Canalization, 251, 258–259, 261
Cancer cells, 108
Cartilage, 276
Cascades
  in brain development, 292–293, 298
  embryonic, 307–308
  regulatory, 127
  signaling, 119–121, 124–127
CDK inhibition, 188
Cell adhesion. *See* Adhesion
Cell aggregation, 36, 208–210, 223–224
Cell autonomy, 122, 126
Cells
  in aggregates, 144–147
  border, 290
  chemistry of, 201–202, 204, 206, 208
  clustering, 203
  discrete types, 203–204
  and immiscible liquids, 149, 222, 224–226
  and mechanical stress, 232
  mobility, 142, 144–147, 159
  morphogenetic behaviors, 120–121, 126
  nuclear control, 77–78

Cells (cont.)
  segregating, 153–154
  self-organization, 140–142, 149, 153–154, 159–160
  signaling between, 77–79, 174
  sorting, 145–149
  stem, 202–209
Cell state numbering, 79–80, 83
Chaotic itinerancy, 206
Character identity, 265
Chemical composition, 201–202, 204
Chemokines, 145
Chemotaxis, 167
Chicks
  embryos, 144, 145, 149
  feathers, 230
  Hox genes, 276
  wings, 276
Chloride, 184
Chondrogenesis, 276
Chromatin, 107
Ciliated larvae, 17
Clocks. *See* Timing
Clustering, 203, 208–210
Cnidarians, 25–26
Collagen, 14
Color, 167, 230–231. *See also Distal-less* gene
Compactedness, 317
Competition, 95
Condensation, 231–232
Connective tissue, 231–233
Connectivity
  boundary patterns, 309–310, 316–317, 321
  morphospaces, 310–317
  quantification, 311–313
  significance, 306–308, 310
  simulation, 314–317, 320
Convergence
  appendages, 40
  definition, 11, 33–35
  and development, 35–38
  vs. homology, 37–38
  intra- vs. inter phyla, 33
  larvae, 42–43
  lophophores, 41–42
  and molecular taxonomy, 43–44
  as open question, 5
  vs. parallelism, 34
  and phenotypic plasticity, 45
  and phylogenetics, 11
  segmentation, 38–40
Crustaceans, 228
Culture systems, 104, 108
Cyclins, 185, 187
Cysts, 226

Designer organisms, 129–130
Determination, 204, 206
  by details, 75–79, 83, 85
Deuterostomes, 23
Development. *See also* Embryos
  of brain, 289–296, 298–299
  cell formation, 81–82
  constraints, 127
  control of, 76–77
  and convergence, 35–38
  of early metazoans, 221, 233–235
  and hierarchy, 127
  and homology, 56, 61, 62
  irreversibility, 208
  and local interactions, 76–77
  morphogenetic fields, 124–125, 126
  noise, 245–246
  of pattern formation, 175, 178
  stabilization, 36–37
  in stressful environment, 260
  of tetrapod limbs, 274–277
Developmental clocks. *See* Timing
Differentiation. *See also* Tissue specificity
  and boundaries, 310
  of brain areas, 289–296
  and chemical composition, 201–202, 204
  intra-inter dynamic model, 198–203
  logic of, 203–210
  phenotypic, 211–217
  plasticity, 103
  rule generation, 204
  spatial structures, 208–210
  and stability, 207–208
  stages, 200
  and transcription factors, 108
Diffusion, 165–173, 176–177, 223–224
Digits, 123, 125, 271–274, 276, 278
Distal-less gene, 40, 170–172
Diversification
  extrinsic factors, 24–25
  genome role, 25–26
  and taxonomy, 43–44
Drosophila
  anal structure, 82–83
  color, 231
  embryonic patterns, 175
  filopodia, 174
  genetic assimilation, 254–255
  Hsp83 gene, 252
  inbreeding, 252
  oocytes, 154–156, 175
  and reaction-diffusion, 167, 168
  stripes, 169, 228–230, 235
  wingless gene, 174
Dynamic reciprocity, 103–104

Echinoderms, 26
Einbahnstrasse model, 185–187
Elasticity, 222, 232
Embryos
  anteroposterior axis, 186
  calcium oscillations in, 183–184
  cell self-organization, 140–142
  chick, 144, 145, 149
  chloride oscillations, 184
  differentiation, 89, 154–155
  environmental factors, 76, 90–98
  and genetics, 89–90
  and life span, 192
  and mechanical stress, 233
  mesenchymal tissues, 231–232
  midblastula transition, 185
  model system approach, 88
  morphological organization, 307–308
  organogenesis, 78–79, 307–308
  physiological approach, 87–88, 90
  potassium oscillations, 184
  self-organization, 138–142
  sex determination, 92
  specification, 78
  syncytial, 228–230
  temperature effect, 192
  yolkiness, 137
Engrailed, 108, 124, 176
Envelopment, 147, 149, 153–154
Environment. *See also* Extracellular microenvironment
  and brain development, 296–297
  and Cambrian explosion, 24–25
  and convergence, 33
  and earliest metazoans, 222–234
  and embryos, 76, 90–98
  and genetic assimilation, 258–259
  genotype sensitivity, 254
  identical, 247–248
  intra-intercell dynamics, 204–209
  nutrition, 93, 97, 98n2
  as open question, 7
  and phenotype differentiation, 212–215
  predators, 95–96
  of reaction-diffusion, 167
  seasonal, 93
  and sex determination, 92–93
  stressful, 253, 258, 259, 260
  transplant experiments, 206
  water, 96–97
Enzymes, 251
Epidermis, 174
Epigenesis
  definitions, 241
  and heredity, 249–251, 259, 261
  and homology, 61, 62–63
  and microenvironment, 107
  as open question, 6, 8
  Waddington landscape, 287, 288(figure)
Epigenetic homology, 287–289
Epigenotype, 91
Epithelia
  and cadherins, 156–157
  and extracellular microenvironment, 104
  liquid behavior, 222
  lumen formation, *225,* 227
  and mesenchymal tissue, 122, 127
  origins, 43
  and soft matter physics, 230–231
Equilibrium shapes, 36
Escherichia coli, 212, 247–248, 250
Eukaryotes, 15–16, 25
Evolution
  of brain, 296–298
  convergent, 33–35
  genetic role, 103, 119, 216–217
  and homologues, 59, 65
  of morphogenesis, 125–126
  morphological, 11–12, 65, 306
  and phenotypes, 211–212, 246–247, 259–263
  of regulatory genes, 45
  theory of, 7–8
Evolvability, 8, 241
Excitable media, 223, 227–230
Extinctions, 25
Extracellular matrix, 232
Extracellular microenvironment (ECM)
  model, 104, 108–110
  molecular cues, 104–108, 110–115
  and natural selection, 103
Extraterrestrial impact, 25
Eyes
  chicken gene in *Drosophila,* 83
  and convergence, 38
  cyclopic mutation, 125
  evolution of, 24
  and homology, 57
  optic nerve, 187
  vision, 292–295, 297–298
ey gene, 83

Feathers, 230
Fenestrae, 319–320
Fins, 242, 267, 279–280
Fish
  apico-ectodermal ridge, 275, 277–278
  basal ray-finned, 280
  calcium oscillations, 184
  and Hox genes, 279–280

Fish (cont.)
  neural tube, 138
  pigmentation, 230
Flax, 249, 252
Follicle cells, 155
Follistatin, 276
Forked mutation, 125
Form, 127, 306–308
Fossils
  Cambrian, 21–22
  early eukaryotes, 15–16
  metazoan, 16–20
  stratigraphics, 20
  trace, 22–23
  of triploblasts, 23
Fringe gene, 40
Frogs, 95, 138
Fungi, 14

Gastrulation, 36, 143, 187, 226
Generic properties, 7, 60, 223, 233
Genes
  bureaucrat, 121, 125, 256
  and cell assemblage, 77
  cell behavior, 122, 126
  controller, 38
  cooptation of, 40
  developmental, 37–38
  differing effects, 124
  duplications, 44, 45
  for eyes, 83
  factors affecting, 91
  5′ region, 105, 187
  and hierarchies, 234–235
  homologous, 57
  large information-containing, 85
  of metazoans, 119
  redeployment, 26
  regulatory, 45, 252, 256, 259
  structural, 256
  and timing, 191
  worker, 121, 256
Genetic assimilation, 254–260
Genetic defects, 81
Genetics
  and body plans, 45
  and brain development, 289–290, 291
  and convergence, 36–37
  and embryology, 89–90
  of epigenic inheritance, 249–250
  and homology, 56–61, 62
  issues, 5–6
  of pattern formation, 176–177
  and phenotypes, 5–6, 62, 91–92, 211–215
  and regulatory signals, 126

  role of, 216–217, 234
  of tetrapod limb, 275–276
Genome
  and body plans, 45
  and diversification, 25–26
  Hox genes, 26, 36
  and microenvironment, 107
  and (pre-)Mendelian phases, 60–61
Gerbils, 249
Golgi apparatus, 77
Gradients
  and cellular aggregation, 223–224
  diffusion, 223–224
  in intra-inter cell dynamics, 209
  in pattern formation, 172–174, 175, 178–179
Gravity, 97
Growth factors, 108, 275, 276
Gulonolactone oxidase, 98

Hairy genes, 188–190, 228
Halkieriids, 41
Hamsters, 293
Heart defects, 249
Heat, 25, 253, 258
Hedgehog, 124, 176
Heparan sulfate, 107
Heredity
  of acquired traits, 261
  of biochemical composition, 202
  at cell level, 82, 202
  epigenetic, 249–251, 259, 261
HES1 gene, 189
Hierarchy
  and developmental constraints, 127
  and evolution, 126
  and intra-inter dynamics, 202
  vs. self-organization, 234–235
  stability, 234–235
  and surface tension, 149, 151
Highly optimized tolerance, 111–112
Homology
  and brain regions, 287
  and character identity, 265
  vs. convergence, 37–38
  definitions, 34, 52
  epigenetic, 287–289
  future study, 66
  morphological, 58–59, 65
  as open question, 5
  as organizational concept, 64–66
  quantification, 59, 63
  semantic issues, 52–59
  significance of, 11–12
  simulation, 320

and structure-function, 62–63
three phase model, 60–64
Homoplasy, 5, 34
Hopf instability, 228
Hox genes
  and brain development, 289
  cluster arrangements, 37
  in *Drosophila,* 82
  duplications, 44
  and fins, 279–280
  and larva, 42
  and limbs, 40, 186, 275–280
  and monophyly, 44
  and segmentation, 39
  and tentacles, 41–42
  and timing, 185–187
Hsp83 gene, 252
Hsp90 protein, 252, 253
Human beings, 95–96, 97
Human immunodeficiency virus (HIV), 108
Hunchback, 178
Hyaladherins, 108
Hyaluronan, 107–108
Hybrids, 216
Hydra, 26, 167

Immiscibility, 149, 222, 224–226, 231
Immune system, 95–96
Inbreeding, 252–253
Incomplete penetrance, 211–212
Inheritance. *See* Heredity
Innovation, 60–61, 233, 265, 274–275
Insects
  butterflies, 170–172, 231
  caste determination, 98
  coloring, 93
  embryonic specification, 78
  filopodia, 174
  nutrition effect, 93
  pattern formation, 175
  predator effect, 95
  season effect, 93
  segmentation, 228
  sex determination, 93
Instructive interactions, 92
Integrins, 107–108
In utero position, 249
Ionic oscillators, 183–184

Jaws, 97
Joints, 307–308, 310, 318

Kidneys, 226–227, 231
Krüppel gene, 176

Larval hypothesis, 17
Larval types, 26, 42–43
Lateral inhibition, 165–168
Limbs
  biramous, 129
  and cell adhesion, 40
  chick embryonic, 144
  from fins, 242, 265–268 (*see also* Autopodium)
  of flies, 124, 127, 129
  and homology, 54–55
  and Hox genes, 40, 186, 275–280
  morphological organization, 308
  and reaction-diffusion, 232
  tetrapod, 242 (*see also* Tetrapod limbs)
  vertebrate, 122–123, 127
Lineage invariance, 123–124
Liquid, behavior of, 149–152, 222, 224–226
Lophophores, 41–42
Lophotrochozoans, 23
Lumen formation, *225,* 226–227
Lungfish, 280

Macromolecules, 77, 78, 81
Mammary cells, 104–105
  malignant, 108
Mammary gland, 109–110, 112
Mass action, 77
Matrix-driven translocation, 231
Maturation-promoting factor (MPF), 185
Mechanical stress, 170, 232–233
Meiofauna, 17
Mesenchymal tissue
  and boundaries, 310
  and condensation, 231–232
  liquid-like properties, 222
  and vertebrate limbs, 122–123, 127
Messenger proteins, 108
Messenger RNA (mRNA), 175, 188
Metabolic control theory, 177
Metazoans
  Cambrian explosion, 20–27
  complexity emergence, 13–14
  forms, 11
  fossils, 16–20
  genes, 119–120
  homology, 60–61
  progenitors, 222–223
  transitional, 14
Mice, 248, 249, 276
Mitosis, 185, 187–188, 201, *229*
Modularity, 5, 126, 208–210, 310
Molecular mechanisms
  binding action, 77
  causality, 3

Molecular mechanisms (cont.)
  of cell adhesion, 152–154
  and diversification, 25–26
  and morphogenesis, 35–36
  and regulatory signals, 236
  stress effect, 253
  of tetrapod limb, 276
  and tissue specificity, 104–108, 110–115
Molecular taxonomy, 43–44
Monophyly, 33, 44
Morphogenesis, 120–121, 125–126
Morphogenetic fields, 124–125, 126
Morphological organization
  boundary patterns, 309–321
  levels of, 306–308
Morphospaces, 308, 310–317
Multicellularity, 208–210
Mutations
  and evolution, 103, 119
  and fly legs, 125
  sensitivity factors, 251–253
  and timing, 191

Natural selection
  conditions, 245
  and details, 75
  and microenvironment, 103
  and morphology, 8, 51
  and parasites, 252
  and phenotypic plasticity, 253–254, 258
  and physical properties, 35–36
  quantification, 80
  and shared functions, 81
  as stress buffer, 251
Nematodes, 26, 123, 191, 212
Nervous system, 188, 242–243
Neural plate, 154
Neural tube, 138, 140, 290
Neuromeres, 289–291
Neurons, 26, 291–296, 298
  optic, 187
Newt, 207
Notch-Delta signaling, 174, 228
Novelty, 5, 6, 60–61, 233, 274–275
Nuclear control factors, 77–78
Nutritional polyphenisms, 93, 97, 98n2

Olfaction, 249
Oligodendrocytes, 187
Orbital fenestrae, 319–320
Organogenesis, 230–231, 307–308
Orientation, 307–308
Oscillations, 183–184, 223, 227–230
Outcrossing, 253
Oxygen, 24

Parallelism, 11, 34
Parasites, 252–253
Pattern formation
  communication of, 167, 168, 169–174
  development, 175, 178
  experimental approach, 168–169
  genetics, 176–177
  models, *171*
  reaction-diffusion, 165–168, 176–177, 228–230
  theoretical approach, 165–168
Pax-6 gene, 38, 57
Permissive interactions, 92
Phenotypes
  differentiation, 211–217
  environmental factors, 91–92
  genetic assimilation, 254–259
  and genotype, 5–6, 62, 91–92, 211–215
  isogenic, 211–212, 247–248, 259
  persistence of, 6
Phenotypic plasticity
  defined, 245
  developmental noise, 245–246
  and evolution, 259–262
  and genetics, 45, 253–260
  isogenic, 211–212, 247–248, 259
  rationale, 261
Phosphorites, 19
Phyla, limitations of, 21
Phyllotactic patterns, 167
Phylogenetics, 4–5, 11
Physics, 159–160, 222–224
Pigment, 167, 230–231
Plasticity, phenotypic, 8, 45, 245–246, 256, 259
p27kip1 gene, 188
Polarity, 226–227
Polarizing activity, 268, 277, 279
Polycystin-1, 226
Polyphenisms, 6, 92–96
Positive feedback, 165–166, 168, 176–177
Potassium, 184
Predators, 95
Prokaryotes, 15
Proteins
  bicoid, 175, 176, 178
  fringe, 40
  Hsp90, 252, 253
  messenger, 108
  in pattern formation, 178
  stress, 108, 253
  TAT, 108
Proteoglycans, 233
Protistans, 14, 15
Protozoa, 248

Quantification
  of cells, 83
  of control elements, 79–80, 84–85
  of homology, 59, 63
  of morphospaces, 311–313
  in pattern formation, 178

Range variation, 247–248, 259
Reaction-diffusion, 165–168, 176–177, 228–230, 232
Reaction norm, 92, 98n2
Recapitulation, 56
Recursive units, 208–210
Redundancy, 6, 81
Regeneration, 124–125
Respiration, 24
Retina, 293, 295, 297–298
Rhombomeres, 289–291
Rhythmic activity, 290

Salinity, 25
Segmentation
  convergence, 38–40
  in *Drosophila,* 82
  physics of, 228–230
  timing of, 188–191
Sensory input, 293–294
Set-aside cells, 17, 42
Sex determination, 92–93, 249
Shape, 127, 306–308
Shared function, 80–81
Shh gene, 277–278, 279
Side effect hypothesis, 61
Signaling
  and brain, 292–293, 294–295
  cascades, 119–121, 124–127
  in culture system, 104
  from ECM, 104–108
  intercellular, 77–79, 174
  intracellular, 107–108
  and molecular changes, 126
  Notch-Delta, 174, 228
  and pattern formation, 170, 172–174
  with small molecules, 173
  and tissue specificity, 103
  with transcription factors, 173–174
Size, 306–308
Skeleton, 21–22, 27, 233. *See also* Bones; Connectivity; Limbs
Skulls, 243, 313–317, 319
Soft matter, 222–223
Somitogenesis, 188
Speciation, 213–216, 260
Sponges, 14, 25, 43
Stability, 207–208
Stem cells, 202–209

Stochastic resonance, 112
Strabismus, 295
Stress
  environmental, 258, 259, 260
  genetic, 251–253, 259
  mechanical, 170, 232–233
Stress proteins, 253
Structure-function relationships, 62–63
Suppressor of forked, 125
Surface contraction waves, 185
Surface tension, 149–152, 154
Switching dynamics, 206
Synapomorphy, 54

TAT protein, 108
Temperature
  and developmental clocks, 192
  heat shock, 253, 258
  ice ages, 25
  and phenotypic plasticity, 248
  and sex determination, 91, 92–93
Tendons, 233
Tentacles, 41–42, 167
Tetrapods, 268–270, 313
Tetrapod limbs
  cell number, 123
  components, 266–267
  development, 274–277
  and fins, 267–268
  genetics, 242, 275–276
  origination, 269–270
  stages, 270–274
Tetrodotoxin, 295
Thalamus, 293–295
Thermodynamics, 159–160. *See also* Temperature
Threshold points, 61
Timing
  and cell cycles, 185, 187, 201
  heterochrony, 191
  hourglass mechanisms, 187–188, 198
  and Hox genes, 185–187
  and intracellular chemicals, 208
  intra-inter dynamic model, 198–203
  ionic oscillators, 183–184
  midblastula transition, 185
  and mutations, 191
  and oligodendrocytes, 187
  for pattern formation, 178–179
  and ratios, 201
  of segmentation, 188–191
  for self-assembly, 26–27
Tissues. *See also* Epithelia; Mesenchymal tissue
  emergence of, 221–235
  enveloping, 147, 149, 153–154
  as excitable media, 227–230

Tissues (cont.)
  and immiscibility, 149, 231
  interaction between, 122–123
  liquid behavior, 149, 222, 224–226
  and mechanical stress, 232–233
  segregation, 158–159
  self-organization, 138–142, 149, 156–158
Tissue autonomy, 122
Tissue organization
  and cadherins, 152–158
  differential adhesion hypothesis, 147, 149
  embryonic, 307–308
  pre-Mendelian
  self-organization, 138–144, 149, 156–158
  surface tension role, 149–152
Tissue specificity
  adaptations, 112–115
  and cadherins, 156–158
  dynamic reciprocity, 103–104
  model, 104, 108–110
  molecular cues, 104–108
  and stochastic resonance, 112
  study requirements, 110–111
Toads, 96–97
Tolerance, highly optimized, 111–112
Tracheae, 167
Transcription factors
  and brain development, 289
  and calcium oscillations, 184
  and differentiation, 108
  and genes, 176–177
  and nuclear control, 77–78
  signaling with, 125, 173–174
Triploblasts, 23
Tumor cells, 204
Turing instability, 228–230

Veins, 167
Vesicles, 77
Viscosity, 222
Vision, 292–295, 297–298
Vitamins, 97, 98n2
von Baer's rule, 35

Waddington effect, 242, 254–256, 287, 288 (figure)
Wingless, 174, 176
Wings, 276
Worker genes, 121, 256

Zebrafish. *See* Fish